VOLUME 3

AUDIOPHILE VACUUM TUBE AMPLIFIERS

Dozens of commercial amplifiers analyzed, more unusual tubes
and ready-to-build designs, and all in a more practical way

IGOR S. POPOVICH

DISCLAIMER & COPYRIGHT NOTICE

The information contained in this book is to be taken in the context of general overview, not specific advice. You should not act on the information contained herein without seeking professional advice. Neither the author nor the publisher (or any other person involved in the publication, distribution or sale of this book) accepts any responsibility for the consequences that may arise from readers acting in accordance with the material given in the book. Professional advice about each particular case / instance should be sought.

Our choice of designs, parts, models and brands was based on their availability and educational value. We were not influenced or induced by anybody in our selection, and their use does not mean we actually endorse or recommend them. You should satisfy yourself that a particular device, component or method is suitable for your intended purpose.

Some circuit diagrams of commercial equipment discussed here were not published by the manufacturers, but were posted online by others, and we cannot confirm their accuracy or authenticity. They are used here for review and discussion purposes as "fair dealing", permitted by international copyright laws. Designs marked with a copyright symbol are intellectual property of their copyright holders and should not be used without their permission. They are discussed here from educational perspective only.

COMMERCIAL USE & INDEMNITY NOTICE

Our designs are available for construction but only for your personal, one-off, noncommercial use. Manufacturers must obtain a license from us for commercial use.

The consequences of any modifications to and deviations from the featured designs are at your own risk, and no responsibility will be accepted by us.

Tube amplifiers involve lethal voltages, high temperatures and other hazards. By purchasing and reading this book you agree to indemnify its author, publisher and retailer against any claims, of any nature, and for any reason!

© **Copyright Igor S. Popovich 2016**

All rights are reserved. No part of this publication may be used, reproduced or transmitted by any means, without the prior written permission from the publisher, except in the case of brief quotations in articles and reviews.

Published by Career Professionals in Australia

P.O. Box 5668, Canning Vale South WA 6155, Australia

Revised & corrected, 2022

Bulk purchases

This book may be purchased in larger quantities for educational, business or promotional use. Please e-mail us at sales@careerprofessionals.com.au

National Library of Australia Cataloguing-in-Publication Data:

Popovich, Igor S.,
    AUDIOPHILE VACUUM TUBE AMPLIFIERS - DESIGN, CONSTRUCTION, TESTING, REPAIRING & UPGRADING, Volume 3

    ISBN: 978-0-9806223-4-8
    1. Electrical engineering 2. Electronics 3. Hi-fi
    I Igor S. Popovich II Title III Index

621.3

AUDIOPHILE TUBE AMPLIFIER BOOKS - "THE TRILOGY" BY IGOR S. POPOVICH

Audiophile Vacuum Tube Amplifiers, Vol. 1:

- BASIC ELECTRONIC CIRCUIT THEORY
- ELECTRONIC COMPONENTS
- AUDIO FREQUENCY AMPLIFIERS
- PHYSICAL FUNDAMENTALS OF VACUUM TUBE OPERATION
- VOLTAGE AMPLIFICATION WITH TRIODES - THE COMMON CATHODE STAGE
- OTHER VOLTAGE AMPLIFICATION STAGES WITH TRIODES
- TETRODES AND PENTODES AS VOLTAGE AMPLIFIERS
- FREQUENCY RESPONSE OF VACUUM TUBE AMPLIFIERS
- IMPEDANCE-COUPLED STAGES AND INTERSTAGE TRANSFORMERS
- NEGATIVE FEEDBACK
- TONE CONTROLS, ACTIVE CROSSOVERS AND OTHER CIRCUITS
- PRACTICAL LINE-LEVEL PREAMPLIFIER DESIGNS
- PHONO PREAMPLIFIERS
- SINGLE-ENDED TRIODE OUTPUT STAGE
- PRACTICAL SINGLE-ENDED TRIODE AMPLIFIER DESIGNS
- PRACTICAL SINGLE-ENDED PSEUDO-TRIODE DESIGNS
- SINGLE-ENDED PENTODE AND ULTRALINEAR OUTPUT STAGES

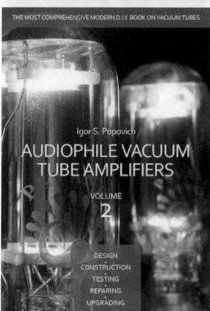

Audiophile Vacuum Tube Amplifiers, Vol. 2:

- PRACTICAL SINGLE-ENDED PENTODE AND ULTRALINEAR DESIGNS
- PUSH-PULL OUTPUT STAGES
- PRACTICAL PUSH-PULL AMPLIFIER DESIGNS
- BALANCED, BRIDGE AND OTL (OUTPUT TRANSFORMERLESS) AMPLIFIERS
- THE DESIGN PROCESS
- FUNDAMENTALS OF MAGNETIC CIRCUITS AND TRANSFORMERS
- MAINS TRANSFORMERS AND FILTERING CHOKES
- POWER SUPPLIES FOR TUBE AMPLIFIERS
- AUDIO TRANSFORMERS
- TROUBLESHOOTING AND REPAIRING TUBE AMPLIFIERS
- UPGRADING & IMPROVING TUBE AMPLIFIERS
- SOUND CONSTRUCTION PRACTICES
- AUDIO TESTS & MEASUREMENTS
- TESTING & MATCHING VACUUM TUBES

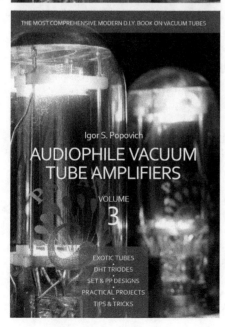

Audiophile Vacuum Tube Amplifiers, Vol. 3:

- THE FRONT-END: SUPERIOR INPUT & DRIVER STAGES
- FROM SHOCKING TO SUBLIME: LESSONS FROM COMMERCIAL LINE STAGES
- DIY LINE-LEVEL PREAMPLIFIERS: $10,000 SOUND ON $500-$1,000 BUDGET
- THE STARS OF THE AUDION ERA: ANCIENT TUBES IN MODERN AMPS
- CHEAP & CHEERFUL: PREAMP & DRIVER TUBES FOR AUDIO EXPLORERS
- SLEEPING GIANTS: OUTPUT TUBES FOR THOSE WHO WANT TO BE DIFFERENT
- THE QUEEN OF HEARTS: SINGLE-ENDED AMPLIFIERS WITH 300B TRIODES
- TRIODES, PENTODES AND BEAM TUBES: MORE SINGLE-ENDED DESIGNS
- BIG BOTTLES: SET AMPLIFIERS WITH HIGH VOLTAGE TRANSMITTING TUBES
- THE WAY IT USED TO BE: VINTAGE PUSH-PULL AMPLIFIERS
- NEW? IMPROVED? MODERN PUSH-PULL AMPLIFIER DESIGNS
- CUTE, CLEVER OR CONTROVERSIAL? INTERESTING IDEAS FROM TUBE AUDIO'S PAST AND PRESENT
- THRIFTY TIPS & TRICKS: TIME & MONEY SAVING IDEAS
- OUTPUT AND INTERSTAGE TRANSFORMERS: FROM COMMERCIAL BENCHMARKS TO YOUR OWN DESIGNS
- MEASUREMENTS VERSUS LISTENING AND OTHER AUDIO DESIGN DILEMMAS

OTHER AUDIO-RELATED BOOKS BY IGOR S. POPOVICH

Available from Amazon, Barnes & Noble, Book Depository and all other major online bookstores

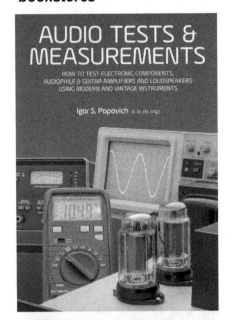

Audio Tests & Measurements: How to Test Electronic Components, Audiophile & Guitar Amplifiers and Loudspeakers Using Modern and Vintage Test Instruments

ISBN: 978-0-9806223-9-3

- TEST INSTRUMENTS, ERRORS, LIMITATIONS & SAFETY ISSUES
- SIGNAL SOURCES, TRACERS, POWER SUPPLIES AND FILTERS
- MULTIMETERS - TYPES, OPERATING PRINCIPLES AND FUNCTIONS
- OSCILLOSCOPES - HOW THEY WORK & HOW TO USE THEM
- TESTING PASSIVE ELECTRONIC COMPONENTS (RESISTORS, CAPACITORS & INDUCTORS)
- TESTING AUDIO AMPLIFIERS AND PREAMPLIFIERS
- DISTORTION MEASUREMENTS
- TRANSFORMER TESTS & MEASUREMENTS
- LOUDSPEAKER TESTS & MEASUREMENTS
- TRANSISTOR TESTERS AND CURVE TRACERS
- TESTING VACUUM TUBES (VALVES)

How to Use, Calibrate, Repair and Upgrade Vacuum Tube Testers

ISBN: 978-0-9806223-7-9

- HOW VACUUM TUBES WORK
- TESTING & MATCHING VACUUM TUBES
- EMISSION TESTERS
- GRID CIRCUIT TESTERS
- DYNAMIC CONDUCTANCE TESTERS
- PROPORTIONAL MUTUAL CONDUCTANCE TESTERS
- HICKOK-TYPE TESTERS
- TRUE MUTUAL CONDUCTANCE TESTERS
- REPAIRING & UPGRADING VINTAGE TUBE TESTERS
- TESTING & MATCHING TUBES WITHOUT A TUBE TESTER

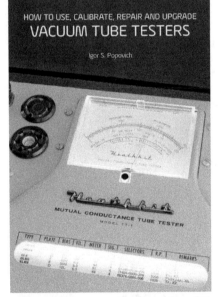

Transformers For Tube Amplifiers: How to Design, Construct & Use Power, Output & Interstage Transformers and Chokes in Audiophile and Guitar Tube Amplifiers

ISBN: 978-0-9806223-8-6

- PHYSICAL FUNDAMENTALS OF MAGNETIC CIRCUITS AND TRANSFORMERS
- FILTERING CHOKES (INDUCTORS WITH DC CURRENT)
- TRANSFORMER MATERIALS, CONSTRUCTION METHODS AND ISSUES
- MAINS (POWER) TRANSFORMERS
- PHYSICAL FUNDAMENTALS OF AUDIO TRANSFORMERS
- SINGLE-ENDED OUTPUT TRANSFORMERS
- PUSH-PULL OUTPUT TRANSFORMERS
- SPECIAL MAGNETIC COMPONENTS: LOW POWER INPUT, PREAMP OUTPUT & DAC OUTPUT TRANSFORMERS, TRANSFORMER VOLUME CONTROL
- INTERSTAGE TRANSFORMERS, GRID & ANODE CHOKES
- OUTPUT AND INTERSTAGE TRANSFORMERS FOR TUBE GUITAR AMPS
- TRANSFORMER TESTS & MEASUREMENTS

CONTENTS OF VOLUME 3:

1. WHY THE NEED FOR VOLUME 3, AND HOW IT DIFFERS FROM THE FIRST TWO	Page 7
2. THE FRONT-END: SUPERIOR INPUT & DRIVER STAGES	Page 13
3. FROM SHOCKING TO SUBLIME: LESSONS FROM COMMERCIAL LINE STAGES	Page 25
4. DIY LINE-LEVEL PREAMPLIFIERS: $10,000 SOUND ON $500-$1,000 BUDGET	Page 47
5. THE STARS OF THE AUDION ERA: ANCIENT TUBES IN MODERN AMPS	Page 57
6. CHEAP & CHEERFUL: PREAMP & DRIVER TUBES FOR AUDIO EXPLORERS	Page 75
7. SLEEPING GIANTS: OUTPUT TUBES FOR THOSE WHO WANT TO BE DIFFERENT	Page 87
8. THE QUEEN OF HEARTS: SINGLE-ENDED AMPLIFIERS WITH 300B TRIODES	Page 99
9. TRIODES, PENTODES AND BEAM TUBES: MORE SINGLE-ENDED DESIGNS	Page 119
10. BIG BOTTLES: SET AMPLIFIERS WITH HIGH VOLTAGE TRANSMITTING TUBES	Page 143
11. THE WAY IT USED TO BE: VINTAGE PUSH-PULL AMPLIFIERS	Page 165
12. NEW? IMPROVED? MODERN PUSH-PULL AMPLIFIER DESIGNS	Page 199
13. CUTE, CLEVER OR CONTROVERSIAL? INTERESTING IDEAS FROM TUBE AUDIO'S PAST AND PRESENT	Page 219
14. THRIFTY TIPS & TRICKS: TIME & MONEY SAVING IDEAS	Page 235
15. OUTPUT AND INTERSTAGE TRANSFORMERS: FROM COMMERCIAL BENCHMARKS TO YOUR OWN DESIGNS	Page 249
16. MEASUREMENTS VERSUS LISTENING AND OTHER AUDIO DESIGN DILEMMAS	Page 269
17. THE END MATTER OF VOLUME 3	Page 281

CONTENTS OF VOLUME 1:

1. WHO WILL BENEFIT FROM THIS BOOK AND HOW	Page 7
2. BASIC ELECTRONIC CIRCUIT THEORY	Page 17
3. ELECTRONIC COMPONENTS	Page 31
4. AUDIO FREQUENCY AMPLIFIERS	Page 49
5. PHYSICAL FUNDAMENTALS OF VACUUM TUBE OPERATION	Page 69
6. VOLTAGE AMPLIFICATION WITH TRIODES - THE COMMON CATHODE STAGE	Page 85
7. OTHER VOLTAGE AMPLIFICATION STAGES WITH TRIODES	Page 99
8. TETRODES AND PENTODES AS VOLTAGE AMPLIFIERS	Page 123
9. FREQUENCY RESPONSE OF VACUUM TUBE AMPLIFIERS	Page 137
10. IMPEDANCE-COUPLED STAGES AND INTERSTAGE TRANSFORMERS	Page 153
11. NEGATIVE FEEDBACK	Page 163
12. TONE CONTROLS, ACTIVE CROSSOVERS AND OTHER CIRCUITS	Page 179
13. PRACTICAL LINE-LEVEL PREAMPLIFIER DESIGNS	Page 193
14. PHONO PREAMPLIFIERS	Page 205
15. SINGLE-ENDED TRIODE OUTPUT STAGE	Page 221
16. PRACTICAL SINGLE-ENDED TRIODE AMPLIFIER DESIGNS	Page 231
17. PRACTICAL SINGLE-ENDED PSEUDO-TRIODE DESIGNS	Page 255
18. SINGLE-ENDED PENTODE AND ULTRALINEAR OUTPUT STAGES	Page 271
19. THE END MATTER OF VOLUME 1	Page 283

CONTENTS OF VOLUME 2:

1. PRACTICAL SINGLE-ENDED PENTODE AND ULTRALINEAR DESIGNS	Page 11
2. PUSH-PULL OUTPUT STAGES	Page 21
3. PRACTICAL PUSH-PULL AMPLIFIER DESIGNS	Page 45
4. BALANCED, BRIDGE AND OTL (OUTPUT TRANSFORMERLESS) AMPLIFIERS	Page 67
5. THE DESIGN PROCESS	Page 79
6. FUNDAMENTALS OF MAGNETIC CIRCUITS AND TRANSFORMERS	Page 87
7. MAINS TRANSFORMERS AND FILTERING CHOKES	Page 97
8. POWER SUPPLIES FOR TUBE AMPLIFIERS	Page 115
9. AUDIO TRANSFORMERS	Page 141
10. TROUBLESHOOTING AND REPAIRING TUBE AMPLIFIERS	Page 169
11. UPGRADING & IMPROVING TUBE AMPLIFIERS	Page 181
12. SOUND CONSTRUCTION PRACTICES	Page 205
13. AUDIO TESTS & MEASUREMENTS	Page 229
14. TESTING & MATCHING VACUUM TUBES	Page 261
15. THE END MATTER OF VOLUME 2	Page 279

WHY THE NEED FOR VOLUME 3, AND HOW IT DIFFERS FROM THE FIRST TWO

- Dozens of commercial amplifiers analyzed, more unusual tubes and ready-to-build designs, and all in a more practical way
- Getting in touch with us
- How to read this book

1

"That is a good book which is opened with expectation and closed with profit."
A. Bronson Alcott, writer and philosopher

Dozens of commercial amplifiers analyzed, more unusual tubes and ready-to-build designs, and all in a more practical way

Like designers of amplifiers and other hi-fi components, authors are fairly thin-skinned creatures. After all, nobody likes criticism, and even the mildest of comments may be enough to deflate egos and make authors and designers miserable. Sometimes, however, the readers' feedback is overwhelmingly positive, and we, the creatives, are our own harshest critics. This has been the case with Volumes 1 and 2 of this book, which have been extremely well received.

Another critical factor is time. It is almost impossible to gain a proper perspective during or immediately after writing a book; a period of time is needed for things to settle down and for the author to start seeing his work in a more impartial light, just as when reading somebody else's book. Such mental and temporal distance is necessary to put things in proper perspective.

The idea for this third tome was born a few months after the first two volumes were published. First of all, we haven't been idle in the eighteen or so months that it took to write and publish the first two books; we kept experimenting, trying out different tubes and circuits, and designing new amplifiers.

The mental and temporal distance already mentioned enabled us to identify topics and issues that were not extensively covered in the previous volumes or were just mentioned in passing and not covered at all. Therefore, the aim of Volume 3 is to devote more time and attention to such issues, methods, and designs.

Most vintage and contemporary books on tube audio technology do not pay much attention to commercial designs and trends. We did analyze a few modern amplifiers in the previous two books, either as practical examples of how certain circuit blocks could be used or as platforms (case studies) the readers could use to practice their skills in analyzing and improving such designs.

Assuming the readers have mastered the basics, we will use more of those practical case studies in this more advanced book and try to emphasize the "detective" or investigative aspect of design and construction.

This means we should try to put ourselves in the shoes of the original designers and builders of audio gear and try to comprehend their design philosophy (if any) and then the logic behind their choices of circuits, tubes, operating points, and other design issues. Again, there is often no valid logic; the manufacturer simply copied a previous design, including its flaws, or financial or marketing people compromised the design.

Our aim is not to criticize or condemn their work but to put them under scrutiny so sadly lacking from the audio reviews in magazines. Our goal is not to dissuade the readers from buying such equipment or to encourage DIY constructors to copy their designs. Although that may not be illegal, it certainly borders on unethical.

We are simply illustrating how to analyze and scrutinize various sources of information to gain a deeper understanding of the field. Such a process happens naturally and intuitively in the minds of most readers, audiophiles, and DIY constructors anyway. All we are doing is outlining it more formally on written pages.

How to become an DIY audio detective

The best way to gain insight into the quality of the design and construction of an amplifier or any audio component, and the only way to experience their sonics, is to have the actual piece of equipment in your hands, on your test bench, and in your listening room. Unfortunately, most are very expensive, so such an approach is not practical or affordable.

Of course, you could borrow a demo amplifier from a local retailer for a few days to hear how it fits with the rest of your audio system and then measure it on a test bench or even reverse-engineer its circuit. It is up to you to decide if such actions would be ethically proper or not.

Some more-or-less accurate circuit diagrams posted on the web were obtained through such a method. We have only done it with amps and preamps we purchased and owned and thus felt that we had the right to do whatever we wanted with them.

The next best thing is to have an accurate circuit diagram, which means one published by the manufacturer. Although in many cases DC voltages are not marked on such basic diagrams, and AC (signal) levels are seldom given, we will teach you how to determine or at least estimate the operating points, amplification factors, output impedances, and other aspects of such circuits.

As we have illustrated in Vol. 1 with Conrad Johnson's "Art" preamplifier, even a few snippets of information and some crucial results of measurements performed by the reviewing magazine may be enough to identify the circuit topology used and to estimate the value of some resistors and capacitors. However, the likelihood of miscalculations and other errors increases significantly when we are removed from the actual physical piece of equipment.

The fewer hard facts we have, the more tentative and inaccurate our analysis becomes. Detectives seldom have all the evidence they need. They often have no evidence at all, so they must base their strategies on speculation, intuition, and educated guesses based on their professional judgment and previous experience. The same applies to audio detectives like us.

DIY audio is about alternatives

Learning from the best commercial designers and builders of tube audio is only a means to an end, not the end in itself. Such knowledge should be applied to your own designs and creations, not merely by copying but by experimenting, evaluating, and applying it in a synergistic way to other circuits, situations and designs. Ultimately, if possible, such learning should be crowned by improving on the ideas and designs of others.

Life, electronics, and tube audio are about alternatives and options, about choices we make to get closer to the goals we set while upholding certain values and principles. All the disagreements in audio (tube vs. transistors, analog vs. digital, single-ended vs. push-pull, and dozens of others) are futile since they are simply choices we make based on what we value more.

If deep, fast, and punchy bass is what makes you happy, you will more likely than not belong to the solid-state camp or perhaps a push-pull school of tube amplification. If midrange magic is what you cherish more than anything else, I would guess that directly-heated triodes (DHTs) in single-ended (SE) circuits are your cup of tea.

If you haven't done so, please study the first two volumes first!

Although some may consider this a separate, stand-alone book, strictly speaking, it isn't. It is Volume 3 in the trilogy of books that should be read and studied together. That is why it is assumed that the reader of this tome is familiar with the concepts discussed in the first two volumes, so most of the previously covered material will not be repeated or explained here.

Invitation to designers and manufacturers

No matter where you fit in the grand scheme of all things audio, if you are a DIY enthusiast, a designer or amplifier maker, or indeed a commercial corporation, we hereby invite you to collaborate with us to make the future editions of all three volumes better, more practical and more relevant to the audiophile community.

Should you wish to share your design philosophy with us in an interview, contribute a design or some of your practical know-how, or even donate an amp or preamp for our analysis and evaluation (and ultimately, its inclusion as a case study), we would welcome your constructive involvement.

Getting in touch with us

If you've liked the book and benefited from it, the best way to repay a favor is to recommend it to your friends and to write an online review. Also, if you spot an error or an omission or have any constructive criticism of the book, I'd like to hear from you to fix it together.

If you want to contribute ideas or projects for the next edition (or Volume 4?), or if you have ideas on making the next edition better, please let me know.

My e-mail is **igorpop@careerprofessionals.com.au**

I hope that this book has answered at least some of your questions about audio design and amplifier building. I also hope that it has also raised a few new questions that should be discussed and the answers sought. I wish you every success on your audio journey in general and DIY projects in particular!

Igor S. Popovich

How to read this book

One way to read a technical book like this one is to immediately go to a section or topic that interests you and then to keep jumping back-and-forth to the related issues and chapters. This will, I suspect, be the way the more experienced designers and constructors will approach this book.

A more systematic approach is sequential, starting from the beginning and reading in order. This is what I would recommend. Although it seems more time-consuming (since you will read about many issues you may already know a lot about), paradoxically, this approach is often faster. You will not miss anything, and you will not waste time flipping forward and backward trying to clarify an issue that you've overlooked and perhaps not fully understood.

Whatever you do, don't treat this book as Holy Scripture. Underline or highlight the important parts, write your thoughts and ideas on its margins, sketch diagrams and circuits in its blank spaces.

MEASURED RESULTS:
- BW: 15Hz - 35 kHz (-3dB, at $10V_{RMS}$ into 8Ω)
- V_{MAX}: $11V_{RMS}$
- $P_{MAX} = 15W$

Most of the circuit diagrams in this book have been tried in practice. When you see this type of frame, you can rest assured that the design is either of a commercial amplifier or one built and thoroughly tested by us.

RULE-OF-THUMB
Load impedance for triodes:
$Z_{AOPT} \approx 3r_I - 4r_I$
r_I = internal resistance of a tube

These Rules-of-Thumb are simple shortcuts for amplifier builders who don't want to bother with high-level maths, models, and similar highbrow concepts. Easy to memorize, they approximate and summarize much more complex formulas, methods, and concepts.

Although each detailed circuit diagram in this book could be a DIY project by itself, small projects are framed and marked with this soldering-iron symbol.

DIY PROJECT

TUBE PROFILE: XXXXX

Each of the tubes discussed or featured in the designs in this book will have its basic parameters and operational data summarized in box such as this one.

Commercial designers and manufacturers wish to protect their practical knowledge and "insider secrets." Framed boxes of this kind will emphasize lesser-known practical tips and tricks. Although a magician's hat and a magic wand are used as symbols, there is nothing magical about these trade secrets; solid scientific and engineering principles underpin all.

TRADE TRICKS

MANUFACTURER'S SPECIFICATION
The checklist symbolizes a list of technical parameters provided by the manufacturer

CRITICAL QUESTION
While there are no stupid questions (only stupid answers!), some questions are far more important than others. These are answered in frames with this symbol.

IMPORTANT FORMULA
The calculator symbol indicates an important or often-used formula.

A WARNING OR A VERY IMPORTANT POINT!
Some issues, myths, and warnings are so important that they warrant being emphasized in a frame of this kind.

KEY FEATURES
The key aspects, strengths or interesting features of certain designs are summarized in a frame with a key.

Currents, voltages and other markings on circuit diagrams

 250V   DC voltage in the marked node (quiescent state, no signal)

 1V   AC signal voltage in the adjacent node (RMS or effective value)

 5mA   DC current through the adjacent branch (quiescent state, no signal)

 A=55   Voltage amplification of the adjacent stage

Abbreviations used (in no particular order)

AC	Alternating current	MAX	Maximum	GND	Ground terminal
DC	Direct current	MIN	Minimum	COM	Common terminal
RIAA	Recording Industry Association of America	BNC	Bayonet Neill–Concelman, video & test equipment connector for coaxial cable	ESR, ESL	Equivalent series resistance and inductance (capacitor)
THD	Total harmonic distortion	IMD	Intermodulation distortion	AWG	American Wire Gauge
PIO	Paper-in-oil (capacitor)	DIN	Deutsches Institut für Normung (German Institute for Standardization)	CCS	Constant current source or sink
F&F	Film and foil (capacitor)	PP	Push-pull (amplifier)	CF	Cathode follower
MF	Metallized film (capacitor)	PP	Polypropylene (capacitor)	SPL	Sound pressure level
MM	Moving magnet (phono cartridge)	PP	Peak-to-peak (AC signal)	NFB	Negative feedback
MC	Moving coil (phono cartridge)	PPP	Parallel push-pull	PFB	Positive feedback
LOG	Logarithmic scale or taper (potentiometer)	NTC, PTC	Negative and positive temperature coefficient (of a resistor or other component)	GBW	Gain-bandwith product of a tube or amplifier
LIN	Linear scale or taper (potentiometer)	SET, PSET	Single-ended triode, parallel SET	EMF, CEMF	Electro-magnetic force, counter EMF
TR, VR	Turns or voltage ratio (of a transformer)	E/S	Electrostatic (field, interference or shield)	RCA	Unbalanced audio connector or Radio Corporation of America
IR	Impedance ratio (of a transformer)	SQ	Special quality (tube)	XLR	Balanced audio connector (3-pins)
RC	Resistor-capacitor coupling between stages	SRPP	Shunt-regulated push-pull (stage or amplifier)	SLO-BLO	Slow blowing fuse, delay fuse
LC	Inductive-capacitive coupling between stages	GOSS	Grain-oriented silicon steel (transformer lamination material)	FET, JFET, MOSFET	Field effect transistor, Junction FET, Metal Oxide FET
CG	Control grid (in a tube)	SG	Screen grid (in a tube)	SP	Suppressor grid
TPV	Turns-per-volt (of a transformer)	CT	Center tap (of a transformer)	DF	Damping factor
CRC	Capacitor-resistor-capacitor filter	CLC	Capacitor-inductor-capacitor filter	UL	Ultralinear (output stage)
AF	Audio frequency	RF	Radio frequency	RFI	Radio frequency interference
PSRR	Power Supply Rejection Ratio	CMRR	Common Mode Rejection Ratio	LCR	Inductance-capacitance-resistance meter or circuit
DHT	Directly heated triode	NFB	Negative feedback	RMS	Root-Mean-Square
KL	Kirchoff's Law	IC	Internal connection (pin of a tube)	TUT	Tube under test
NOS	New Old Stock	BW	Bandwidth	EM	Electromagnetic

Symbols used

≈ APPROXIMATE
≡ EQUIVALENT
|| PARALLEL CONNECTION

MAGNETIC COMPONENTS

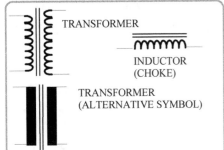

CAPACITORS

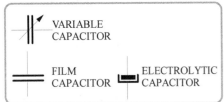

AC AND DC SOURCES

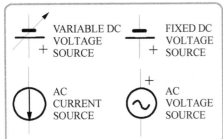

MISCELLANEOUS SYMBOLS

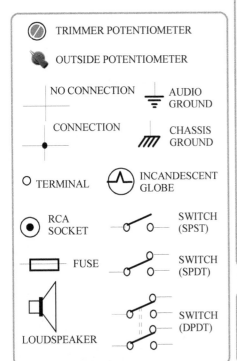

RESISTORS

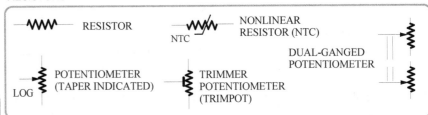

SEMICONDUCTORS

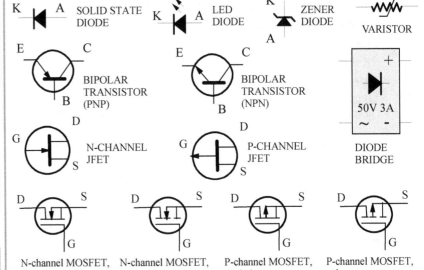

ELECTRON TUBES

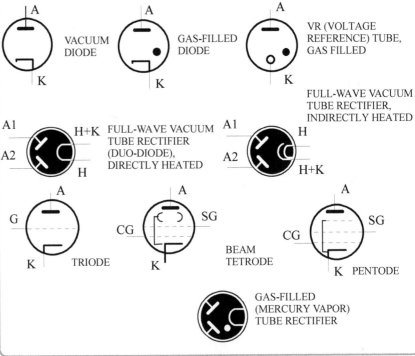

TEST INSTRUMENTS

THE FRONT-END: SUPERIOR INPUT & DRIVER STAGES

- ECC88 + VT-25 driver
- 6N30P + VT-25 driver
- KT61 (KT81) - EL33 (CL33) - 6P25 driver or line stages
- Bel Canto's modified SRPP stage
- Triode -strapped C3m and C3g driver stages

"Music can change the world, because it can change people."
Bono

2

ECC88 + VT-25 (10Y) driver

This preamp/driver circuit topology is highly regarded in hi-end circles. Most of its gain comes from the first common cathode stage using a high perveance triode such as 5687, ECC88, 6N30P, but almost any other triode can be used, whatever lights your fire.

The first stage is directly coupled to the grid of the driver stage, a directly-heated power triode (DHT) called 10Y or VT-25. Again, there are many choices - VT-25, 45, 2A3, 300B; the possibilities are limited only by your imagination, or in the case of AD1 and similar rarities, by the depth of your wallet.

The anode load of the DHT is anode choke with 50H minimum inductance; 100h or higher is better. The DC resistance of such chokes is usually 1-2 kΩ. The output is inductively (LC) coupled to the load for three reasons.

To increase gain, since the choke impedance is much higher than any resistance we would usually use there. The load line is almost horizontal (very high impedance), so current swings are minimized, and the distortion is reduced. Demands on the anode voltage supply must be considered.

Due to DC coupling between the two stages, the cathode of VT25 is at a high DC potential of 130-150 Volts. Adding the required anode-to-cathode operating voltage of say 250V means the anodes are at 380-400 V_{DC}. With say 20 mA of anode current through DHT, the voltage drop on a 22k anode resistor would be 440V, meaning our anode power supply would need to produce, say 380+440 = 820 Volts!

Such high voltage would be available if we were driving high voltage tubes such as 211, 845, or GM70, which would use 900-1,100V power supplies. Still, other output tubes operating at 300-400V would necessitate a separate high voltage power supply, with filtering elcos in series, bypassed by equalizing resistors. This would complicate things and increase the output impedance of such power supplies.

Since the DC resistance of anode chokes is around 2kW, or ten times lower, the voltage drop due to the 20mA anode current is also ten times lower, or around 40V, which keeps the HV supply under the 450V limit (common voltage rating of a single elco)!

TUBE PROFILE: ECC88 (6DJ8)

- Indirectly-heated duo-triode
- Noval socket, 6.3V, 365 mA heater
- Maximum anode voltage 130V_{DC}
- V_{HKMAX}=150V_{DC}
- P_{AMAX}=1.8 W, I_{AMAX}=25 mA
- TYPICAL OPERATION:
- V_A=90V, V_G=-1.3V, I_0=15 mA
- gm = 12.5 mA/V, μ=33, r_I = 2.6 kΩ

KEY FEATURES:

- DC-coupling between stages
- LC output coupling
- Only one coupling capacitor in the signal path
- Low distortion

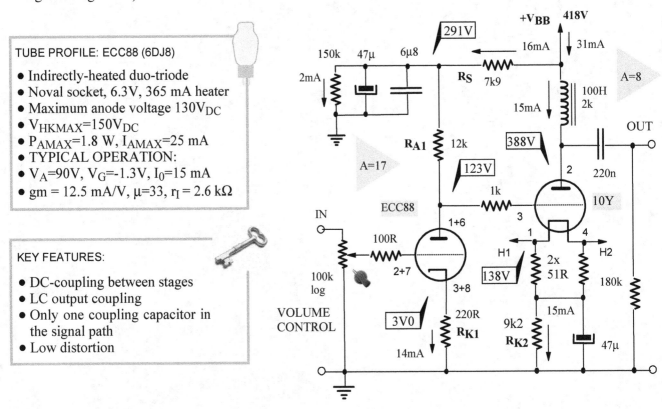

Design sequence

The values of the components on the schematic above have been determined through a relatively simple process that we will outline in a moment. Once we have decided on the topology, the aim is to determine the values of the four resistors (in bold) and the anode supply voltage +V_{BB}. All other values are chosen based on previous experience and are not critical.

This procedure assumes that we can have any supply voltage we need. That is usually not the case in practice, we have a specific power transformer that provides a certain secondary AC voltage, and our options are limited. However, we still have a few choices regarding the power supply, what rectifier tube to use (voltage drop across it will vary with its type), LC or CLC filtering, DCR of the filtering choke, etc.

THE FRONT-END: SUPERIOR INPUT & DRIVER STAGES

Those choices can take the final value of the DC supply voltage up or down, allowing the designer a certain degree of freedom. ltimately, if the required V_{BB} is too high, we can go back to our design sequence, reduce the anode currents in one or both stages, and recalculate the voltages in significant points (those marked on the schematic).

STEP 1: Positioning the quiescent point Q1 of the 1st stage

Since we are using two ECC88 triodes in parallel, the current through this stage should be between 10 and 20mA (5-10mA per triode). Higher current stages sound better. The anode resistance should be between 10 and 33k. However, the higher the current and the resistance, the higher the voltage drop on this anode resistor, so don't go crazy here. Let's choose 14mA and 12kΩ for the anode resistor. Now draw the load line AB and position its middle (Q1-point) in a reasonable spot, such as $V_G=-3V$.

The input voltage swing is from 0V to -6V or 6V in total. We can read the anode voltage swing by drawing vertical lines down to the VA axis, from 35V to 195.

Since the positive voltage swing $\Delta V_+ = 85V$ and the negative voltage swing $\Delta V_- = 75V$ aren't equal, harmonic distortion will be present. The second harmonic distortion is $D_2 = (\Delta V_+ - \Delta V_-)/2\Delta V_A = 10/320 = 3.1\%$, not a great result, but not bad for such a large voltage swing either.

The 1st stage operating parameters are $R_{A1}=12k$, $V_0=120V$, $I_0=14mA$, and $P_{IN}= I_0 V_0 = 1.68$ W.

With maximum anode dissipation (both triodes together) of 3.6 W, that is very conservative, resulting in long and stress-free tube life.

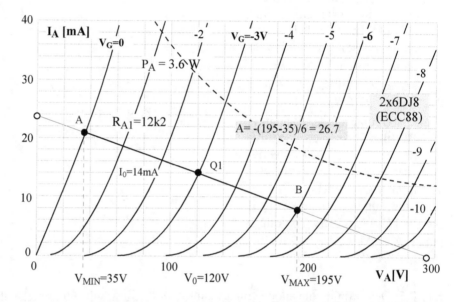

The input voltage swing is DVG = 6V, and with the output voltage swing of $\Delta V_A = 195-35 = 160V$, the voltage gain is $A_1 = -\Delta VA/\Delta VG = -160/6 = -26.7$ or 28.5 dB. However, this formula assumes that the cathode resistor is fully bypassed for all audio frequencies by a capacitor; ours isn't. We will have mild local negative feedback here, and the gain will be lower. We'll come back to that later.

STEP 2: Determining the DC voltages and cathode resistance of the 2nd stage

Now we know that the grid must be -3V with respect to the cathode, and since the grid is at ground potential (zero), the cathode voltage must be +3.0V. We know the current (14 mA) and the voltage drop across the cathode resistor R_{K1}, so its resistance must be $R_{K1}=V_{K1}/I_{K1} = 3/0.014 = 214$ Ω, so we can use a standard value of 220 Ω!

From the anode characteristics, we know that anode-to-cathode voltage must be $V_0=120V$, and since the cathode is at 3V with reference to ground, the anode must be at 123V. An Anode current of 14mA will create a voltage drop of 14*12 = 168V on R_{A1}, so the supply voltage for the 1st stage must be 123+168 = 291V!

STEP 3: Determining the quiescent point Q2 of the 2nd stage.

Looking at the anode characteristics of 10Y triode (next page), we choose a moderate current of 15 mA. Again, don't go too high with anode voltage since that will increase the supply voltage significantly. Let's stay at around $V_0=250V$.

The impedance of the 100H anode choke at 1kHz is 628k ($X_L= \omega L = 2\pi f L$), which the signal sees in parallel to the 180k grid resistor of the output tube. The impedance of the 220n coupling capacitor is only 723Ω at 1 kHz; such a small impedance compared to 180k and 628k can be considered a short circuit.

Since the voltage across the anode choke is 90o ahead of the voltage on the 180k resistor, the two impedances should not be added algebraically as simply being in parallel but should be added using Pythagora's theorem: $Z=\sqrt{(Z_{L2}+R^2)}$ = √(658² + 180²) = 653kΩ!

On IA-VA characteristics, such a high impedance is almost a horizontal line! We positioned V_0 at 250V, so from the graph, we read the required grid bias in Q2 (intersection with $I_A=15mA$), which is around -15V, meaning the cathode must be at grid potential +15V!

STEP 4: Determining the DC voltages and cathode resistance of the 2nd stage

The grid of the 2nd stage is at the same DC voltage as the anode of the 1st stage, and that is 123V, so the cathode of the 10Y tube must be at 123+15 = 138V! Now we can calculate the required cathode resistor: $R_{K2}=V_{K2}/I_{K2}$ = 138/0.015 = 9.2kΩ, perfect!

However, 10Y is a directly-heated triode, and there is a DC heating voltage of 7.5V between points H1 and H2. We don't know exactly how that voltage will be distributed, but we can assume that H1 will be approximately 3.75V above point X, so the voltage in point X needs to be 3.75V lower, or 138-3.8=134 V and $R_{K2}=V_{K2}/I_{K2}$ = 134/0.015 =8,947 Ω, or a standard value of 8k9!

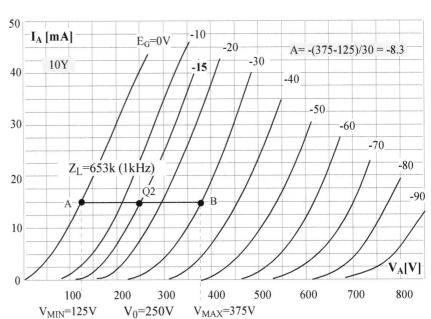

Once the amp is made and energized, we can fine-tune that voltage by changing the cathode resistor or by installing a 9k2 resistor and then, if needed, adding a suitable resistor in parallel to reduce the resistance by 300Ω.

2nd stage: V_0=250V, I_0 =15mA, P_{IN}= $I_0 V_0$ = 3.75W

With maximum anode dissipation of 10W, that is very conservative for 10Y/VT25, so these lovelies will live deep into their old age. The input voltage swing is ΔV_G = 30V, the output voltage swing is ΔV_A = 375-125= 250V, so voltage gain is A= $-\Delta V_A/\Delta V_G$ = -250/30 = -8.3 or 18.4 dB.

STEP 5: Determining the anode voltage supply of the 2nd stage and the decoupling resistor R_S

Since the cathode is at 138V and the anode must be 250V above the cathode, the anode voltage is 138+250 = 388V. 15mA current on a 2kΩ choke will cause a 30V voltage drop, so we finally know that the anode power supply must provide 388+30 = 418V_{DC}! This is a bit too close to the 450V limit of common electrolytic capacitors, but we can use 500V_{DC} units or, even better, use 630V_{DC} rated film capacitors instead.

We have a bleeder resistor of 150k to ground, so 291V across it will cause a current of I_3= 291/150 = 2 mA to flow. That current plus the 1st stage current I_1=14mA flow through R_S, so R_S = (418-291)/(2+14) = 127/16 = 7k9!

Voltage gain and output impedance of the 1st stage

To estimate the voltage gain of the first stage and its output impedance, we need to consult the graph that links three basic parameters (μ, gm, and r_I) for the ECC88 tube.

We have 7mA through each triode. Once we mark that operating point Q on the graph, we get r_I=4.1kΩ and μ=30! With two triodes in parallel, μ stays the same, 30, but gm is doubled, and the internal impedance is halved, so r_I=2.05kΩ.

The voltage gain is **A_1=−μR_L/[r_I+R_L+(1+μ)R_K]**

A_1= -30*12/(2.05+12+31*0.22) = -30*0.575 = -17.25

The local NFB caused by the un-bypassed cathode resistor has reduced the gain of the 1st stage from -26.7 to -17.25 or from 28.5 to 24.7 dB, a reduction of -3.7dB.

Z_{OUT1} = R_L[r_I + (1+μ)R_K]/[R_L+ r_I + (1+μ)R_K] = 12*(2.05+31*0.22)/(12+2.05+31*0.22) = 12*8.87/20.87 = 12*0.425 = 5k1. The output impedance of the 1st stage is approximately 42% of its anode resistance (12k).

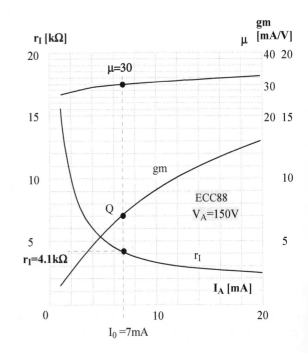

THE FRONT-END: SUPERIOR INPUT & DRIVER STAGES

6N30P + VT-25 driver

6N30P is a Russian duo-triode with original name 6H30π in Cyrillic alphabet. There are a few varieties that all seem to have identical electrical parameters, but have slightly different internal construction, such as 6N30P-DR and 6N30P-I. The amplification factor is low (μ=15), the transconductance very high (gm=18mA/V), so the internal resistance is low, around 800Ω. Anode currents of up to 40mA are possible. Some commercial manufacturers are full of praise for this tube, even naming it a "supertube", while most have been ignoring it completely.

Instead of paralleling two ECC88 to get around 2k internal impedance we could use a single 6N30P with 0.8k internal impedance. However, the gain would be halved, but we could bypass the cathode resistor and all would be well again! One physical tube can be shared between two channels of a stereo amp.

With an 18k load and 15mA anode current the quiescent anode-to-cathode voltage is low, only 115V, but the bias is higher than for ECC88, -6V. A 10k anode resistor would be a low load and take the voltage swing down towards the nonlinear part of the curves.

The voltage gain of this stage is around 14 times, just over half of the gain achievable with two ECC88 in parallel. The distortion situation? Since $\Delta V_+ = 89V$ and $\Delta V_- = 80V$, $D_2 = (\Delta V_+ - \Delta V_-)/2\Delta V_A = 9/338 = 2.7\%$, slightly better than with two ECC88 in parallel.

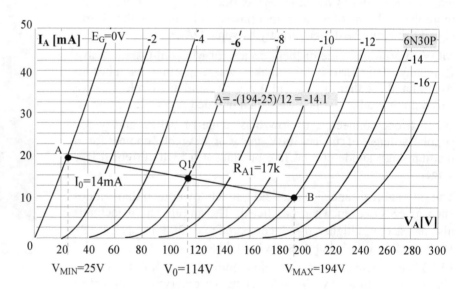

The cathode resistor must be $R_{K1} = V_{K1}/I_{K1} = 6/0.015 = 400\ \Omega$, so we can use a standard value of 390 Ω.

From the anode characteristics we know that anode-to-cathode voltage must be $V_0=115V$, and since the cathode is at 6V with reference to ground, the anode must be at 121V.

The anode current of 15mA will create a voltage drop of 15*17 = 255V on R_{A1}, so the supply voltage for the 1st stage must be 115+255 = 370V.

Let's keep the quiescent point Q_2 of the 2nd stage the same as in the previous case, $V_0=250V$, $I_0=15mA$ and $V_G=-15V$.

The cathode must be 15V above the grid at 115V, so $V_K=130V$ and $V_A=130+250 = 380V$, almost the same as before (388V).

The current through the bleeder resistor is now 370V/150k = 2.5mA and the total current through R_S is 17.5mA.

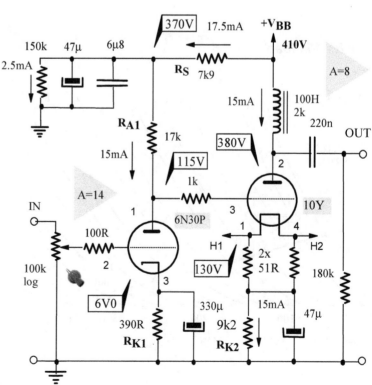

TUBE PROFILE: 6N30Pi

- High perveance indirectly-heated duo- triode
- Noval socket, heater: 6.3V/800mA
- V_{AMAX}= 250V, P_{AMAX}= 4 W
- I_{AMAX}= 40mA
- gm=18mA/V, µ=15, r_I = 830Ω

Now there is only 410-370 = 40V across R_S, so R_S = 40/17.5 = 2.28k, so a standard value of 2k2 should be used.

With 330µF cathode bypass capacitor, the 1st stage time constant is $\tau=R_K C_K$ = 0.1287 seconds, which means the lower -3dB frequency will be f_L = 1/(2πτ) = 1.23 Hz!

Since R_K of the 2nd stage has a high value (almost 10k), even a 22µF bypass cap will ensure a low f_L of that stage.

KT61 (KT81) - EL33 (CL33) - 6P25 driver or line stages

First produced in 1939, KT61 was a high sensitivity audio output tetrode ("KT" stands for "kinkless tetrode"), capable of producing more than 4 Watts of audio output power with only 3 volts of grid signal - it could be driven directly by some DACs!

MOV imprint on some of them stands for Marconi-Osram Valve Company. This British tube maker was exporting its tubes to Australia, those that local brand AWV (Australian Wireless Valve Co) did not produce, which were then branded AWV Radiotron.

When triode-connected, its low bias range of 0-10V and anode supply voltage of under 300V make it a viable choice for a medium-high anode current line preamp or a high current driver stage.

The internal resistance is under 3kΩ, much lower than duo-triodes such as 6SN7, even when paralleled. The amplification factor is also at least 50% higher, µ=33, compared to µ=20 for 6SN7. KT81 is identical to KT61, but it uses a Loktal base.

EL33 is not identical but a very close equivalent of KT61. There is also CL33, with a 33V, 200mA heater, but otherwise identical to EL33.

Top view of Mullard CL33. Notice the elliptically-shaped anode.

EL33 with 15k load

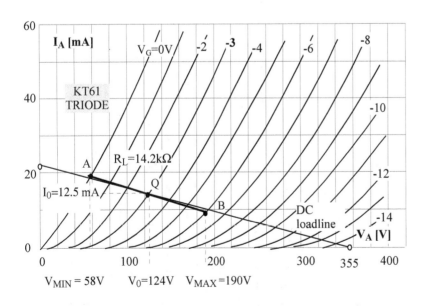

TUBE PROFILE: KT61

- Power audio tetrode
- Octal socket, 6.3V/1A heater
- P_{SCREEN}= 2W, P_{AMAX}= 10 W
- V_{SMAX}= V_{AMAX}= 275V
- I_{AMAX}= 40mA
- AS A TRIODE: V_{AMAX}=300V
- gm=12mA/V, µ=33, r_I=2.8 kΩ

GRAPHIC ESTIMATION

V_{bias}= -3V, V_0=124V, I_0 =12.5mA
VOLTAGE SWING:
$\Delta V = V_B - V_A$ = 190-58 = 132V
$A_V = -\Delta V_A/V_G$ = -132/6 = - 22
DISTORTION (2nd harmonic):
ΔV_+=66V, ΔV_-=66V
$D_2 = (\Delta V_+ - \Delta V_-)/2\Delta V *100$ =
0/(2*130)*100 = 0 %

The -3V bias is a fortunate value; you can use battery or LED bias. Here we use the standard cathode bias. Instead of one 15k anode resistor, use two 33k or three 47k resistors. For some reason, two or three resistors in parallel sound better than a single resistor of the same equivalent value.

Remember, the AC and DC load line are not identical; thus, our analysis of the gain and distortion is only approximate. The AC load is 15kΩ in parallel with 330kΩ grid leak resistor of the next stage, or 14k2, but that is close enough to 15kΩ for estimation purposes.

With three 47k resistors in parallel, the effective load would be 15.7kΩ in parallel with 330kΩ, or exactly 15k! The gain of 22 times with such a low value of anode resistor (15k) is respectable. With a choke-loading or if interstage transformers are used, the gain will approach 30, so with a 2 V_{RMS} input signal, up to 60 V_{RMS} will be available at the output, enough to drive many directly-heated triodes to full power.

Mazda 6P25

Mazda 6P25 is recognized by its sandy gray coating on the bottom half. While it is pin-compatible with EL33 and KT61, as a triode, it has a much lower mju of around 17 and a slightly lower mutual conductance. It would work well as a line stage (a preamplifier output tube), but as a driver for a power stage, the voltage gain would be around 10, which is too low for all output triodes and even many pentodes.

The metallic outside coating acts as an electrostatic shield connected to pin 1 of the base, which should be connected to the audio ground.

ABOVE: Mazda 6P25 can be recognized by its metallic outside coating that acts as an electrostatic shield

Bel Canto's modified SRPP stage

Bel canto is an Italian phrase for "beautiful singing" or "beautiful song." The Minnesota, USA-based company released their model SET40 amplifier in the late 1990s, with a retail price of US$5200.- + US$295.- for optional tube covers (in 1999).

A spartan black powder-coated chassis, four 12AX7 tubes and two 845 output tubes. The power and output transformers are top-mounted under their enclosures, while the audio section and the power supply are on one large PCB. Bel Canto claimed that PCB-based amplifiers sound better (?!?)

The continuous power output of 37 watts, and peak power output of 70 watts, with -3 dB bandwidth from 6 Hz to 35 kHz, are claimed for Bel Canto SET40. The total gain is specified as A=21 dB, and the input level needed for 37 Watts output is 1.5 VRMS. As an educational exercise, let's check if the two claims match.

How to analyze and scrutinize equipment specification figures

At the rated power level of 37 Watts the output voltage on 8Ω speaker terminals must be $V_{OUT}=\sqrt{(P_{OUT}*Z_L)}=\sqrt{(37*8)} = \sqrt{296} = 17.2$ V_{RMS}. Now we can calculate the overall voltage gain of the whole amplifier as A=17.2/1.5=11.5, or in decibels A = 20log11.5= 21dB, just as claimed.

The signal-to-noise ratio (SNR) at 1 Watt output level is specified as 96 dB (A-weighted). Since SNR [dB]= 20log (V_S/V_N), and 1Watt on 8Ω requires a voltage signal of V_S= 2.83V, this means the noise voltage at the 8Ω speaker terminals would be $V_N=V_S/(invlog(SNR/20)) = 2.83/63,096 = 0.045$ mV. If true, that is an incredibly low noise level for a tube amplifier!

Likewise, the harmonic distortion at 1 Watt output level (using a 1 kHz test signal) is under 0.1%, which is an excellent result for an amplifier without global NFB. However, the damping factor is low (DF = 3), in the ballpark typical of such designs.

The "modified SRPP" input/driver stage

The single input/driver stage is unusual. SET40 owner's manual says, "The philosophy behind the design of the SET 40 is based on the observation that simplicity of the circuit and the reduction of the number of active devices in an audio amplifier will permit the greatest quantity of information to get through the amplifier." Sure, there is only a single driver stage in SET40, but has the number of active devices really been reduced? With four triodes in this "modified SRPP" stage, it hasn't.

The promotional blurb continues, "This chameleon-like performance is achieved through an elegantly simple circuit." A simple circuit? Again, a driver stage with four triodes cannot be called simple.

Bel Canto gives us the essential parameters in their circuit description: 30 dB gain (31.6 times), 1kΩ output impedance, and over $100V_{RMS}$ of output voltage swing. Superior PSRR (Power Supply Rejection Ratio) is claimed, making the stage less sensitive to power supply variations and modulation.

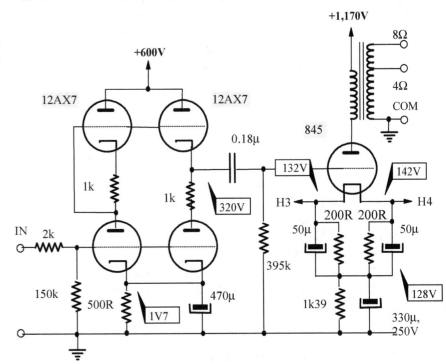

RIGHT: The in-principle circuit of Bel Canto SE40 amplifier. The actual component values used may vary from those marked.

Ordinary 12AX7 SRPP stage

Since Bel Canto claims that this "modified SRPP" has superior properties compared to the ordinary SRPP circuit, we started our experiment there, using a 12AX7 SRPP stage as a benchmark.

Since 12AX7 is a very low bias triode, it can only take low input signals, generally under $1V_{RMS}$. With 100mV at the input, the gain of the SRPP stage was 70, but as the input signals rose, the stage gain dropped, so with 1V input, the gain was only 35!

Furthermore, look at what happened to the frequency range. With 100mV input, the -3dB bandwidth was modest, 15 Hz to 51 kHz. With a $1V_{RMS}$ signal, the bandwidth narrowed to 163 Hz - 20 kHz, which is unacceptable for hi-fi use.

In the 2nd part of this illuminating experiment, using the same tubes made by Philips Australia, on the same testing platform, using the same regulated $400V_{DC}$ supply voltage, we expanded the ordinary 12AX7 SRPP into a "Bel Canto-style" circuit.

ABOVE: 12AX7 SRPP stage

MEASURED RESULTS:

V_{IN} [Volts AC]	V_{OUT} [Volts AC]	A (Gain)	BANDWIDTH [-3dB]
0.1	7	70	15 Hz - 51 kHz
1.0	35	35	163 Hz - 20 kHz

Bel Canto - style 12AX7 SRPP stage

For small input signals, the gain was 70, the same as that of the ordinary SRPP stage. At higher input levels, the gain was reduced, although not as much as before. The input voltage range was wider, up to $1.5V_{RMS}$, after which the drop in gain became significant (only 55 with a 2V input).

The frequency range also improved compared to the ordinary SRPP circuit, but with 1V input, the -3dB bandwith was barely acceptable 25 Hz - 40 kHz.

The output impedance was relatively high, around 12kΩ! We did not measure other parameters like the PSRR (Power Supply Reject ratio) or harmonic distortion. One rainy day, when you feel inclined to do so, by all means, conduct those further tests yourself.

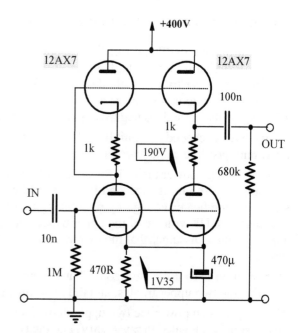

ABOVE: Bel Canto-style SRPP with 12AX7

MEASURED RESULTS:

V_{IN} [V_{AC}]	V_{OUT} [V_{AC}]	A (Gain)	BANDWIDTH [-3dB]
0.1	7	70	17 Hz - 115 kHz
0.5	35	70	17 Hz - 88 kHz
1.0	69	69	25 Hz - 40 kHz
1.5	95	63	30 Hz - 35 kHz
2.0	110	55	30 Hz - 35 kHz

"Suono più bello" driver stage

We have been raving about triodes such as E86C and ECC40 throughout Volumes 1 and 2 of this book, so it was only natural that in the next step of this experiment, we check how would ECC40 perform in Bel Canto style circuit.

Unfortunately, E86C is a single triode, so four would be needed per channel for this circuit, which would complicate construction (eight sockets) and violate the Simplicity Rule, so we'll proceed with ECC40 lovelies.

Since ECC40 is a higher bias tube, it can take much higher input signals than 12AX7 (just like 6SN7 and 12AU7), we changed the values of cathode and anode resistors to suit, but the circuit remained the same.

One glance at the tabulated test results is enough to conclude:

1. The gain (around 25) did not drop with the increasing amplitude of the input voltage.

2. The frequency range was incredibly wide, 6Hz - 760 kHz, and narrowed slightly with increasing signals. Even with 2V at the input, it was 6Hz - 175 kHz. Compare that to 30 Hz - 35 kHz with 12AX7! And the output impedance? It was ten times lower, only 1.2kΩ!

Wait, you will say, but the gain is almost three times lower than with ECC83?! Sure, this circuit cannot be used to directly drive transmitting tubes such as 845, but it can drive all pentodes and many power triodes. Try it with lovelies such as EL156, EL153, LS50, F2a or EL12 - bliss!

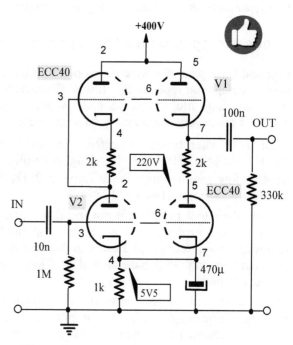

ABOVE: Bel Canto-style SRPP stage with ECC40

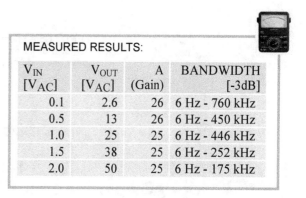

MEASURED RESULTS:

V_{IN} [V_{AC}]	V_{OUT} [V_{AC}]	A (Gain)	BANDWIDTH [-3dB]
0.1	2.6	26	6 Hz - 760 kHz
0.5	13	26	6 Hz - 450 kHz
1.0	25	25	6 Hz - 446 kHz
1.5	38	25	6 Hz - 252 kHz
2.0	50	25	6 Hz - 175 kHz

The maximum heater-cathode voltage issue in totem-pole circuits such as SRPP

In both ordinary SRPP and Bel Canto-style SRPP stages the cathodes of the upper triodes are at an elevated DC potential, typically around half of the anode supply voltage. In the last SRPP circuit with ACC40, the voltage between the upper triodes' cathodes and the heaters (assuming they are at ground or zero DC potential) is approx. $220V_{DC}$! The maximum heater-cathode voltage for ECC40 if the cathode is positive with respect to the heater is $175V_{DC}$. If the cathode is negative with respect to the heater, the allowed DC voltage is only -100V!

So, we have exceeded the limit by $45V_{DC}$ or around 25%! Most other duo-triodes (except 12AX7) have 100V as the maximum positive cathode voltage allowed, which is much less than the limit for ECC40.

Some designers power their SRPP stages with up to $250V_{DC}$, since most quality tubes will tolerate up to $125V_{DC}$ on the cathodes (the limit exceeded by 25%). Others elevate the DC voltage on the secondary transformer winding supplying the heater voltage by referencing them to a voltage divider usually powered from the HV supply. This decreases the HK voltage of the upper triode and increases the HK voltage of the lower triode by the same amount. This, of course, assumes that the two triodes of such vertical or "totem pole" arrangement are the two halves of one physical tube.

One solution is to use a "horizontal" arrangement of triodes instead of the "vertical" one. One physical tube (V1) contains the two upper triodes, and the other tube (V2) contains the two lower triodes.

That way we can power the two upper duo-triodes (one for each channel of a stereo amp) from their own secondary heater supply, elevated to say +200V, and the two lower duo-triodes would have their own 6.3V secondary winding simply grounded at one end or at the CT if available. That is the best solution, but the price to pay is the need to have two separate heater circuits for the preamp tubes.

Triode-strapped C3m and C3g driver stages

C3g and C3m pentodes were developed for German Post and used in telephone repeater amplifiers. Characterized by high reliability, long life, low noise, and low microphonics, these pentodes make very linear and good sounding triodes.

With µ=19 when triode connected, C3m and C3o are in the 12AU7 and 6SN7 class; C3g has a higher amplification factor (µ=40), low internal impedance (around 2kΩ) but a much higher mutual conductance of 17 mA/V.

Despite their looks, these are not metal but glass tubes; the black metal housing is just an external shield that can be removed. Even the loktal base can be removed and pins soldered onto an octal base should you have the time and patience to do so.

Triode connected, they are sensitive and powerful enough to be used in single-stage headphone amplifiers. Plus, their low output impedance makes them suitable as drivers for Class A_2 triode output stages.

TUBE PROFILE: C3g

- Pentode for wideband amplifiers
- Loktal socket, heater: 6.3V, 370 mA
- $V_{AMAX}=250V_{DC}$, $V_{SMAX}=220V_{DC}$
- $P_{AMAX}=3.5W$, $P_{SMAX}=0.7$ W
- $I_{KMAX}= 30$ mA, $V_{HKMAX}=120V$
- TYPICAL OPERATION (triode):
- $V_A=V_S=200V$, $R_K=180\Omega$, $I_0=17mA$
- gm=17 mA/V, µ=40, $r_I = 2.3$ kΩ

TUBE PROFILE: C3m - C3o

- Indirectly-heated pentode
- Loctal socket
- Heater C3m: 20V/125mA
- Heater C3o: 6.3V/400mA
- $V_{AMAX}= V_{G2MAX}=300V$,
- $P_{AMAX}= 4$ W, $P_{G2MAX}= 1$ W
- As a triode:
- gm=7 mA/V, µ=19, r_I = 2k7

LEFT: C3m pentode after the removal of its metal shield, Loctal base and padding rings.

THE FRONT-END: SUPERIOR INPUT & DRIVER STAGES

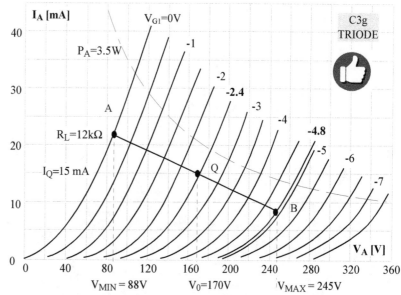

GRAPHIC ESTIMATION

V_{bias}= -2.4V, V_0=170V, I_0 =15mA
VOLTAGE SWING:
ΔV_G = 4.8V, ΔV_A=245-88 = 157V
A_V=-ΔV_A/ΔV_G= -157/4.8= -32.7
DISTORTION (2nd harmonic):
ΔV_+=82V, ΔV_-=75V
$D_2 = (\Delta V_+ - \Delta V_-)/2\Delta V *100 =$
7/314 = 0.77 %

BELOW: C3m pentode in triode conenction and with battery bias. It can be used as a driver stage in power amps or as a line stage preamplifier.

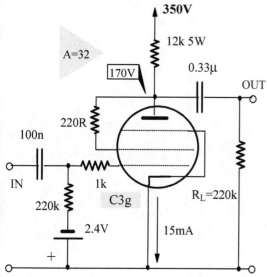

One advantage of battery bias is that it is stable and not affected by the operating conditions of the stage. An even more significant sonic benefit is that the unfortunate cathode bypass capacitor (electrolytic!) has been eliminated. There is no more time constant with cathode resistance and the associated attenuation of low frequencies.

However, the input coupling capacitor must be added; otherwise, the bias battery would discharge through the volume control pot or the output impedance of the previous stage or CD player, whatever is driving this stage. However, that is a film capacitor, and its sonic signature isn't nearly as bad as that of the cathode bypass electrolytic capacitor.

The output impedance is $Z_{OUT} = R_A \| R_L \| r_I$ = 12k||220k||2k7 = 1k9, not a bad result at all!

C3m common cathode stage

Let's estimate the voltage gain and 2nd harmonic distortion from the anode characteristics of a triode-strapped C3m pentode. The distortion is lower than in the C3g case, but the voltage gain is also halved, down to A=15.

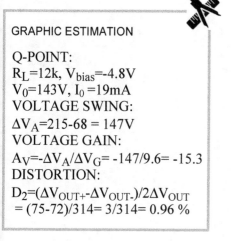

GRAPHIC ESTIMATION

Q-POINT:
R_L=12k, V_{bias}=-4.8V
V_0=143V, I_0 =19mA
VOLTAGE SWING:
ΔV_A=215-68 = 147V
VOLTAGE GAIN:
A_V=-ΔV_A/ΔV_G= -147/9.6= -15.3
DISTORTION:
D_2=(ΔV_{OUT+}-ΔV_{OUT-})/2ΔV_{OUT}
= (75-72)/314= 3/314= 0.96 %

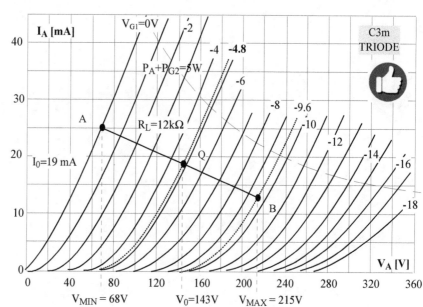

FROM SHOCKING TO SUBLIME: LESSONS FROM COMMERCIAL LINE STAGES

- Audio Research Corporation LS7 and LS8
- BEZ Q6A μ-follower line stage
- BEZ Q3B-1 transformer-coupled line stage
- Opera Consonance Cyber 222 line stage
- Sheer Audio DHT-01A line stage
- Yarland PM7 line preamplifier
- Xiang Sheng 708B headphone amplifier & line stage
- Mingda MC-7R line stage
- Dared SL2000-A line preamplifier

"People admire complexity. They don't trust something that looks too simple. But only the simple ideas will work. The more powerful ideas have an elegant simplicity about them. Less is more."
Al Ries and Jack Trout, in their book "Horse Sense: The Key to Success Is Finding a Horse to Ride"

Audio Research Corporation LS7 and LS8

Both LS7 and LS8 (MkI and MkII) are relatively simple, unbalanced (single-ended) line stages using four 6DJ8 (6922, ECC88) duo-triodes. According to the manufacturer's specifications, LS8 provides 12dB of voltage gain. The frequency range is specified as 0.5Hz to 100kHz (±0.5dB) or 0.1Hz to 250kHz (-3dB).

The maximum allowed input voltage is 3.5V, while the output impedance is 200Ω. Since its circuit is somewhat simpler, let's start with the later model, LS8 MkI.

To gain a global view of a particular design, draw its block diagram. This may seem a waste of time to more experienced, but it will provide a broader perspective, which often gets lost in the sea of detail.

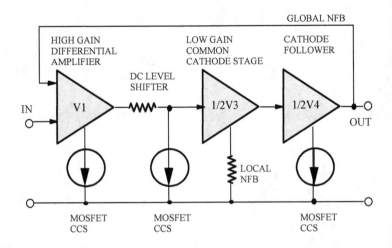

LS8 MkI: The block diagram of the audio section

DC analysis

The output of the first stage passes through a high resistance of 76.3 kΩ, which acts as a DC level shifter. The CCS (constant current sink) with an enhanced MOSFET transistor makes sure the current to ground is constant, and since the grid of the second stage does now draw any current, the DC voltage drop on that series resistor is also constant.

Thus, the DC voltage is dropped from +55V at the output of the 1st stage to +13V at the input to the 2nd stage. That is why the anode resistors on the differential amplifier are unequal. The right one (47k5) passes a lower current of 2.2mA, the one on the left (lower resistance of 37k4) passes a higher current (2.8mA) which is 2.2mA plus 0.6mA that gets diverted through the CCS. That ensures that both anodes (pin 1 and 6) are at the same DC potential of 55V!

The first stage is a differential amplifier used in a single-ended mode (unbalanced input and unbalanced output), so the second triode (pins 6,7,8) acts only as an input for global negative feedback signal through the resistive β network. The circuit is a simple voltage divider. The lower that ratio, the weaker the feedback. This feedback ratio is $\beta = R_2/(R_1+R_2) = 19k6/(2 \times 35k7 + 19k6) = 0.215$ or 21.5%. This is very strong feedback!

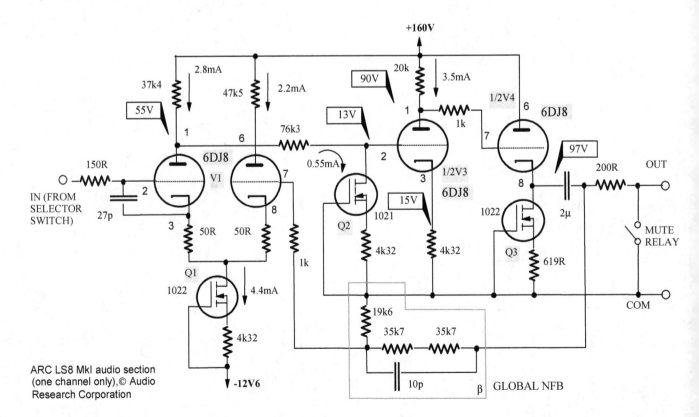

ARC LS8 MkI audio section (one channel only), © Audio Research Corporation

Graphical analysis

Instead of a simple resistor, the 1st stage uses a CCS in the cathode circuit, powered from the source of negative DC voltage (-12.6V). This means the load line of that stage (A-B) is practically horizontal, since the dynamic resistance of the CCS is very high, while its DC resistance is very low (kind of like an electronic choke). The distortion of the 1st stage is very low since the positive (Q1-A) and negative (Q1-B) voltage swings are approximately equal (55-10=45V, 100-55=45V).

The cathode follower at the output also has a MOSFET CCS in its cathode circuit.

Stage 1 parameters are V_0=55V, I_0 =2.8mA, input voltage swing: ΔV_G = 3V, output voltage swing: ΔV_A = 94-14= 80V, so the voltage gain is A= $-\Delta V_A/\Delta V_G$ = -80/3 = -26.7. The graph below is imprecise; the tube's µ in this operating point is around 26, so let's assume the gain of -24.

The gain of a differential amp (for the differential signal) is the same as for the common cathode stage, assuming a differential output is used. However, ARC did not use the differential output (from both anodes) but only from one. The gain of a differential amplifier with a single-ended output is thus halved to -12.

Stage 2 parameters are V_0=75V, I_0 =3.5mA, input voltage swing: ΔV_G = 4V, output voltage swing: ΔV_A = 122-22= 100V, so the voltage gain is A= $-\Delta V_A/\Delta V_G$ = -100/4 = -25.

However, that result is only valid for a common cathode amplifier with a bypassed cathode resistor, which is not the case here. We need to calculate the local NFB.

The rule-of-thumb formula for voltage gain of a CC stage with local current NFB due to the un-bypassed cathode resistor is

$A_F = A/(1+A*R_K/R_A)$,

where R_K is the cathode resistance, R_A is the anode load and A is the gain when R_K is bypassed (without NFB). Now we have A_F= -25/(1+25*4.32/20) = -25/6.4 = -3.9!

Due to DC-coupling R_K is high (4k32), creating a strong local negative feedback - that is why the gain of the 2nd stage is so low.

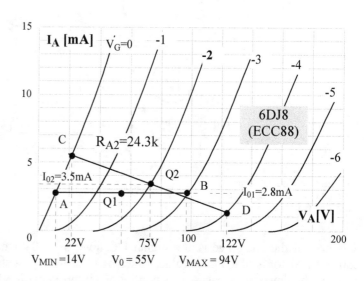

ARC LS8 MkI: operating points of the two amplifying stages

VOLTAGE GAIN OF CC STAGE WITH UN-BYPASSED CATHODE RESISTOR

$A_F=A/(1+A*R_K/R_A)$
R_K is the cathode resistance, R_A is the anode load and A is the gain when R_K is bypassed (gain without NFB).

Alternative analysis using small signal models

The gain of a differential amp (for the differential signal) is A=$-\mu R_A/[r_I + R_A]$ = -26*37/(8+37) = -21, and the single-ended topology reduces that gain to a half or -10.5 times!

The 2nd stage is common cathode amplifier with un-bypassed cathode resistor, whose gain is A_A= $-\mu R_L/[r_I+R_L+(1+\mu)R_K]$ = -26*20/[7.5+20+(1+26)4.3] = -28*0.13 = -3.6

So, after two inverting stages, the output of the 2nd stage is also DC-coupled to the input of the final stage, a cathode follower, which is non-inverting, making the whole preamp non-inverting.

Estimating the gain of cathode follower to be around 0.8, the overall gain without feedback is approximately A_0=A1*A2*A3 = (-10.5)*(-3.6)*0.8 = 30 or 20*log30 = 29.5 dB.

ARC specifications say, "Input sensitivity: 500mV for a rated output of 2V", meaning the voltage gain of 4 times or 12dB, while the "maximum" gain is specified at 12.2 dB.

The overall gain (with feedback) is $A_F=V_{OUT}/V_{IN}$= $A/(1+\beta A)$, and since we know β=0.215, we can calculate the gain without feedback, or A= $A_F/(1-\beta A_F)$ = 4/(1-0.215*4) = 29! So, our initial estimation of the overall gain is very close to reality.

LS8 MkII dispenses with the DC level shifter and the associated constant current sink. As a result, the two anode resistors in the first stage are identical.

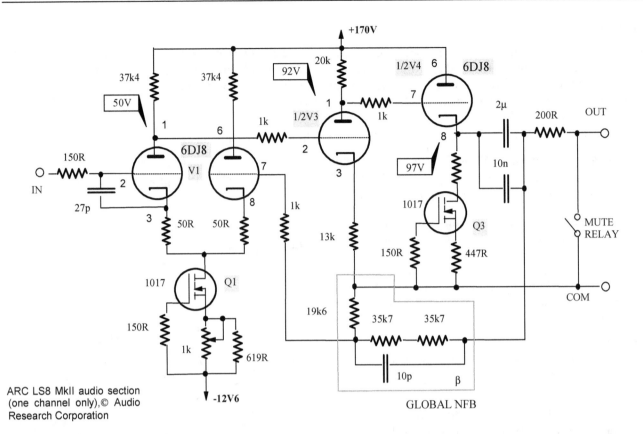

ARC LS8 MkII audio section (one channel only), © Audio Research Corporation

The grid of the 2nd stage is at a higher DC level of +50V (the same as the 1st stage anode), and the 2nd stage's cathode resistor is of a higher value, 13k. The CCS in the 1st stage is adjustable, always a good idea so that a sweet spot can be tuned in by ear. ARC's continuous improvement efforts are evident!

LS7 (below) uses a simpler, J-FET-based CCS in the first stage and no CCS in the last stage. Notice a very unusual design; the anode of the other triode in the differential amplifier is not connected via its anode resistor to the +225V anode supply but is connected to point X instead. Since it needs to be the same as DC voltage on pin 1 of the first tube, its DC voltage level is much lower, determined by the current through the second and last stage (CF) and the resistive divider (15k - 1k22 - 10k).

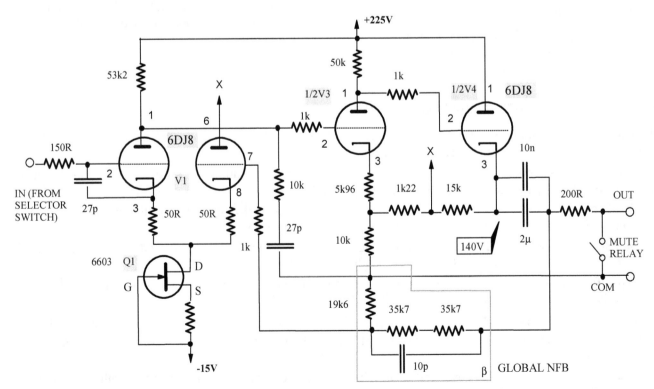

ARC LS7 audio section (one channel only), © Audio Research Corporation

BEZ Q6A µ-follower line stage

An interesting line stage with only two inputs (?) and two paralleled outputs. Assuming you have studied Volume 1 of this book, you should recognize this circuit as a µ-follower. The first stage in this totem-pole (vertical) arrangement is a common cathode amplifier with an un-bypassed cathode resistor, whose load is a cathode follower. In a µ-follower, the same anode current flows through both stages, in this case, a relatively high current of 12.2mA. Both stages use two paralleled 6SN7 triodes, or, more precisely, their copy, 6N8P.

Despite paralleled triodes and the CF, the output impedance is still relatively high, around 2kΩ! This is because the 6SN7 triode has a relatively high internal resistance.

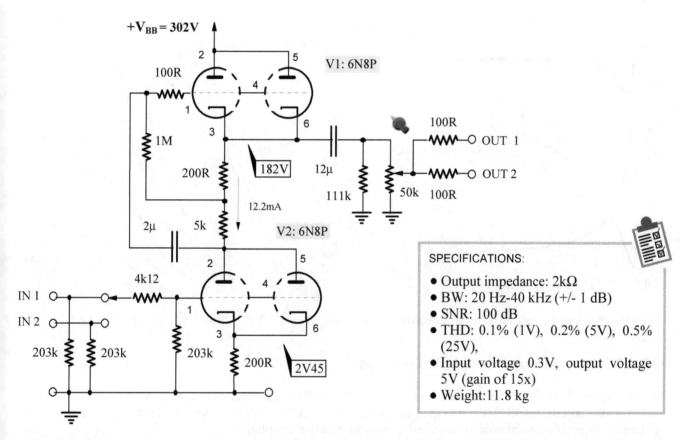

SPECIFICATIONS:
- Output impedance: 2kΩ
- BW: 20 Hz-40 kHz (+/- 1 dB)
- SNR: 100 dB
- THD: 0.1% (1V), 0.2% (5V), 0.5% (25V),
- Input voltage 0.3V, output voltage 5V (gain of 15x)
- Weight:11.8 kg

The second interesting aspect of this design is positioning the volume control potentiometer at the output of the stage instead of the much more common location at its input.

The power supply (not shown) features tube rectification in the anode voltage supply (5Z4P rectifier tube), followed by a CLC filter and a simple shunt voltage regulator with two OD3 VR diodes in series, the output of which (+302V) is taken to each of the two channels through its own 200Ω resistor and 33µF decoupling capacitor to ground.

One 6.3V heater secondary winding feeds two bridge rectifiers, one for each channel, and then each of the four 6SN7 duo-triodes has its own CRC filter (2x1,000µF and 0.15Ω series resistor.

The maximum HK (heater-to-cathode) DC voltage for 6SN7 is 200V (heater negative with respect to cathode), and here the cathodes of the upper triodes are at 182V. The BEZ designer did not reference their heater grounds to a higher DC level despite having four separate DC power supplies where that would be easy to do on two of them, with the other two grounded.

WHAT ABOUT THOSE STRANGE COMPONENT VALUES?

A few commercial amplifier designers & builders, including ARC and BEZ, seem to like specifying exact values of resistors, such as 111 kΩ, 4k12, and 203 k here. Just in case you may be tempted to spend hours trying to find a single resistor or their combination that will give you those nonstandard (read "weird") values, these are not critical at all! Use whatever the closest standard value you have!

There are a few instances where component values are critical for the proper operation of a circuit, and these are usually marked (with an * or #) by the designer or manufacturer on the circuit diagram.

The gain of the cathode follower is $A_K = \mu R_K/[(1+\mu)R_K + r_I]$, and since in the chosen operating point (see the graph below), the internal resistance for two paralleled 6SN7 triodes is $r_I=4k\Omega$ and $\mu=20$, with $R_K=5,200\Omega$ we get a gain of $A_K= 0.92$!

The dynamic load that the bottom tube (V2) "sees" is $r_A=R_L/(1-A_K) = 5.2/(1-0.92) = 5k2*12.5 = 65k$! We can estimate the gain of that tube using a formula for common cathode amplifier with un-bypassed cathode resistor, whose gain is $A_A = -\mu R_L/[r_I+R_L+(1+\mu)R_K] = -20*65/[3.8+65+(1+20)0.2] = -20*0.89 = -17.8$!

The cathode resistor is of very low value (only 200Ω), meaning the local negative feedback is very mild, so the amplification factor has only been reduced from around -20 down to -17.8.

Finally, our estimated overall gain is -17.8*0.92 = -16.3 times, which comes reasonably close to the specified voltage gain of 15x.

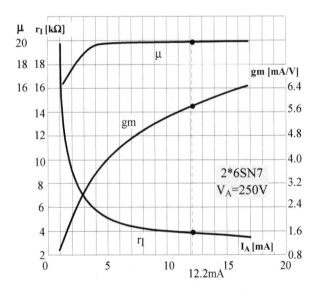
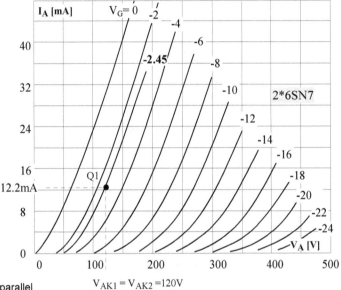

ABOVE: How the three tube parameters vary for two 6SN7 triodes in parallel

$V_{AK1} = V_{AK2} = 120V$

BEZ Q3B-1 transformer-coupled line stage

This line stage offers three unbalanced inputs and switchable RCA (unbalanced) and XLR (balanced) outputs. Priced for sale on eBay at AU$475 + AU$130 postage in Nov. 2015, it represents an excellent value-for-money, considering its point-to-point construction and transformer-output coupling.

The audio section uses one 6DJ8 duo-triode per channel. The first is a CC cathode stage with a bypassed cathode resistor (to maximize gain). The second stage is a cathode follower with its load (the output transformer's 19k primary impedance) in the cathode circuit. Two 600Ω secondaries are terminated by 2.03kΩ resistors.

Different tubes can be substituted, ECC85, 6N11, 6H30, 6N1, 6N6 and 6H23.

The audio circuit

Referring to the circuit diagram on the next page, at 5.5mA ECC88 has an internal impedance of 5kΩ and μ=30 , so the voltage gain of the first stage is $A_1 = -\mu R_L/(r_I+R_L) = -30*7.3/(5+7.3) = -30*0.6 = -18$!

The load $R_L = 7k3$ is the parallel combination of the second stage's 8k35 anode resistor and 56.5k grid leak resistor. Normally, the grid leak resistor is much larger than the anode resistance, so in the 1st approximation, we can disregard its impact on gain, but that is not the case here.

The impedance ratio of the output transformer is IR=19,000/600 = 31.67 and the voltage ratio is VR=5.63. The 2k03 terminating resistance will be reflected to the primary as 2k3*31.67 = 64k3.

SPECIFICATIONS:

- Output impedance: 600Ω
- Output voltage: 2V (3V maximum)
- BW: 30 Hz-100 kHz (+/- 0.5 dB)
- SNR: 100 dB
- THD: under 0.2%
- Gain: 10 dB (3.2x)
- Weight: 7.3 kg

The voltage amplification of the cathode follower is then
$A_K = \mu R_K/[(1+\mu)R_K + r_I] = 30*64.3/[(1+30)64.3 + 4] = 0.966$!

The total voltage gain is a product of three voltage gains, that of the 1st stage (A_1), cathode follower (A_K), and the transformer (A_3). A_3 is a reciprocal of the voltage ratio VR or $A_3 = 1/5.63 = 0.178$!

Finally we have $A = 18*0.966*0.178 = 3.1$! The specified gain is 3.2 times, so our estimation is spot on!

The designer chose to connect the output transformer in the cathode circuit of the 2nd stage instead of its anode. Obviously he didn't need the additional gain that would have provided.

If you remember the material we covered in Volume 1 of this book, a cathode follower is an "electronic" impedance transformer with a voltage ratio of just under 1 (0.8-0.95, depending on the tube and resistance values). Its unique combination of very high input and very low output impedance is not possible with output transformers since they reflect impedance both ways.

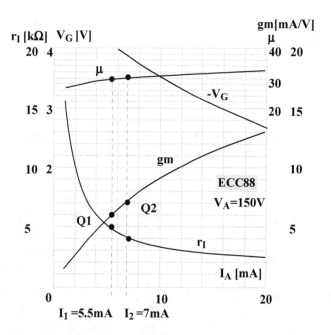

ABOVE: How the three static tube parameters vary for ECC88 triode

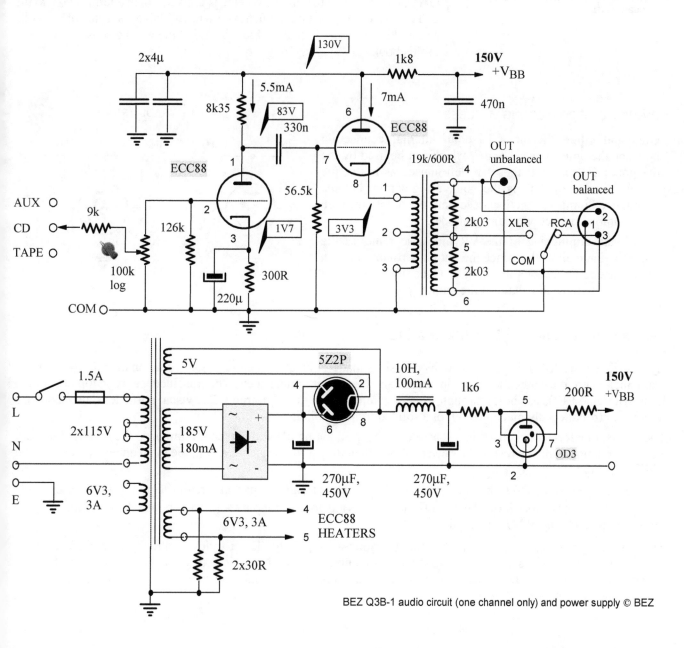

BEZ Q3B-1 audio circuit (one channel only) and power supply © BEZ

Solid-state rectification is used, with a Chinese 5Z2P rectifier tube connected as a series diode before the filtering choke. Other octal rectifiers, both directly- and indirectly- heated, such as 5Z4P, 5Z3P, 5U4, 5Y3, and 5V4, can be substituted.

CLC filter is followed by a simple shunt regulator using the VR150 (OD3) cold cathode gas tube. Audio tubes are heated by raw $6.3V_{AC}$ from one transformer secondary; the other identical 6.3V 3A secondary is unused.

Connecting the unused secondary winding in series with the primary

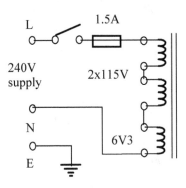

Adding an unused secondary winding in series with the primary to reduce secondary voltages

If you live in a country such as Australia, where the mains voltage is nominally $240V_{AC}$ but is often up to $250V_{AC}$, connect any unused low voltage secondary windings in series with the $230V_{AC}$ primary winding(s). The best would be an unused 12.6V winding, but even 6.3V (as in this case) or even 5V for unused tube rectifier heating would help.

Since there is no high current load on the 6.3V heater winding, its no-load voltage is usually above 7 Volts.

Now we don't have $240V_{AC}$ feeding a primary designed for $230V_{AC}$, but a primary designed for $237V_{AC}$. This will reduce all secondary voltages back closer to their nominal values. This method also works with 100V equipment from Japan or amps designed for 110V working on the actual 115-120V mains voltage in the USA.

Of course, if your actual mains voltage is $250V_{AC}$, then adding $7V_{AC}$ to the $230V_{AC}$ primary won't help much, you will still have a surplus of 13 Volts and will need to rectify that problem in one of the ways discussed in Vol. 2 of this book.

TWO WAYS OF CONNECTING VR TUBES   TRADE TRICKS

OD3 and other VR tubes have an internal connection between the unused pins 3 and 7, which can be used by the designer/constructor. If these are left unused as in (a), the removal of the tube stops the voltage regulation, but the unregulated input voltage is passed onto the output through the common connection at pin 5 (anode).

In (b), the unplugging of the VR tube also removes the output voltage altogether since the output is routed through the connection between pins 3 and 7. Once that connection is broken, there is no voltage at the output (pin 7).

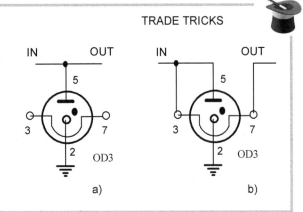

Opera Consonance Cyber 222 line stage

Cyber 222 is a sleek-looking line stage. Its chassis is made of black-coated mild steel, clad on all sides (except the rear panel where the connections are) in natural anodized aluminium sheets. The translucent acrylic top decoration lit up with blue LEDs has holes through which the tops of five tubes protrude. The visual attractions also include a remote control allegedly machined out of a solid aluminium billet.

The mains transformer is housed in a separate el-cheapo-looking black metal case (similar to the one used in Cary SLP-98 preamp). Still, the rest of the power supply, including a filtering choke, is in the main enclosure, next to the audio circuitry.

Cosmetics aside, what about the quality of its audio circuit design? A partial circuit diagram posted on the web, whose accuracy, of course, hasn't been confirmed, shows a µ-follower input stage, DC-coupled to a cathode follower output stage, all using 6SN7 duo-triodes.

The first thing you should notice is the strange choice of topology. The µ-follower already contains a cathode follower (the upper tube's circuit). This assumes the output is taken from pin 6 of the upper triode (top of the 2k resistor). However, here the output is taken from the high impedance output (the anode of the lower triode) and then fed into another cathode follower!

There are two possibilities. Either the designer had a secret reason for this choice and knew something we don't, or he did not understand the first thing about µ-follower operation.

The upper tube in the CF (pins 5,4 & 6) is the active element, the lower triode is a simple constant current sink.

Speaking of currents, the other major issue that needs mentioning is that both stages were designed to operate at very low (for 6SN7 triode) DC currents, only 2mA for the 1st and 1.5mA for the output stage. This is wrong from two aspects.

At such low currents, the internal resistance of 6SN7 is very high, and its amplification factor is low.

Secondly, and more importantly, class A, single-ended preamplifier (and power) stages sound much better when biased to work with as high DC currents as possible (respecting the tube's anode power dissipation limits). Stages with low currents sound flat, lifeless, and "constipated" in comparison.

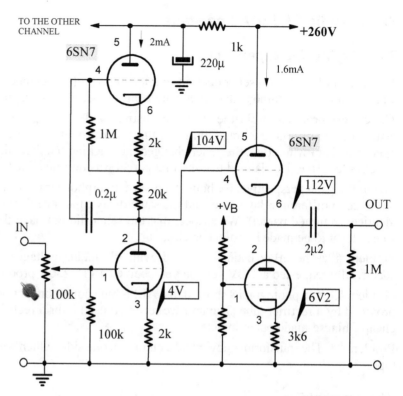

The alleged Opera "Consonance Cyber 222" audio section, © Opera Audio

The cathode follower's gain is $A_K = \mu R_K/[(1+\mu)R_K + r_I]$, and since in the chosen operating point, the internal resistance of the 6SN7 triode is $r_I=10k\Omega$ and $\mu=17$, with $R_K=22k$ we get $A_K = 0.92$!

The dynamic anode load that the bottom tube (V2) "sees" is $r_A = R_L/(1-A_K) = 22/(1-0.92) = 275k$! The gain of that tube can be estimated using a formula for common cathode amplifier with un-bypassed cathode resistor, $A_A = -\mu*r_A/[r_I + r_A + (1+\mu)R_K] = -17*275/[10+275+(1+17)2] = -14.5$!

Finally, the overall gain is approx. $-14.5*0.92 = -13.3$ times. Minus means the preamp inverts the absolute phase.

Mk II changes

Describing the MkII model they were selling, one German website refers to it as a "Röhren-Vorverstärker in Serial-Regulated Pull-Push (SRPP) Schaltung". Does that mean the circuit was redesigned and changed from the µ-follower in MkI to SRPP in MkII?

Meanwhile, Opera's own website claims that the new version (Mk II) uses "Class A power supply" and that the "sound and reliability have been improved significantly". Class A refers to a specific mode of operation of active (amplification) components such as tubes and transistors, where one component conducts the whole signal, not just a part of it as in classes B and C. Referring to a power supply as class A makes no sense whatsoever, since power supplies don't amplify audio signals, but simply provide AC and DC voltages required for audio circuits.

As for the improved sound and reliability, is that a polite way of saying that the previous version didn't sound that good and there were reliability issues?

Strangely, the Opera website does not specify the gain at all, and even the user's manual stays silent on the issue! Voltage gain of a line stage is its most important parameter, akin to the output power of a power amplifier. However, the German website says (for MK II): "Eingangsempfindlichkeit 280mV, Ausgangsspannung Pre-Out 2,2V". Since *Eingangsempfindlichkeit*, a typical German sausage word, means "input sensitivity", and *Ausgangsspannung* means "output voltage", the gain is $A_V=2.2/0.28 = 7.86$ times or 17.9 dB.

Since 6SN7 triodes have $\mu \approx 20$ and the gain of a SRPP stage is slightly lower than half of tube's mju, around 9, once the signal is attenuated by the output cathode follower (say 15% down), one could quickly estimate the overall gain at $A \approx 9*0.85 = 7.65$, close to the specified value of 7.86, so it seems that Mk II is indeed based on a SRPP input stage. A minus sign needs to be added here, of course, since a SRPP is also a phase-inverting stage.

Sheer Audio DHT-01A line stage

The eBay buying experience

This Taiwan-based eBay seller deals in smaller audio parts, attenuators, binding posts, audio connectors, cables, and all sorts of related componentry, even whole circuit boards and, as in this case, fully-made phono and line stages.

There were problems from the start. The listing said, "This preamp will be shipped by EMS air express with insurance and on-line tracking." Initially, the seller provided a fake tracking number. Four weeks later, when we opened a claim with eBay, the seller changed the tracking number and finally sent the product, which took another ten days to arrive. He claimed that the first parcel was returned due to a postal error.

To put things in perspective, we bought a pair of amorphous transformer cores from this seller around the same time but under a different eBay name, and the same delay of almost four weeks happened. Even the seller's excuse was identical (returned parcel). We suspect that the seller did not have the preamplifier in stock and was waiting four weeks for it to be made by someone else.

Problem #2: The eBay listing specified available mains voltages as 100V, 120V, 200V, 230V, or 240V. We specifically requested a 240V version yet received a 115/230V product.

Problem #3: The photographs showed that ALPS blue potentiometer was used and that high voltage rectification was provided by a rectifier tube. The unit we received used hybrid rectification (two silicon diodes were added) and a cheap Chinese-made potentiometer.

Problem #4: The transformer cover had a cutout on one side, which someone covered with black electrical insulation tape.

The preamplifier

As with many Made-in-China and Made-in-Taiwan preamps, the preamp has only two inputs (??) and one pair of output RCA connections. It was priced at AU$550 plus AU$100 shipping. The pricing in AU$ instead of more common US$ indicates that the seller was specifically targeting Australian buyers.

The preamp is small; it measures only 26 x 26 x 19 cm. Its 5.5 kg weight (not 7kg as specified) comes mainly from its dark-colored hardwood base, supposedly made of "very expensive Millettia laurentii wood," a tropical tree classified as endangered. Its timber, known as wenge or African rosewood, is used in instrument making (guitars) and archery. The dust produced when cutting or sanding irritates the eyes and skin, causing dermatitis, respiratory problems, and drowsiness. Its splinters are septic.

The tube complement includes a 25Z3 rectifier tube and two 112A directly-heated triodes. Type 01a, 201a, and 301a tubes can also be used, according to the seller.

SPECIFICATIONS:
- Voltage gain: 6 (15.6 dB)
- BW: 7 Hz - 40 kHz (+/- 1 dB)
- SNR: 90 dB
- Weight: 7 kg

MEASURED RESULTS:
- Voltage gain: A=7 (16.9 dB)
- BW: 8 Hz - 63 kHz (-3dB)
- Weight: 5.5 kg

The audio tubes

112A (or 12A) are directly heated triodes that hail from the 1920s. Both the amplification factor and transconductance are low, while the internal resistance is high, so they are low merit tubes. A more modern 201A and 301A retained the same anode dissipation of 2.5 Watts and mju of around 8, but had a lower gm and much higher resistance of around 10 kΩ!

Its ancestor, UV201, made by General Electric (GE), was followed in 1922 by UV201A, a thorium filament version. In 1925 a new 4-pin base (UX4) and standardized construction was introduced and the name was changed to UX201A. Do not confuse this tube with Western Electric 201A, a totally different type. To add to the confusion, Cunningham, another manufacturer, named the tube CX301A.

TUBE PROFILE: UX-112A

- Directly-heated triode, 4-pin socket
- Heater: 5V/250mA
- V_{AMAX}= 180V, P_{AMAX}= 2.5 W
- gm=1.8 mA/V, μ=8.5, r_I = 4.7 kΩ
- TYPICAL OPERATION:
- V_A=180V, V_G=-13.5V, R_A=10.65kΩ, I_0=7.7 mA

TUBE PROFILE: UX-201A (CX301A)

- Directly-heated triode, 4-pin socket
- Heater: 5V/250mA
- V_{AMAX}= 150V, P_{AMAX}= 2.5 W
- gm=0.8 mA/V, μ=8, r_I = 10 kΩ
- TYPICAL OPERATION:
- V_A=135V, V_G=-9.0V, I_0=3 mA

The 25Z6 tube rectifier

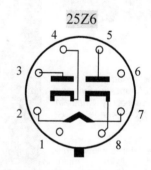

25Z6 is an indirectly-heated tube rectifier. It differs from the more common types such as 5Y3 not just due to its unusually high heater voltage (25V) and nonstandard pinout, but also by its cathode arrangement. The cathodes are completely separate, and the maximum AC voltage per anode is only 260V, which indicates that this rectifier tube was designed for voltage doubler circuits.

Since no commercial amp or preamp uses this obscure tube and has no replacement, it is cheap to buy, $5-$10 for USA-made NOS tubes.

TUBE PROFILE: 25Z6

- Indirectly-heated twin diode
- Octal socket
- Heater: 25V/300mA
- V_{AMAX}= 260V_{AC}, I_{AMAX}= 83mA
- V_{HKMAX}= 385V_{DC}

Topology

All connections are mounted at the back of the top plate, which means there was no need to drill and mount anything on the preamp's timber frame, a smart decision (internal photo on the next page). Notice two copper strips that connect the aluminum bottom plate to the grounded top plate (the photo of preamp's internals on the next page). The power supply, built on a large PCB, occupies one side, while the audio section was built in a point-to-point fashion on the other half of the top plate.

The audio section

From the two inputs, the signal goes via unshielded twisted pairs to the selector switch (1), whose output feeds the grid stopper resistors (2). The grid leak resistor is connected between the grid and the ground (3). The two large anode resistors, one for each channel, (4), are connected to the small PCB that also mechanically supports the ground bus, stretching from the output RCA sockets up to the volume control potentiometer (6).

The coupling capacitors (5) are connected between the anodes and another small PCB, supporting the ground bus. From there, ordinary hookup wires carry the output signals to one side of the dual volume control pot, whose wipers are hooked up directly to the output RCA sockets via a pair of shielded cables.

The cathode elco (7) and a trimmer potentiometer between the two heater pins (8) are mounted on another pair of small PCBs, joined by another small module, on its tiny PCB (9). These two modules are constant current sinks, one for each triode consisting of a transistor, LED, and three resistors.

The power supply

Since the 220-110V voltage selector switch (1) was wired in, that means the preamp's power transformer has a dual primary winding. Each of the tubes has its heater supply circuit. The AC input voltage (2) is rectified by a diode bridge, filtered, and regulated by a 7805 voltage regulator on a small heatsink (3). Notice low inductance air coils, one in each of the 5VDC supply legs (4), made by wrapping transformer winding wire around some thin cylindrical base, a nail, for instance.

The high voltage circuit starts with a hybrid rectifier bridge, two silicon diodes, and the 25Z6 tube rectifier together (5). The DC voltage goes through two CRC (π) filters (6) and is then regulated by IRF830 Hexfet power MOSFET transistor (7) and LM31T7 voltage regulator (8).

ABOVE: The preamplifier came with NOS 12A tubes by Sylvania and 25Z6 rectifier by Chatham Electric, both USA-made.

RIGHT: The internal view of the preamplifier with major components and points of interest numbered, as discussed in the text.

Volume control placement - at the input or at the output of a preamplifier?

Most audiophile preamplifiers and power amplifiers have their volume control potentiometers at the very input, between the signal source and the first amplification stage. Some have it after the first stage, and a few (preamplifiers only) have the gain potentiometer at their output, as is the case here, with Sheer Audio DHT-01. Why would a designer make such a decision?

Input volume control adjusts the magnitude of the signal voltage brought to the input stage's control grid, so the anode AC (signal) voltage swing will vary depending on the position of the potentiometer's wiper arm. That swing will be small at low volumes, and the corresponding harmonic distortion will be very low. As the volume control is turned clockwise, a larger signal feeds the grid, and the anode voltage swing increases, as does distortion.

With a potentiometer at the output of the stage, the stage permanently operates with a maximum anode voltage swing and maximum distortion. Reducing the volume at the output does not reduce the distortion level. This is the first main drawback of output-placed volume control.

To discover other consequences of such potentiometer placement, we need to consider a typical single-stage line preamplifier, connected to a tube power amplifier (next page).

For the circuit with input potentiometer, the stage's output impedance is $Z_O = r_I \| R_L$, where load is the parallel combination of R_A, R_P, and R_{IN}, or $R_L = R_A \| R_P \| R_I = 28k$.

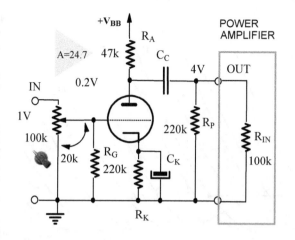
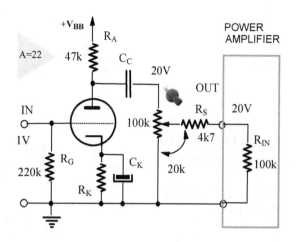

a) Input potentiometer circuit

b) Output potentiometer circuit

The output impedance is $Z_O = r_I \| R_A \| R_P \| R_{IN}$. Assuming the same tube's internal resistance of 6kΩ, we get Z_O = 5k5. The constant load means the output impedance is also constant, as is the amplification factor of the stage.
$A = -\mu R_L/(r_I+R_L) = -30*28/(6+28) = -24.7$

In circuit b) the load is 80kΩ in series with the parallel combination of 20k lower pot's section, which is in turn paralleled by a series combination of 4k7 resistor and 100k input resistance of the power amp, so the load resistance is $R_L = 80 + 20 \| 104.7 = 16{,}8$, so $Z_O = r_I \| R_A \| R_L = 4k$! The output impedance is lower than in the previous case, but the gain is also reduced: $A = -\mu R_L/(r_I+R_L) = -30*16.8/(6+16.8) = -32* 0.78 = -22$

In this circuit, neither the gain nor the output impedance is constant; both change with the setting of the volume control potentiometer!

The circuit

The stage operates with a constant current sink in its cathode, which not only fixes the anode current and makes the dynamic load line horizontal, but also keeps the cathode voltage and thus the grid bias (-6.4V in this case) constant. Since the current is independent of the anode load, by using CCS, the designer has compensated for one drawback of the output-located volume control, namely the variable gain of the stage. The CCs operation keeps the gain constant.

Notice that the LED and the transistor in CCS are both in the path of the audio signal! Sure, it may be argued that the 100μF electrolytic capacitor bypasses all audio frequencies around the CCS and that CCS effectively isn't in the signal path, but that is the theory; in practice, they do impact the sound!

The anode current is not specified, but we can deduce it from the DC voltage drop on the 15k anode resistor, $I_O=(200-100)/15 = 6.67$ mA!

Let's estimate the gain and distortion from the anode curves of the 112A tube. The characteristics deviate from the specified values on the circuit diagram, which says that the anode-to-cathode DC voltage is 100-6.4 = 93.6V. The characteristics show that with the 200V anode supply voltage, 13kΩ load, and the nominated operating point Q (6.67mA and -6.4V bias), the anode-to-cathode voltage should be around 115V!

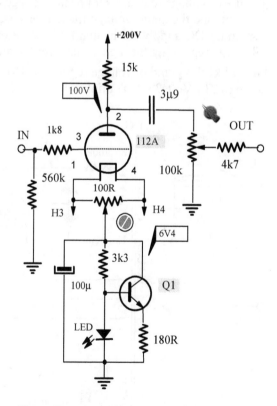

The audio section of Sheer Audio 12A line stage

Why a 13kΩ load when the diagram shows a 15k anode resistor? Remember, the coupling capacitor is considered a short circuit for AC signal, which effectively connects the volume control potentiometer in parallel with the 15kΩ anode resistor. Assuming the volume control at maximum, the full 100kΩ is in parallel with 15kΩ, making the total AC load 13kΩ. As we have seen in the previous analysis, the load will reduce with any other volume control setting.

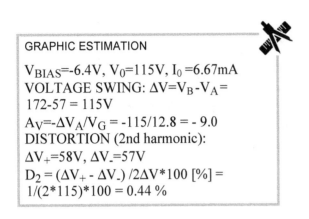
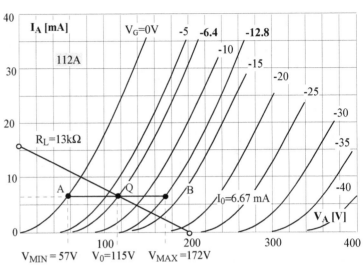

GRAPHIC ESTIMATION

V_{BIAS}=-6.4V, V_0=115V, I_0=6.67mA
VOLTAGE SWING: $\Delta V = V_B - V_A$ =
172-57 = 115V
$A_V = -\Delta V_A/V_G$ = -115/12.8 = -9.0
DISTORTION (2nd harmonic):
ΔV_+=58V, ΔV_-=57V
$D_2 = (\Delta V_+ - \Delta V_-)/2\Delta V * 100$ [%] =
1/(2*115)*100 = 0.44 %

The power supply

The rectified voltage is filtered by two RC filters (no choke was used) and regulated, firstly by a simple series regulator with a bipolar transistor driving an N-channel power MOSFET as a series pass element, followed by the LM317 voltage regulator. Both solid-state components passing the supply current are mounted on identical heatsinks, as seen in our visual review of the preamp's topology.

The audio tubes are DC heated; the 7805 voltage regulator keeps the required 5V level. Notice a couple of low-pass LC filters in each heater supply line, which shunt RF interference to the ground.

The complexity and design of its power supply, however, are nonsensical. Since the high voltage regulator decouples the rectifier from this line stage's audio load, there are no sonic benefits from using a rectifier tube, especially not in a hybrid arrangement.

The use of voltage regulators for heater voltages can also be questioned. Heater filaments have a relatively large time constant, and the mains voltage fluctuations usually aren't that wide anyway. A sufficiently large filtering capacitor (assuming DC supply for heaters) will store enough charge to make up for any instantaneous current shortfalls. Likewise, regulating the high voltage for the anodes does not always result in a superior sound.

I don't like the sound of voltage regulators. Does it matter if the anode supply voltage changes from 222 up to 224 or down to 219 Volts? I don't think so!

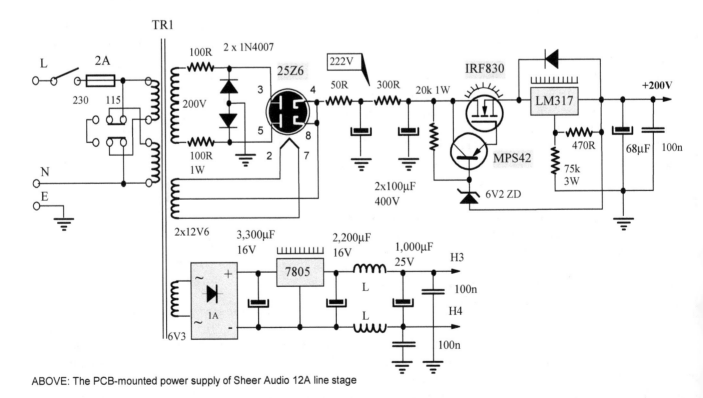

ABOVE: The PCB-mounted power supply of Sheer Audio 12A line stage

Listening tests and a final verdict

The tubes were highly microphonic. Just touching the volume control or top aluminium plate would be audible from the loudspeakers.

The sound was OK, but there was none of that expected DHT magic. Our line stage with #46 or #47 tubes (also featured in this volume) sounds significantly better. A DHT line stage priced under US$380 would be a bargain of the century if it sounded like a directly heated triode. Unfortunately, for some reason, this one doesn't.

Yarland PM7 line preamplifier

The preamp looks impressive with its solid timber fascia in a high gloss finish. The chassis is hefty and rigid, finished in high gloss black. The booklet that came with the preamp specifies that PM7 has four line-level and one MM/MC input; however, this preamp came with only line-level preamps marked. The confusion became apparent once the topside cover was removed.

The whole preamp, including its power supply, is on a large PCB, designed to accommodate two 12AX7 (one per channel) as a phono stage and one 12AU7 as a line stage, shared between channels. Component values and symbols are screen-printed on the board, but most of such components are missing (1) due to the different circuit used.

The factory modified the PCB by drilling some tracks out (breaking the connection) and rewiring the circuit. Each 12AX7 works as a voltage amplifier, driving one-half of 12AU7, which is now wired as a cathode follower. Large output capacitors were added in, hanging in the air. There are a dozen or more jumper wires (2) and modifications to the PCB. What a shocker!

The promotional spiel on various websites calls this "2009 New improved edition with Full Music vacuum tubes, Class A legendary circuit, vacuum tube rectification, ALPS volume controller (JAPAN)." Tubes are indeed marked "Full Music," but the volume control is not ALPS but the cheapest possible Made-in-China potentiometer.

What that "legendary" circuit is, heaven knows, we did not want to waste the hours required to re-engineer the circuit from this messy PCB.

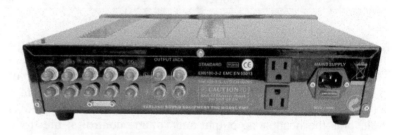

ABOVE: Bloopers galore! The engraving reads "Headphone tube amplifier". Also, the switched mains outlets are American type, yet the preamplifier is a model dedicated to the European market !?!

SPECIFICATIONS:

- Tubes: 1x12AU7, 2x12AX7, 1x6Z4
- Bandwidth: 25 Hz - 20 kHz – 1.0 dB
- 2 outputs, 5 inputs
- Input impedance 100 kΩ
- S/N Ratio: 91db
- Dimensions : 400 x 280 x 80mm
- Net weight: 9.0 kg

MEASURED RESULTS:

- Voltage gain: A=20 (26 dB)
- Output impedance: 188 Ω

The rectifier tube is soldered onto the PCB (3) without a socket. When it fails (it is not a matter of "if" but "when"), it will be a very messy and fiddly job to replace it. The whole PCB will have to be removed.

The board is bolted onto a dark gray Perspex backing (4). For some reason, a large rectangular opening was cut out of the bottom of the thick metal chassis, and the Perspex (acrylic) cover was bolted onto it from the outside. I guess that the whole PCB introduced a high level of microphony, as another Chinese preamp suffered from (Music Angel).

A steel divider (5), which is a great idea, welded to the chassis, screens the power supply from the audio circuit.

Notice a thick self-adhesive tar tape, used in a few places (under the mains transformer and on both sides of the divider) most likely to dampen chassis vibration and resonances (6).

Instead of installing an extension shaft and mounting the selector switch at the back (close to the input RCAs), there is a huge bunch of thick shielded cables (7) running across the whole depth of the chassis never a good idea.

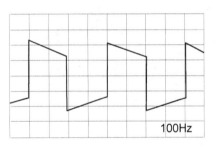

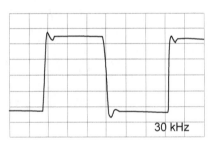

LEFT: Preamp's reproduction of a 100Hz test signal (square wave). Not the greatest of results (high sloping edges that ideally should be horizontal).

The 1 kHz output signal was perfect, and so was the 10 kHz reproduction. Only above 30 kHz a slight overshoot is noticeable, but it settles very quickly, so no prolonged oscillations are present, a very good result!

Xiang Sheng 708B headphone amplifier & line stage

The earlier version of this line stage-cum-headphone amplifier (pictured below) had a window at the front. An array of LEDs and a dummy tube that only gets heated were supposed to make the preamp glow nicely in the dark and look cute, but it only made it look cheesy and silly. There is only one pair of line-level inputs, one pair of outputs at the back, plus the headphone output and volume control at the front. The retail price is around US$200, including airmail worldwide.

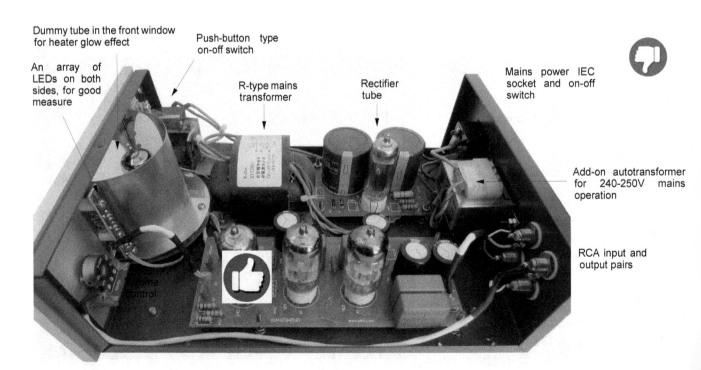

FROM SHOCKING TO SUBLIME: LESSONS FROM COMMERCIAL LINE STAGES

The 6N3 input tube is shared between the channels. It can be replaced with 5670, whose copy it seems to be. 6N11 can be replaced with 6DJ8 (ECC88, 5692), there is one for each channel.

Notice that 6Z3 is a Chinese version of this rectifier, which has a different pinout from the usual 6Z4 rectifier tube. Thus it cannot be replaced by a western 6Z4 without a significant modification of the printed circuit board or a special adapter!

All the voltage gain comes from the first stage, which amplifies 13.6 times (measured on our specimen). The second stage is a White cathode follower, which provides the low output impedance, measured around 100Ω.

Since the mains transformer was designed for $220V_{AC}$, all voltages were way too high on our $250V_{AC}$ mains. The heaters were at 7.2V instead of 6.3V, so we added a small transformer, whose two 6.3V secondaries were connected in series with its primary winding, making it operate as an autotransformer, a trick covered in Vol. 2 of this book.

We did not test 708B as a headphone driver, but as a line stage it sounded ordinary, I guess that is all one could hope for from a $200 device. Although genuine WIMA film capacitors were used by the factory, perhaps the sound would improve after an upgrade of a few resistors and coupling capacitors.

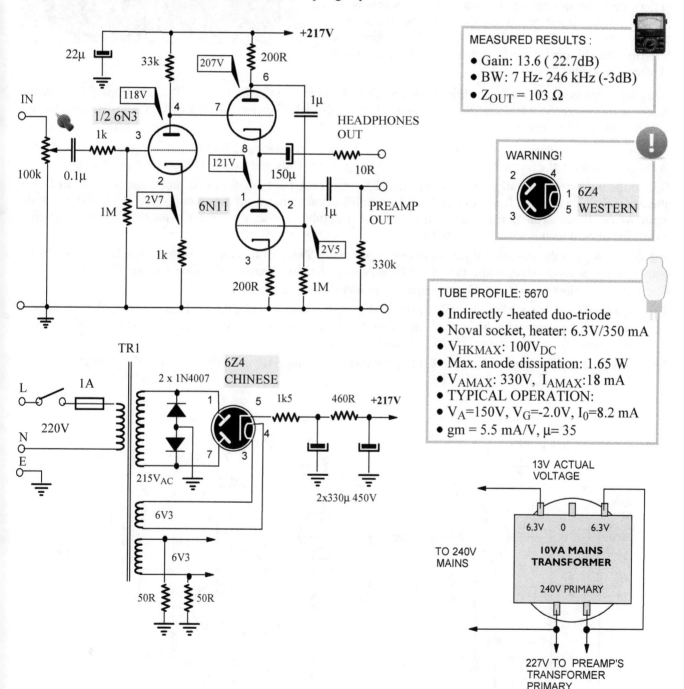

ABOVE: How to connect a small low voltage transformer to run 220V preamplifiers on 240V mains

Mingda MC-7R line stage

MC-7R is a Chinese-made line-level preamp. The 220V rated power transformer burned out most likely due to a very high mains voltage in Perth (250V or higher) or tube rectifier shorting the secondary windings. The primary fuse was of the wrong size and unable to protect the transformer.

Unfortunately, the transformers were potted in their steel enclosures, so they had to be thrown out, too. In general, we noticed that Chinese-made transformer winding wire often has poor insulation. The laminations are usually of acceptable quality, but the winding wire fails, not the laminations.

Sellers describe the circuit in various ways, some as an "improvement circuit based on the classical Hidomushi circuit."Others talk about "adopts Shigeru Wada circuit" or "Hetian Maoshi circuit giving clear, smooth and warm playback in sounds."

Googling supposedly "classical" Hidomushi circuit only comes up with four results, all from this preamp's descriptions. So much for Hidomushi, whoever or whatever that is! Likewise, " Shigeru Wada" and "Hetian Maoshi" search terms yield no results.

The online circuit diagram shows the first stage's high anode resistance (270kΩ), but our specimens had 150k installed. Likewise, the anode resistors of the second stage were 100k instead of 180k on the circuit diagram.

Due to the very strong negative feedback (1k and 22k resistors), the first stage does not amplify at all. In fact, it attenuates the signal by the factor of 0.33! All the amplification comes from the second stage. With 100 mV at the input, we got 30 mV at the first anode, 700 mV on the second tube's anode, and 700mV at the preamp's output, which means the overall gain is around 7 or 16.9 dB.

The output stage is the so-called White Follower, analyzed in Vol. 1 of this book. Its upper triode is driven by the output of the second stage, while the lower triode's amplification factor and anode resistor (3k3 in this case) determine the reduction in Z_{OUT}. Output impedances under 10Ω are possible.

The stage is effectively a push-pull amplifier where the upper triode provides the load current when the input signal is positive, and the lower tube drives the load during the negative periods of the voltage on the upper grid. Higher linearity (lower distortion) than ordinary CF is also claimed.

As in most totem-pole circuits, the cathode of upper triodes is at high DC potential, 160-170V in this case.

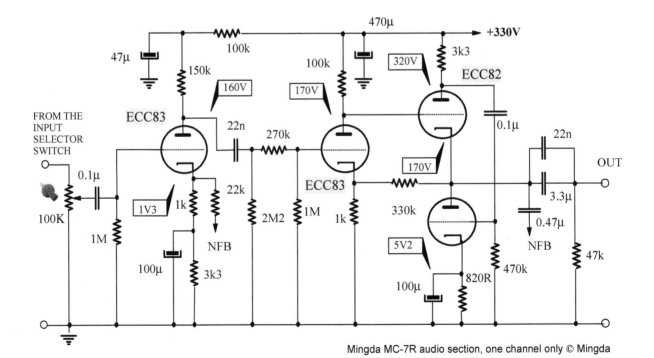

Mingda MC-7R audio section, one channel only © Mingda

For most tubes (except 12AX7), to avoid the breakdown in H-K insulation, the heaters of these tubes must be referenced to a higher DC voltage point. This was done here for the output duo-triode only (ECC82), raising its heater reference to 50V. However, that is not enough; the heater-cathode voltage is still at 120V, above the 100V limit! Don't you just hate when designers and manufacturers do the right things but in a slipshod or haphazard manner? You see many such instances in Chinese-made amplifiers, the right intention but poor implementation.

Notice the local negative feedback via the 339k resistor, from the cathode of the upper triode back to the cathode of the 2nd stage triode (that is why the 1k cathode resistor is un-bypassed).

The power supply

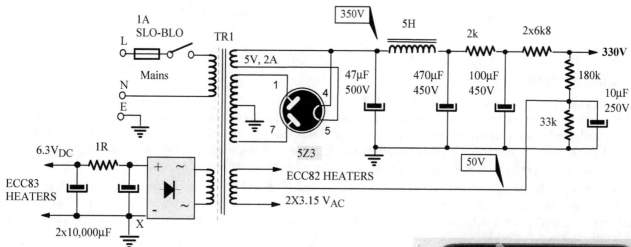

Upon the inspection of the two failed preamps, one had a 7A fuse installed; the other one had a 3A fuse. The circuit diagram did not even show any of the three primary components - the on-off switch, the mains voltage selector switch, or the fuse!

Instead of installing a new choke, a 1k2 (5W) resistor was substituted without any effect on the power supply's performance. Please note that the measured DC voltages marked on the circuit diagram are not the original figures but were obtained after installing a different replacement mains transformer.

The filtering choke and two power transformers of the same size were partially epoxy-encapsulated inside the transformer cover. The circuit diagram shows only one power transformer.

The 5Z3 rectifier (a copy of 5U4 beast that can power up large power amplifiers) is an incredible overkill for a preamp of such a feeble current draw, but it looks impressive, so we left it in place.

The view inside the transformer cover. The choke (1) tested OK, but the upper transformer showed signs of overheating around the area where wires come out of the coil (2).

Tube rolling

Due to the already mentioned heater-cathode insulation limit, no other duo-triode should be used in the first stage except ECC83. Of course, you could add a resistive voltage divider, just as it was done for the ECC82 heaters and choose the resistor values to get around 80V_{DC} at their junction. Disconnect point X from the ground and connect it to that junction. That will ensure that the heater of the input tube is at 80V_{DC}, so the cathode-heater DC voltage of the upper triode is 160-80 = +80V, and the cathode-heater DC voltage of the lower triode is 1.3-80 = -78.7V, both within the 100V limit for ECC81 and ECC82, which can then be substituted.

The tube used in the White cathode follower can be ECC81, ECC82 or ECC83. The output impedance is not significantly affected, but the sonics are, so experiment to determine which voicing you prefer.

Test results

The -3dB frequency bandwidth was extremely wide, in fact, the widest of any line stage we designed & built or any commercial line stage we have tested or repaired, from 1Hz to the incredible 530 kHz! While the output impedance was specified as 100Ω, we measured 85Ω, an excellent result. However, ordinary CF with 12AU7 can easily achieve 180 Ω, so the reduction of Z_{OUT} by the use of the more complex White follower is not significant.

MC-7R's reproduction of a 100Hz square wave was OK. The 1 kHz output signal was almost perfect, with a tiny glitch at the top rising edge. As illustrated below, that glitch becomes a slight overshoot at higher frequencies, an acceptable result for such a high frequency.

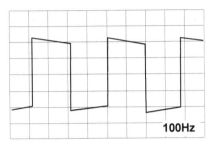
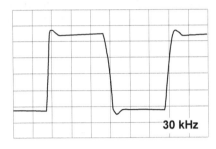

MEASURED RESULTS:
- BW: 1Hz - 530 kHz (-3dB)
- Gain: A=7.6 (16.9 dB)
- Z_{OUT} = 85 Ω

ABOVE: Preamp's reproduction of a 100Hz and 30 kHz test signals (square wave).

Construction details

The inputs (1) are very close to the selector switch (2) and volume control pot (3), a very wise practice. However, the on-off switch mounted on the narrow front panel required a metal shield (4) and two pairs of very long twisted cables (5). Normally only one pair would be needed, but a double-pole mains switch was used. Had this switch been mounted on the side in position (6), both measures would have been avoided, and the two front knobs could have been symmetrically positioned at the front. The manufacturing cost would be lower as well.

Since the two paralleled outputs are at the rear (7), the global NFB cables had to be of a shielded type (8). Tube sockets are mounted on a black powder-coated steel mounting plate. Notice that the photo below was taken after the failed original transformer was taken out and before its replacement was wired in through the holes in the chassis (9). That is why the rectifier bridge has no wires connected to it.

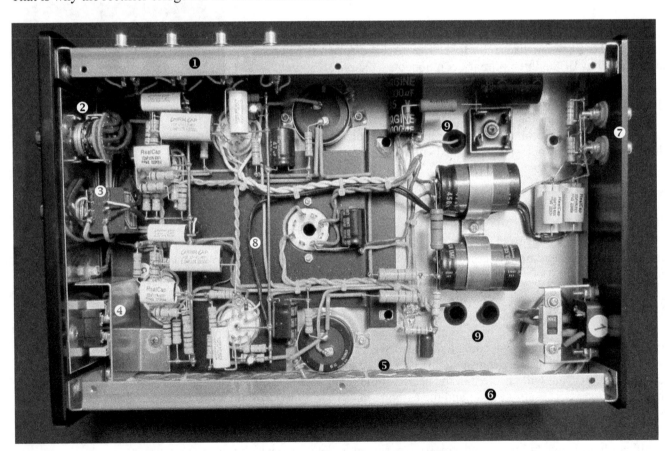

Listening impressions

With NOS American and European tubes installed, the preamp sounded confident and dynamic. While it did color the sound a bit (it was not absolutely transparent, but then what preamp is?), its preservation of details and microdynamics seemed above average. It certainly made CD players in the low-to-medium price range sound less digital by removing glare and digital artifacts. So, if your power amp is a bit sterile and anemic sounding, or if you want to get your CD player to sound more like a turntable, try this budget performer.

Dared SL2000-A line preamplifier

With solid timber front and side trim, this tiny line stage looks cute. Well, apart from that silly bent bar above the three exposed tubes. As in many Made-in-China amps and preamps, the minuscule chassis compromised everything. Only two inputs and one output could be accommodated, and the insides are overcrowded, filled by a large PCB, so significant modifications are impossible.

The circuit is Marantz 7a inspired; only 12AT7 was used instead of 12AX7 in Marantz's vintage design. Notice three film capacitors in the signal path and very strong feedback, with β=R2/(R1+R2)= 2k2/(2k2+20k) = 0.1 or 10%.

A line stage design with two gain stages, three coupling caps, and a very strong negative feedback is passé these days. If you get one of these on a secondhand market cheaply, and if two inputs are enough for you, the best option would be to ditch the PCB and build a better, simpler, no NFB circuit inside using point-to-point wiring.

How to do a DC analysis of the circuit when DC voltages aren't marked

DC voltages are not marked on the diagram, but we could draw the bias lines and determine DC currents through each stage, which would enable us to calculate the voltage drops on the two anode resistors and thus the DC voltages on the anodes.

To draw the DC bias line for the second stage, we need two points. For example, one mA of anode current on 330Ω cathode resistor will produce a voltage drop (bias) of 0.33V, 2mA will create a drop of 0.66V (point X), and 4mA will result in a bias of -1.32V, (point Y).

We have chosen the 2nd stage for this analysis since the 1st stage bias line cannot be drawn due to the extremely low current and the imprecise scale of the anode curves.

The intersection of the bias line and the DC load line gives us the quiescent operating point of the second stage, Q2, from which we estimate the cathode voltage of around 1.1V (since the grid is at -1.1V). This means the anode current is 1.1V/0.33k = 3.3mA.

We could try to graphically estimate the voltage gain, but that would be very imprecise from this graph.

To check the accuracy of our conclusions, the DC voltage drop on the anode resistor is 3.3mA*47k = 155V.

Since the anode-to-cathode voltage is V_{AK}=90V, the anode supply voltage should be V_{BB}=90+1+155=246V. Since 250V is specified, it seems that our estimation is spot-on!

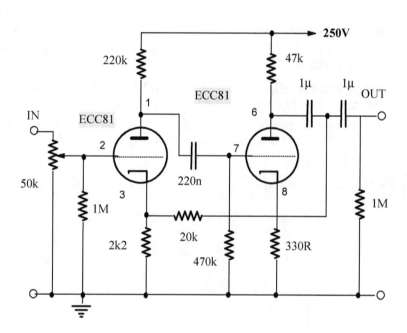

ABOVE: The audio section of Dared Sl 2000-A preamplifier (one channel only).

BELOW: The load line of the 1st stage is just visible at the bottom of the anode curves, but any further graphical analysis of that stae is impossible!

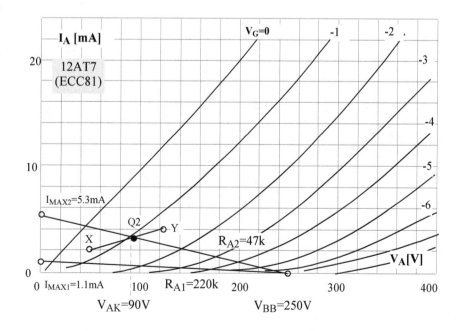

How to calculate AC voltage gains and negative feedback figures (feedback ratio, feedback factor and NFB in decibels)

When RK is bypassed to ground (for AC signal) by a capacitor, the voltage gain (without the local negative feedback) is $A=-\mu R_A/(r_I+R_L)$, and with local NFB through the un-bypassed RK, the gain is $A_F= -\mu R_A/[r_I+R_A+(1+\mu)R_K]$ or $A_F= A/[1+(R_K/R_A)A]$.

Since the first stage operates at a very low DC current of, say 0.5mA, the published average parameters are not valid, so we have to read them from the graphs. At such low currents, 12AT7's internal resistance is very high, around 60k, and mju is very low, around 28! So, we have $A_1=$ $-\mu R_A/[r_I+R_A+(1+\mu)R_K] = -28*220/[60+220+(1+28)2.2] = -28 *0.64 = 18!$

For the 2nd stage, operating with around 3.3mA of anode current, the internal resistance is lower, around 20k, and mju is higher, around 48! Thus, $A_2= -\mu R_A/[r_I+R_A+(1+\mu)R_K]=$ $-48*47/[20+47+(1+48)*0.33] = -48*0.57 = 27!$

Thus, the overall voltage gain (without the global NFB) is $A=A_1*A_2 = -18*-27 = 486$, which is way too high for a line stage. We know that feedback ratio is $\beta= R_2/(R_1+R_2) \times 100\% = 2k2/(2k2+20k) = 0.1$, so the "closed loop" voltage gain is $A_F = A/(1+\beta A) = 486/(1+0.1*486) = 486/49.6 = 0.02026*486 = 9.8!$

Factor $1+\beta A$ is called a feedback factor (FF), which tells us how much the gain has been reduced. In this case FF is 49.6 times or $20*\log(49.6) = 39dB!$

You can see that the global NFB (through the 20k feedback resistor) reduced the overall gain from 468 (or 53.7 dB) down to 9.8 (or 19.8 dB), meaning the NFB is 53.7-19.8 = 33.9dB, which is incredibly high.

DIY LINE-LEVEL PREAMPLIFIERS: $10,000 SOUND ON $500-$1,000 BUDGET

- Design project: 12SN7 line stage with a separate power supply and transformer output
- Design project: 5670 (WE396A) line stage
- Design project: 12SN7 line stage
- 6V6, 6F6, 6K6 and 6Y6 in line stages
- The ECL series: ECL80, ECL82 (6BM8), ECL83, ECL84, ECL85, ECL86 (6GW8)

"Pass the music through too many gain stages, and you'll no longer recognize the genius and beauty of your favorite recordings."
From Balanced Audio Technologies Rex II line-stage preamplifier data sheet

Design project: 12SN7 line stage with a separate power supply and transformer output

Commercial benchmark: Raphaelite CR7 preamplifier

Raphaelite CR7 preamplifier is made in China. It seems that most amplifiers under this name are sold as kits, then assembled by individuals and sold on eBay and elsewhere. Various Raphaelite components are also sold separately, such as mains and output transformers, chokes, and MC step-up transformers.

Its eBay price in Sept. 2015 was steep for a China-made piece of gear, US$2,220 + US$199 transport!

The technical data specified: input impedance of 100kΩ, 600Ω output impedance, input sensitivity of 230mV, distortion under 1% at $3.8V_{OUT}$, frequency range 6Hz-67kHz (-3dB), and S/N ratio of 75dB (not a great result). The tube complement comprises of 2x12AT7, 2x LS7, and one 5Z3P rectifier (equivalent to 5U4G).

LS7 or CV1660 is an ancient directly-heated triode made by GEC in the UK, dating back to 1925. Compared to modern triodes, its maximum anode voltage and its mutual conductance are very low, while its internal impedance is a moderately high 5kΩ.

So, based on parameters only, this seems an unwise choice. However, based on our experience with various directly heated triodes, perhaps LS7 sounds good, which is why the designer of this line stage chose it.

TUBE PROFILE: LS7 (CV1660, VT82)
- Directly-heated triode, 4-pin socket
- Heater: 4V/150mA
- V_{AMAX}= 150V, P_{AMAX}= 2.5 W
- gm=2.4 mA/V, µ=12, r_I = 5 kΩ
- TYPICAL OPERATION:
- V_A=150V, V_G=-4.0V, I_0=21 mA

Commercial case study: BAT Rex II transformer-coupled line stage

BAT claims that Rex II is "the world's preeminent preamplifier," and, indeed, in early 2016, it was priced accordingly at US$25,000! However, measurements by John Atkinson of Stereophile magazine (review in Jan 2016 issue) revealed a few faults and drawbacks.

Although an output impedance of 200Ω was claimed, it varied between 700Ω at 20Hz and 1,770Ω at 20kHz.

The frequency response of the two channels was also significantly different. One channel started to roll off around 10kHz and was down 1.6dB at 20kHz, while the other was down -0.5dB at the same frequency.

Paralleling four triodes (which can never be perfectly matched) and coupling them by a transformer to the output without any feedback means the variations in tube and transformer parameters will cause such unavoidable discrepancies between channels.

Finally, the spectral signature showed that the 3rd harmonic was higher than the 2nd, and the 5th above the 4th, meaning the discordant (unpleasant) sounding artifacts are more prominent than the euphonic even harmonics. This is obviously a consequence of the preamp's balanced design, which reduces pleasant-sounding even distortion products, leaving the harsh sounding odd harmonics to dominate, just like in push-pull output stages.

Rex II uses four 6H30 duo-triodes per channel, meaning four and four triodes are paralleled in a symmetrical arrangement. Since the internal resistance of those triodes is 800-900Ω, four in parallel would achieve 200–225Ω. Had capacitive coupling been used, that would be the output impedance of the whole preamplifier since there is only a single stage. However, MkII uses output transformers, which increase the output impedance!

Its predecessor, model Rex, with capacitive output coupling, cost US$18,500 in 2008. Is transformer coupling worth the US$6,500 price premium (35% price increase, while the inflation in the USA has been very low during those seven years)? We'll never know!

The non-technical reviewer, Fred Kaplan, commented on Rex II aesthetics: "Finally, the two boxes are pug-ugly: plain, black, bluntly rectangular, with no design flair. Even the buttons on the faceplate are too dark to read without a flashlight - which heightens the need for diagrams in the manual. But these are minor qualms." Minor at a $1,000 price level, perhaps, but at 25 times that, you'd expect much, much better!

The "power module" contains 2x 5AR4 rectifiers, 2x 6C45 (1/2 of 6C33C-B monster triode) or 2x 6H30 in a dual shunt regulator (two options are switch-switchable), and 4x 6C19 as current sources. How is that for complexity?

Finally, for good measure, some serious criticism of the review side of equation. Neither Rex nor Rex II reviews in *Stereophile* featured oscillograms of the preamps' square wave reproduction, which is essential for judging their fidelity.

Since it is a transformer-output type of line stage, and those usually exhibit all kinds of resonant peaks and related amplitude and phase distortion, I would love to see Rex II's square wave reproduction. A serious oversight by such a reputable audio authority as *Stereophile*.

12SN7 line stage - audio circuit

The first step in designing a transformer-coupled line stage is to ascertain the maximum allowed DC current through the output transformer's primary winding. In this case (nondescript China-made transformers bought on eBay) that was specified as 20mA.

It is always a good idea to reduce the anode current to below half that value. Parameters of audio transformers, especially those with an air gap, vary widely, depending on how wide the gap is, how well packed the laminations are, and how tightly was the transformer assembled and bolted together.

In short, do not place too much trust in the published maximum current figures and don't get too close to them, or you may be unpleasantly surprised by the results.

If you calculate the anode (and cathode) DC current through two 12SN7 triodes in parallel (as per our schematics) you will get $I_A = 1.7V/330\Omega = 5.15$ mA or just over 2.5 mA per triode. This is quite a low value for such a beefy tube and could be doubled if you wish.

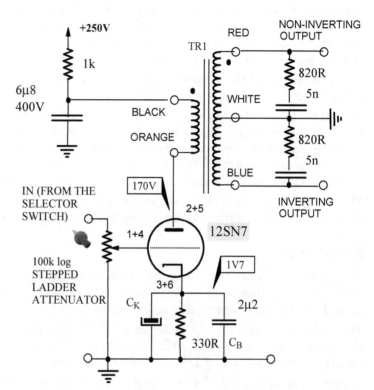

The audio section of our capacitorless transformer coupled line stage. The input selector switch is not shown.

Choosing the value for the cathode bypass capacitor

The cathode resistor can be left un-bypassed, but the stage's voltage gain will be reduced due to the local negative feedback thus introduced.

The capacitance of the bypass capacitor will determine the lower -3dB limit of our line stage, so to determine the minimum capacitance required, we need to decide what limit we'd be happy with. Let's say we would like $f_L = 7$Hz. This is an approximate calculation, but it does the job. The angular frequency is $\omega_L = 1/\tau = 1/(R_K C_K)$. Remember, as we have seen in Volume 1, τ (tau) is the time constant of the cathode circuit. Since $\omega_L = 2\pi f_L$, the minimum capacitance required is $C_K = 1/(R_K 2\pi f_L) = 1/(330*2*3.14*7) = 7.6*10^{-4}$ or 760 µF!

The required capacitance is relatively large due to the low value of the cathode resistance. You can use the standard common value of 1,000 µF. Make sure you bypass it with a good quality film capacitor, 220nF to 10µF in value.

The output wiring

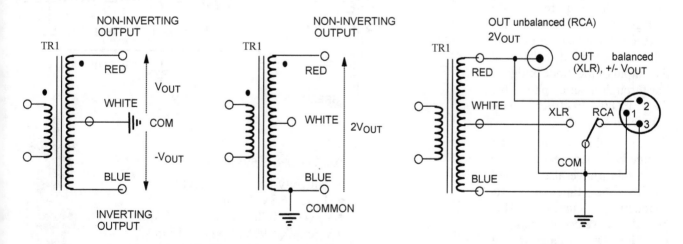

LEFT: Both inverting and non-inverting single-ended (unbalanced) outputs are available, making phasing of the whole audio system very easy.
MIDDLE: If the whole secondary winding is used, the output voltage is doubled compared to the previous case. The inverting output is obtained as pictured; the non-inverting output would be taken from the blue terminal (+), while the red terminal would be grounded (-).
RIGHT: A switchable arrangement between RCA (unbalanced) and XLR (balanced) outputs. The unbalanced output is the same as in the middle illustration, while the balanced output is the same as in the first case on the left.

Our output transformer has a dual secondary with a CT (center tap). This is handy if you want to use it as an interstage transformer (phase splitter) in a push-pull amplifier, but it brings us two benefits in this application. For an unbalanced operation, you can ground the CT and use either half of the winding. The blue wire will give you an inverted signal while the red wire carries the output signal in phase with the input signal (non-inverting output).

If both single-ended (RCA connector) and balanced (XLR connector) outputs are needed, a simple switch makes selecting between the two output modes easy.

Pin 1 of the balanced output is permanently grounded. The positive (pin 2) is wired to the RED secondary terminal. The negative output (pin 3) is wired to BLUE secondary terminal #6, making this a non-inverting arrangement. By swapping the red and blue terminals, a balanced non-inverting output is obtained. When the switch is in the "RCA" position, the center tap (WHITE) of the output transformer is disconnected from the ground, and a whole secondary (between RED and BLUE terminals) is connected across the unbalanced output.

The two RC networks (Zobel networks) across the transformer's secondaries improve the amplitude response by flattening the resonant peaks and widening the frequency range. However, the gain is reduced as a consequence.

Power supply

The power transformer had two identical coils so that all secondary windings could be connected in series or parallel. For the EZ80 rectifier, two 6.3V secondaries were paralleled to double their current capacity, while the 12.6V heaters for audio tubes were connected in series and powered from two 22V secondaries in parallel.

The audio tubes are slightly underheated (22V instead of 25.2V), reducing noise and prolonging their life. Connecting our mains voltage of 250V to the 240V primary tap raises all secondary voltages by 250/240=1.042 times or 4.2%, so instead of 22V, we had 23V, perfect.

A high heater voltage (22-23V here compared to the standard 6V) means low heater current, which reduces radiation from heater wiring since the strength of the radiated field is proportional to the current, not voltage! The basic heater wiring precautions still need to be followed.

Twist the heater wires, avoid looping them around tube sockets and run them parallel with enclosure walls, away from the audio circuit. This, plus the fact that our mains transformer was of low radiation type (PU-laminations) and was removed entirely from the audio chassis, resulted in a very quiet preamplifier and a very high S/N (signal-to-noise) ratio!

To get the required anode voltage, two 125V secondaries were connected in series. The full-wave rectified voltage at the output of the diode bridge is fed through a series pass tube, the EZ80 rectifier with paralleled anodes.

Since this is an indirectly-heated rectifier, the audio tubes have also warmed up by the time it starts conducting current. This provides a welcome delay, so the high DC voltage is not immediately applied to the audio tubes. It also improves the sonics.

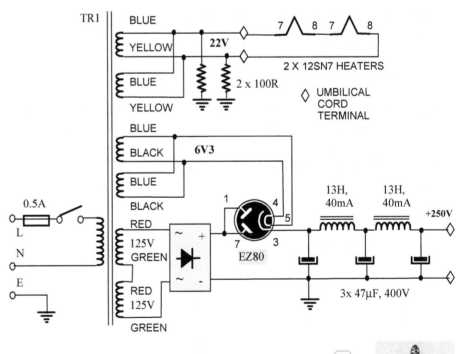

TUBE PROFILE: 6V4 (EZ80)

- Indirectly-heated dual rectifier
- Heater: 6.3V/435mA
- V_{AMAX}= 300V, P_{AMAX}= 2.5 W
- V_{HKMAX}= 90V, I_{KMAX}= 15 mA
- TYPICAL OPERATION:
- V_A=250V, V_G=-2.3V, I_0=10 mA
- gm=5.9 mA/V, μ=57, r_I = 9.7 kΩ

Topology and enclosures

The power supply is in its own enclosure, connected to the audio section by an umbilical cord. A minimum of 4-cores are required, two for the heaters and two for the anode supply (+V_B and ground). The cord is permanently wired to the audio chassis, while the power supply side is terminated in a 5-pin DIN connector. Since high voltage is present, the socket on the power supply side must be a female type (pins are recessed to be touch-safe and pass the "finger test") while the connector on the cable (audio side, the load) is a male type (with protruding pins).

The "enclosures" are large cigar-cases made of cedar in burl finish, with gold hinges at the rear. These aren't usually cheap but can be found at reduced prices online and at clearance sales.

The top and bottom halves are held together by the screw-on decorative front panels in gold mirror-finish. To open either enclosure up and access the components, the front panels are simply unscrewed (self-tapping wood screws) and the top panels hinge up, so the preamp does not have to be turned upside down.

A very elegant and classy-looking preamp, which sounds even better than it looks!

The only drawback of non-metallic enclosures such as these is their lack of electrostatic and electromagnetic screening. In this case, with line-level signals and a separate power supply, we didn't have to shield anything, but using them for phono stages is not recommended.

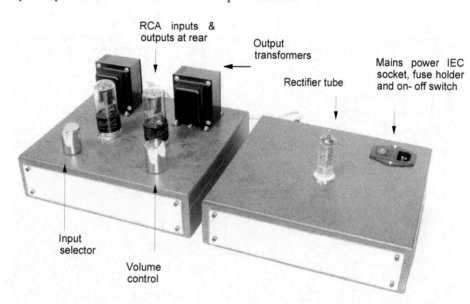

Sonic impressions

Apart from being very quiet, the preamp carries that unmistakable sonic signature of 6SN7 or 12SN7 triodes. It is also free from the frizziness and other artifacts associated with even the best of coupling capacitors. However, the output transformers do introduce their own subtle coloration, so the eternal question remains. Is transformer coupling really more pleasant-sounding than capacitive coupling or cathode follower outputs? Many audiophiles think so, but I am not convinced. Not all coupling capacitors sound the same, and not all output transformers sound better than all capacitors.

Design project: 5670 (WE396A) line stage

This cute little (in size only) duo-triode comes in quite a few varieties, from Western Electric's 407A and 396A, through American industrial versions, 6385 and 6854, to Russian 6N3P variant. A very linear, reliable, and good-sounding tube, 5670 has a decent amplification factor of around 35 and a (relatively) low internal impedance.

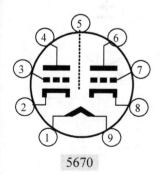

5670

TUBE PROFILE: 5670 (6N3P)

- Indirectly-heated duo-triode
- Noval socket, heater: 6.3V/350 mA
- $V_{HKMAX}=100V_{DC}$
- Max. anode dissipation: 1.65 W
- V_{AMAX}: 330V, I_{AMAX}:18 mA
- TYPICAL OPERATION:
- V_A=150V, V_G=-2.0V, I_0=8.2 mA
- gm = 5.5 mA/V, μ= 35, r_I=6k3

Operating point, load line and voltage gain estimation - Version 1

Since this minimalist design uses battery bias, the operating point needs to be positioned at $E_G=-2.4V$ because that is the nominal voltage of the battery used. VARTA, part number 55608602059, is a rechargeable Nickel Metal Hydride battery of 80 mAh capacity. It's a button cell type with PCB pins, 15.5 mm in diameter. This battery costs under $10 and is perfect for this application. A charge will last 2-3 years.

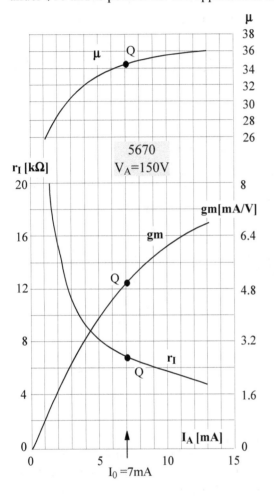

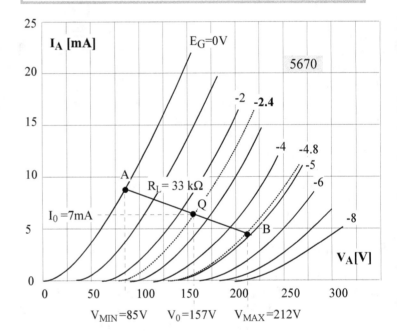

GRAPHIC ESTIMATION

$R_L=33k$, $V_0=157V$, $I_0=7mA$
$P_{IN}= I_0 V_0 = 1.1W$
$\Delta V_G = 4.8V$, $\Delta V_A = 212-85 = 127V$
$A = -\Delta V_A/\Delta V_G = -127/4.8 = -26.5$ (28.5 dB)
$\Delta V_+=72V$, $\Delta V_-=55V$, $D_2 = (\Delta V_+ - \Delta V_-)/2\Delta V *100$ [%]
$= 17/(2*127)*100 = 6.7\%$

With maximum anode dissipation of 1.65W, our Q-point is at 1.1Ω, a conservative positioning. 5670's parameters in that point are 7kΩ internal resistance, μ=34.5, and gm=5 mA/V!

However, the distortion is very high, and the gain of 25 times is great for a driver stage in an amplifier, but it's way too high for a line stage. Transformer loading could be used to reduce the output voltage if used as a line stage.

Due to the relatively high anode current of 7mA, the voltage drop on the 33k anode resistor is 33*7=231 V, so a power supply of 390V is needed.

However, to use a shunt regulator, we need a small tube that can pass at least 7mA (same as the audio stage) of current and operate with 400 Volts on the anode. Most duo-triodes have a maximum anode-cathode voltage of only 250-200V, some even less. Even many smaller output tubes such as EL84 cannot work on 400V!

So, we either forget about shunt regulation and build the circuit as it is or find a tube that can work on 400V between its anode and cathode.

The third option is to reduce the anode current to say 4 mA. That would also reduce the voltage drop on the anode resistor to 33*4 = 132V. The required supply voltage is now $V_{BB}=157+132 = 289V$, still too close to the 300V limit of most small tubes.

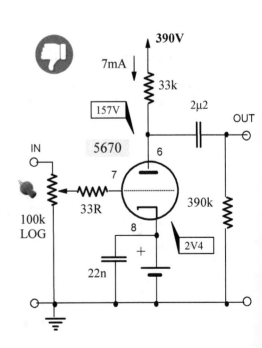

Using a more powerful tube here is possible but would violate the Simplicity Rule. More importantly, in terms of sonics, amps and preamps with regulated power supplies sound different from the unregulated ones. The tube used in a shunt regulator is in the signal path, so ideally, it should be the same tube used in the audio circuit, and in this case, it should be the other half of the 5670 duo-triode!

Version 2: transformer coupling and shunt regulation

One solution to our high voltage troubles is to use an output transformer, such as the one we designed in this volume, 9k primary and 150Ω secondary, which can take up to 20mA of the primary current. With 7mA through the load, the DC voltage drop on its primary is only 10V, so we need to regulate our voltage at around 167V.

LM431 is a 3-terminal adjustable precision shunt regulator integrated circuit in a TO-92 case, basically a programmable Zener diode. By changing the referent voltage, output voltages of +3V to +36V can be obtained. The maximum output current is 100mA.

V_{REF} is the referent or cathode voltage that we want as the tube's grid bias voltage. It is set by a simple voltage divider R_1-R_2 and equals $V_{REF} = [R_2/(R_1+R_2)]V_{OUT}$

We need to find the referent voltage of the programmable shunt regulator. First, the 5670 tube's anode characteristics must be consulted and its operating point determined. To get a 5mA anode current at 167V anode-to-cathode voltage, around -3.3V of bias is required, and that is our V_{REF}!

We need a regulated output voltage of 167V, so V_{OUT}=167V and V_{REF}=3.3V.

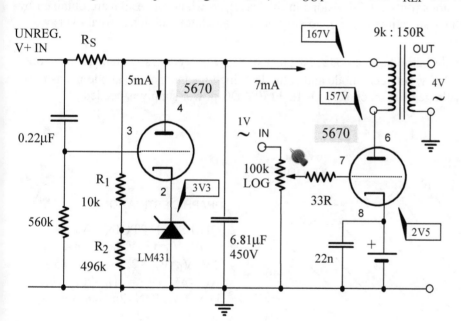

$V_{OUT}=(1+R_1/R_2)V_{REF}$
$167=3.3(1+R_1/R_2)$
$R_1=163.7R_2/3.3$ or $R_1=49.6R_2$

If we choose R_2 to be 10k, R_1 must be 496k.

The current through the series resistor R_S is $I=I_{LOAD}+I_{VR} = 7+5 = 12$mA. The voltage drop across it will depend on the unregulated output voltage of the power supply $+V_{BB}$ coming in.

With 240V_{DC} coming in, the voltage drop across R_S is $\Delta V=240-167 = 73$V. Its value needs to be $R_S =\Delta V/I = 73/0.012 = 60.1$ kΩ, so use a standard value 62k resistor.

The power dissipated on R_S is $P=V^2/R_S = 73^2/62,000 = 2,500/1,200 =0.09$W, so even a 1/2 Watt resistor should be fine.

With the maximum voltage swing, the 2nd harmonic distortion is only 1.2%; with normally-used levels, it will stay below 0.5%.

The gain is (267-62)/6.6 = 31, but the output transformer attenuates VR=√IR = √(9,000/150) = 7.75 times, for overall gain of 31/7.75 = 4 times!

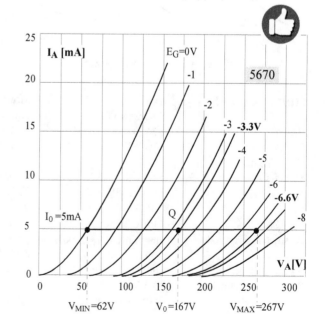

KEY FEATURES:

- Battery bias in the cathode circuit
- No coupling capacitors (transformer output)
- Shunt-regulated high voltage power supply

6V6, 6F6, 6K6 and 6Y6 power tubes in line stages

Although 6V6 is a beam tetrode and 6K6 and 6F6 are pentodes, the three tubes are relatively similar in appearance, parameters, and performance. 6Y6 is a low voltage-high current tube, its max. VA is almost half that of the other three, but its anode current is doubled, and its required heater power is also much higher than the other three! Since it is not interchangeable with the other three types, it is included here for comparison purposes only.

All four tube types sound very good, especially when triode connected, and since millions were made in the golden era, they are still plentiful and relatively cheap to buy. NOS are still available as pairs or even quads.

While all three have internal resistance around 2.5kΩ, 6V6 has the highest amplification factor. 6Y6 also has a very low internal resistance (under 800Ω); if your line stage needs to drive long interconnect cables or low-impedance solid-state power amplifiers, and you don't want to use a cathode follower to reduce Z_{OUT}, that is the way to go.

Let's see what can be expected when used as a voltage amplifier. Assuming you'd want to use anode chokes of 100-200H inductance, we will choose a relatively low current of 12mA since most available anode chokes were designed to operate with less than 20mA of magnetizing DC current.

The higher the DC magnetization, the lower the permeability of its magnetic core and the lower the choke's inductance. Lower inductance means higher lower -3dB (corner of cutoff frequency) frequency and weaker bass.

The anode characteristics of most tubes, especially power types, aren't very precise in the region of low anode currents, since in regular service, such tubes aren't meant to be operated in that region. Plus, the scale resolution isn't very high. There aren't any minor divisions (only 50V divisions in this case), so whatever results are obtained by such coarse estimation, they are just that - an estimation. So, go easy on the calculator and don't sweat over two or three decimal places!

With 12mA of current and 10V bias on the cathode, a cathode resistance of 833Ω is needed, so a standard resistor of 820Ω is a perfect choice. Q point is at 135V between anode and cathode, which means that the anode needs to be 145V above ground, and with a 25V voltage drop on choke's DCR, +170V DC power supply is needed.

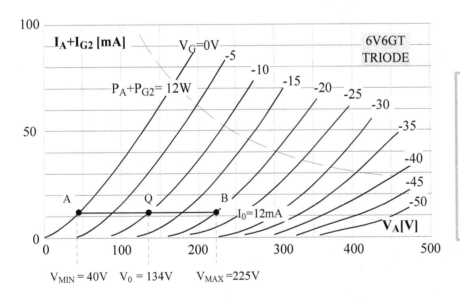

SIDE-BY-SIDE	6V6	6K6	6F6	6Y6
	Beam tetrode	Power pentode	Power pentode	Beam tetrode
Heater	6.3V/0.45A	6.3V/0.4A	6.3V/ 0.7A	6.3V/ 1.25A
P_A / P_{G2} max.	12W / 2W	8.5W / 2.8W	11W / 3.75W	12.5W / 1.75W
V_A / V_{G2} max.	315 / 285 V	315 / 285 V	375 / 285 V	200 / 200 V
SE class A_1 pentode P_{OUT}	5.5W	4.5W	4.8W	6.0W
V_A, V_S, V_G	315V, 225V, -13V	315V, 250V, -21V	285V, 285V, -20V	200V, 135V, -14
R_L	8.5kΩ	9kΩ	7kΩ	2.6kΩ
I_{A0}, I_{S0}	34mA, 2.2mA	26mA, 4mA	38mA, 7mA	61mA, 2.2mA
Parameters as a triode	gm= 4mA/V, r_I=2400Ω, μ=9.6	gm= 2.7mA/V, r_I=2,500Ω, μ=6.8	gm= 2.6mA/V, r_I=2600Ω, μ=6.8	gm= 7.3mA/V, r_I=750Ω, μ=5.5
V_A max. triode	300V	315V	350V	250V
P_A + P_{G2} max. total triode	12.5W	7W	10W	14W
V_A, V_G, I_0, R_L	250V, -20V, 39mA, 4.8kΩ	250V, -18V, 38mA, 6kΩ	350V, -38V, 48mA, 6kΩ	250V, -42V, 50mA, 5kΩ

DIY LINE-LEVEL PREAMPLIFIERS: $10,000 SOUND ON $500-$1,000 BUDGET

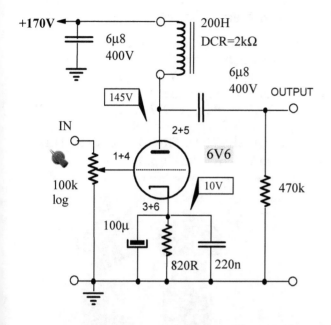
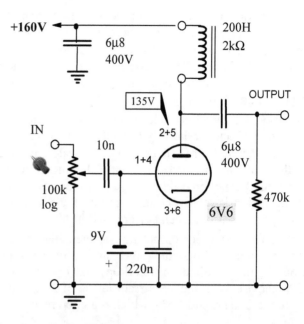

Option 1: cathode bias. The input can be DC coupled, but an electrolytic cathode bypass capacitor is needed, otherwise local NFB will reduce gain. If lower gain is acceptable, eliminate cathode capacitors.

Option 2: battery bias. No cathode RC network, but an input coupling capacitor is now required. However, that can be a low value, low voltage quality film&foil or PIO unit.

Not a bad result. As expected, voltage gain close to tube's mju, and distortion around 1% for the maximum voltage swing. Under normal listening levels the 2nd harmonic should stay below 0.5%!

A standard 9V alkaline battery could be used to bias the stage, obviating the need for cathode resistor and bypass capacitors, but now an input coupling capacitor must be added, otherwise the battery would be discharged through the output resistance of the source (CD player, DAC, phono stage, etc.) However, a low-value low voltage film or PIO capacitor degrades the sound much less than a high value (100µF+) elco.

The ECL series: ECL80, ECL82 (6BM8), ECL83, ECL84, ECL85, ECL86 (6GW8)

The ECL80-86 series of tubes feature a triode and output pentode, so a complete push-pull power amplifier could be built by using only two tubes. While ECL82, ECL83, and ECL86 were developed for audio applications, ECL84 and ECL85 were used in TV sets as frame output tubes, frame oscillators, sync-separation, and sync-amplification.

The triodes have a relatively high mju, ranging from 50 to 100, and a low anode power rating, under 1 Watt (except ECL83, whose triode section is rated at 3.5 Watts!) See the comparison table below.

Triode connected pentode sections have mju ranging from 7 to 36, and anode power ratings between 4 and 9 Watts, so, among the five tubes, you are bound to find one combination that will suit your needs or application.

ECL82 (6BM8) is a triode-pentode developed for audio use in radios, tape recorders/players, and low-cost stereo systems. The 300mA heater version, PCL82, was used in the audio stages of domestic TV sets. The successor, ECL86 (6BW8), was released in 1962.

SIDE-BY-SIDE	ECL82	ECL83	ECL84	ECL85	ECL86
Also known as	6BM8	CV8534	6DX8	6GV8	6GW8
Heater	6.3V/0.78A	6.3V/0.6A	6.3V/0.72A	6.3V/0.9A	6.3V/0.7A
TRIODE: P_A / I_A max.	1 W, 15 mA	3.5W, 15 mA	1W, 12 mA	0.5W, 15 mA	0.5 W, 4 mA
gm (mA/V)	2.5	2.5	4	5.5	1.6
r_I (kΩ)	28	34	16	9	62
µ	70	85	65	50	100
PENTODE: V_A / V_{G2} max. (V)	300/300	250/250	250/250	250/250	300/300
PENTODE: P_A / P_{G2} max. (W)	7, 1.8	5.4, 1.2	4, 1.7	9, 1.5	9, 1.8
PENTODE: triode connection					
gm (mA/V)	8.5	6	11	9	12
r_I (kΩ)	1.1	1.7	3.2	0.8	1.7
µ	9.5	10	36	7	21

The amplification factors of the triode section are high, 70, and 100, respectively, but their internal resistances are also high, in the order of 30kΩ and 60 kΩ, respectively. With a 100k anode load, the triode section easily achieves a voltage gain of 65 with 3.8% distortion, not a bad result.

The triode-strapped pentode section has a decent mju or around 21 and a relatively low internal resistance of between 1.5 and 2 kΩ (depending on anode current). With an 8k2 anode load, the gain of this high current (I_0=22mA) common cathode stage is around 15, but its D_2 distortion is relatively high, around the 5% mark. Perhaps that's why it sounds warm and lush?

With an interstage or output transformer, biased at a modest level of 14 mA, the stage gain approaches mju, around 20-21, while distortion drops to 2.1% (loadline X-Y above). With higher current levels of 20-25 mA D_2 reduces to around 1.5%, which is a good result.

With the same 9k:150Ω line output transformer as in the 5670 project, the overall gain would be 20/7.75= 2.6 times or 8.3dB, perfect for a low gain line stage.

TUBE PROFILE: ECL86 (6GW8)

- Indirectly-heated audio triode-pentode
- Noval socket, heater: 6.3V/700 mA
- TRIODE: P_{AMAX} = 0.5W, I_{AMAX}= 4mA
- gm = 1.6 mA/V, r_I=62kΩ, μ=100
- PENTODE: P_{AMAX}=9W, P_{SMAX}= 1.8W, I_{KMAX} = 55mA, V_{HKMAX}= 300V
- PENTODE SECTION TRIODE CONNECTED:
- gm=12 mA/V, r_I=1.7kΩ, μ=21

ECL86 from two angles

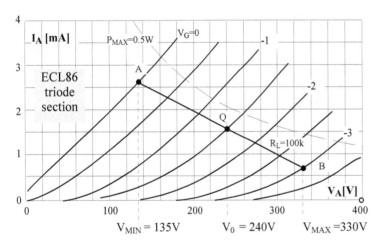

GRAPHIC ESTIMATION

QUIESCENT POINT: V_G= -1.5V, V_0=240V, I_0=1.5mA
$\Delta V = V_B - V_A$ = 330 - 135 = 195V
$A_V = -\Delta V_A / V_G$ = -195/3 = -65
DISTORTION (2nd harmonic):
ΔV_+=105V, ΔV_-=90V
$D_2 = (\Delta V_+ - \Delta V_-)/2\Delta V$ = 3.8 %

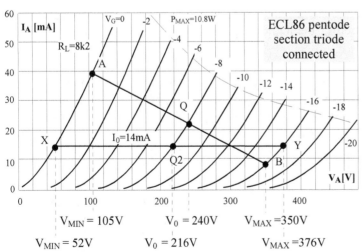

GRAPHIC ESTIMATION

QUIESCENT POINT: V_G= -8V, V_0=240V, I_0=22mA
$\Delta V = V_B - V_A$ = 350 - 105 = 245V
$A_V = -\Delta V_A / V_G$ = -245/16 = -15.3
DISTORTION (2nd harmonic):
ΔV_+=135V, ΔV_-=110V
$D_2 = (\Delta V_+ - \Delta V_-)/2\Delta V$ = 5.1 %

THE STARS OF THE AUDION ERA: ANCIENT TUBES IN MODERN AMPS

- AD1 power triode
- Design project: VT-25 line stage
- Design project: VT-25A SET amplifier with EC8010 driver stage
- 1626 (VT-137) triode
- Design project: 1626 line stage
- Design project: PSET 1626 power amplifier
- #46 dual grid triode
- Design project: Line stage with #46 or #47 tubes
- Design project: #46 or #47 PSET power amplifier

"Research is simply the manner in which men solve the knotty problems in their attempt to push back the frontiers of human ignorance."
Paul D. Leedy, "Practical Research: Planning and Design"

AD1 power triode

AD1 was the most linear power triode ever produced. Just look at those straight and equidistant lines long enough, and your eyes will get all teary, and you'll start salivating just imagining how magical those babies would sound. However, back to earth, matched pairs are rare (even single triodes are hard to find) and thus very expensive! If this book sells well (please ask all your friends, relatives, and their pets to buy a few copies each), perhaps I'd be able to afford one in a few years!

The filaments were coated by a thin layer of evaporated barium during tube activation, which made AD1's filaments very efficient. The heater power is only 3.8 watts, compared with 6.25W for 2A3, so AD1's heater consumption is 40% lower for 20% more audio power. This makes the AD1 triode the most efficient triode of all time.

However, there is a drawback, too. The filaments are sensitive to and intolerant of high heater voltages, which quickly result in grid current and loss of emission. Testing these lovelies on primitive tube testers whose heater voltages can easily exceed the nominal figures is a recipe for disaster.

Many don't even have a 4V setting, or, if they do, the voltage is the same as for the 5V setting because, in both positions, the selector switch is connected to the same tap on the power transformer!

We bought a very weak Tungsram AD1 for a few bucks and tried various rejuvenating it, but without success. It seems that its filament was permanently damaged in the past, in the manner just described. At least we could take a photo of it, reproduced at the bottom of this page.

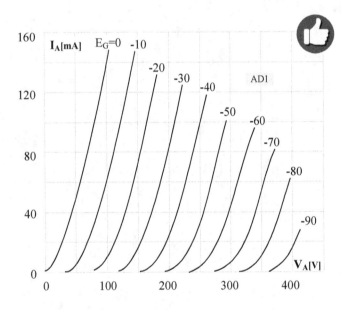

TUBE PROFILE: AD1

- Indirectly-heated power triode
- Heater: 4V, 0.95A (3.8W)
- Maximum anode voltage 250V
- Maximum anode dissipation: 15W
- TYPICAL OPERATION (Z_L=2.3 kΩ):
- V_A=250V, V_G=-45V, I_0=60 mA
- gm=6mA/V, μ=4, r_I=670Ω
- P_{OUT}= 4.2 W @ THD = 5%

LEFT: AD1 was the most linear power triode ever produced. Just look at those straight and equidistant characteristics!
BELOW LEFT: AD1 made by Tungsram in Hungary
BELOW RIGHT: Due to darkened glass the internal construction on most AD1 tubes isn't visible as it is on this specimen.

As a comparison, the modern production by Emission Labs uses barium oxide coated filaments. Their heater current is 1.5A, 55% higher, but reliability and longevity are (allegedly) improved.

Notice how similar 2A3 is to AD1, with the same plate dissipation and bias voltage. 2A3 has a slightly higher internal impedance and thus needs a slightly higher load, but its output power is lower, 3.5W versus 4.3W for AD1.

Unfortunately, the sockets and heater voltages are different; otherwise, we could have the A-B shoot-out of the millennium AD1 versus 2A3, in the same SET amplifier, of course.

Some audiophiles claim that directly-heated tubes with 4 Volt heaters (such as AD1) sound better than their 2.5V or 5V counterparts due to the optimal filament geometry, particularly the ratio between filament diameter and its length.

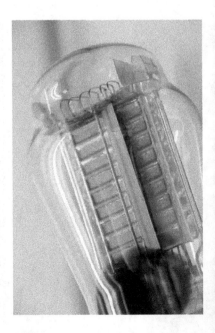

Design project: VT25 line stage

VT25 and VT25-A

Succeeding UV210, UX210 hails from 1925. Several US tube manufacturers manufactured it until the early 1930's when it was replaced by type 10 triode. Western Electric used its own naming convention and called it VT-25. The 10Y version soon followed, from which 801A or VT-62 variety was developed, with an increased anode dissipation of 20 Watts.

While VT25 has a thoriated tungsten filament that produces a bright glow, the uprated VT25A uses an oxide-coated filament that glows very dimly, just as 2A3 and 300B tubes do. VT25A has a higher anode power rating of 15 Watts.

TUBE PROFILE: VT-25 (10Y)

- Directly-heated triode
- 4-pin socket, heater: 7.5V/1.25A
- V_{AMAX}=425V, P_{AMAX}= 10 W
- gm=1.5 mA/V, µ=8, r_I = 5.3 kΩ

The German competitor: Da triode

Da directly-heated triode was developed for telephony applications and first released in 1928. The glass bulb shape evolved over the decades, from pear-shaped to the most modern straight version pictured (by Siemens). Even the internal structure varied so widely that just by looking at the anode and grid structures, you would never guess it was the same triode.

Compared to VT25, it has a 30% higher anode power rating of 13 Watts (halfway between VT-25 and VT-25A), higher transconductance, and almost four times lower internal impedance, which seem promising. However, the amplification factor is less than half that of VT-25, so higher grid driving voltages are needed.

Notice that despite its higher power rating, the heater power consumption of Da triode is only P_H=5.8*1.1 = 6.4 Watts, compared to 9.4 Watts for VT-25, meaning this German masterpiece is a much more efficient design than its American counterpart!

TUBE PROFILE: Da

- Directly-heated triode
- European telephone 5-pin socket (G5K)
- Heater: 5.8V/1.1A
- V_{AMAX}=230V, P_{AMAX}= 13 W
- gm=2.5 mA/V, D=27.5 (µ=1/D = 3.6), r_I =1.45 kΩ

VT-25 and VT-25A as output tubes

VT25 and its family are small transmitting tubes, and that fact is reflected in their anode characteristics. A very high anode supply voltage is needed to get even a Watt or two of audio power out of the SE output stage, complicating the power supply design. The high internal resistance (around 5kΩ, compared to 800Ω for 300B!) requires output transformers with high primary impedances, at least 10kΩ, preferably 12-15kΩ.

Since the heaters are rated at 7.5V, AC heating is out of the question. Remember, even 5V_{AC} heating of 300B is a borderline case, making it very hard, if not impossible, to keep the hum down to the level demanded by the horn and other high-sensitivity speakers.

On the other hand, DC heating brings its own issues to the fore. Various types of DC supplies (regulated voltage supply, multiple LC filters, or regulated constant current supply) affect the sonics, usually negatively. Plus, there is always a voltage drop (gradient) across the filament with DC heating, which wears out unevenly, thus reducing the tube's life. Sonically, DC heating is inferior to AC heating, so another example of the "Solve one problem, create two or three new ones" rule!

The oxide-coated variety (VT25A) is more forgiving when it comes to heater supplies. Still, most audiophiles prefer the sound of the thoriated tungsten versions, so that is not much of a consolation.

Due to the AC heater hum issue and the required high anode supply voltages, VT25 and VT25-A are demanding triodes, so only experienced constructors should consider using them in their projects. The sonic rewards, however, are immense. These lovelies sound incredible; in a word, they sound right!

VT-25A line stage with resistive load (R_L=15k)

We have chosen 400V anode supply voltage under the 450V limit of common filtering elcos. If you opt for polypropylene film capacitors, which usually go up to at least 600 V, your load line can be made more horizontal while avoiding the low anode current region below 5 mA!

The DC load line has anode and cathode resistances in series (15k+1k2=16k2), but for the AC signal, the cathode resistor is bypassed; the 180k output resistor has to be added in parallel to the anode resistance, making the anode AC load 13k8. Biased at -15V and 11.5 mA (Q1), the voltage gain is around -5, and the 2nd harmonic distortion is very close to zero! Although there is no need for an anode choke to reduce distortion, let's see what that option (Q2) would give us.

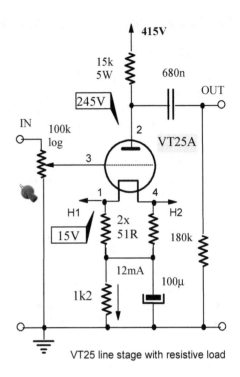

VT25 line stage with resistive load

GRAPHIC ESTIMATION

A-Q1-B OPERATION:
V_{BB}=400V, V_{G0}= -15V, V_o=230V
ΔV_G = 30V, ΔV_A=320-140 = 180V
A_1=-ΔV_A/ΔV_G= -180/30 = -5
$V+$ = 90V, $\Delta V-$ = 90V
D_2= (ΔV_+- ΔV_-)/2ΔV_A= 0%

X-Q2-Y OPERATION:
V_{G0}= -15V, V_o=250V
ΔV_G=30V, ΔV_A=375-125=250V
A_2= -ΔV_A/ΔV_G= -250/30 = -8.3
$\Delta V+$ = 125V, $\Delta V-$ = 125V
D_2= (ΔV_+- ΔV_-)/2ΔV_A= 0%

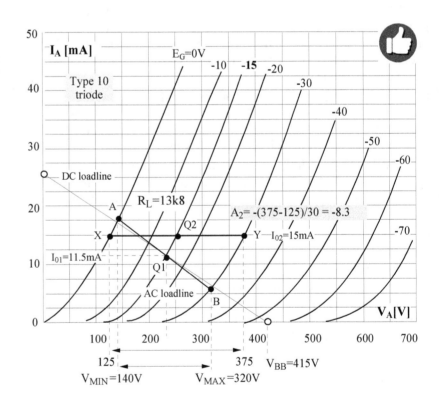

VT-25A line stage with inductive load (100H anode choke)

The impedance of the anode choke at 1kHz is around 628kΩ (X_L=ω_L = 2πfL), which the signal sees as parallel to the 180k grid resistor of the output tube. The impedance of the 220n coupling capacitor is only 723Ω at 1 kHz; such a small impedance compared to 180k and 628k can be considered a short circuit and neglected.

Since the voltage across the anode choke is 90 degrees ahead of the voltage across the 180k resistor, the two impedances should not be added algebraically as simply being in parallel but should be added as vectors. The modulus of the impedance can be calculated using Pythagora's theorem: $Z=\sqrt{(Z_L^2+R^2)} = \sqrt{(658^2+180^2)} = 653k\Omega$! On I_A-V_A characteristics, such a high impedance is almost a horizontal line.

We positioned V_{02} at 250V, so from the graph, we read the required grid bias in Q2 (intersection with I_A=15mA), which is around -15V, meaning the cathode must be at grid potential +15V! The voltage gain of the LC coupled stage is much higher, around -8.3 times, an increase of 66% compared to RC coupling.

Design project: VT-25A SET amplifier with EC8010 driver stage

We had a pair of large SE output transformers with 12k primaries that we wound for GM70 beasts, so let's see what could be expected from the VT25-A power stage with such a load (graphic estimation below right). To get any decent output power, the load line must be positioned to maximize both the voltage and the current swing.

2.2 watts with only 0.55% distortion is very promising, better than a 45 triode, for instance.

Assuming a cathode bias, for +40V cathode voltage with 24mA current, the cathode resistor needs to be 40/24= 1.67k. If a standard value 1k8 resistor is used, the actual bias will be $V_K=1.8*24 = 43.2V$. The power dissipated into heat on the cathode resistor is $P=V_K*I_K = 43.2*24 = 1W$, so, as a minimum, a 3W rated resistor is needed.

Since the output tubes are biased at approx. -40V, a driver stage with a voltage gain of around 30 is needed, which means the driver triode must have mju above 40 or 45 and high driving power. 5687, 12BH7, 6SN7, and 6CG7 duo-triodes could drive VT25A's grids, but their mju is less than half required.

EC8010 is a virtually unknown and rare SQ tube. It has an anode power rating of 4.5 Watts, low internal resistance, and a high mju (60), a very rare combination indeed! More details on pages 83-84.

Starting with Q at 13mA and 135V, a very conservative choice, far from EC8010's maximum power dissipation limits, a 12k load would give us a voltage gain of -46 with 1.5% distortion, which isn't a bad result. To get a lower distortion we could try moving the quiescent point towards higher anode voltages and currents (Q2).

Not just a slightly lower distortion, but the maximum voltage swing is also increased (higher headroom), two additional benefits. The input power is $P_{IN}=I_0*V_0= 165*0.017 = 2.8$ Watts, still way below the maximum allowed of 4.5 Watts. These beauties can be pushed even harder towards Q3, but we are not sadists, so let's stay with Q2!

GRAPHIC ESTIMATION

V_{G0}=-40V, V_0=500V, I_0=24mA
V_{MIN}=270V, V_{MAX}=725V
I_{MIN}=5mA, I_{MAX}=44 mA
ΔI_A=39mA, ΔV_A=455V
$P_{OUT}= \Delta V_A*\Delta I_A/8$
$P_{OUT}= 455*0.039/8 = 2.2W$
$D_2=(V_+ - V_-)/2\Delta V_A$
$D_2= (230-225)/2*455= 0.55\%$

GRAPHIC ESTIMATION

A-Q1-B OPERATION:
V_{G0}= -1.75V, V_0=135V
$\Delta V_A=V_{MAX}-V_{MIN}= 216-55 = 161V$
$A_V=-\Delta V_A/V_G = -161/3.5 = -46$
ΔV_+=80V, ΔV_-= 75V
$D_2 = (\Delta V_+ - \Delta V_-)/2\Delta V *100 [\%] =$
$5/(2*161)*100 = 1.5 \%$

X-Q2-Y OPERATION:
V_{G0}= -2V, V_0=165V
$\Delta V_A=V_{MAX}-V_{MIN}= 260-65 = 195V$
$A_V=-\Delta V_A/V_G = -195/4 = -48.75$
ΔV_+=100V, ΔV_-= 95V
$D_2 = (\Delta V_+ - \Delta V_-)/2\Delta V *100 [\%] =$
$5/(2*195)*100 = 1.3 \%$

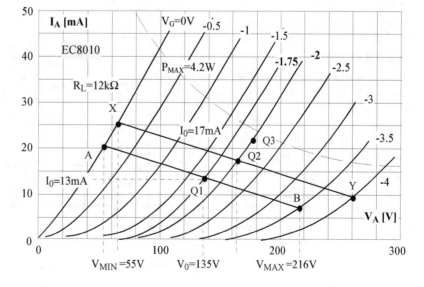

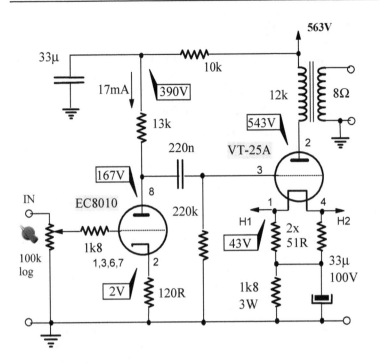

An observant reader will notice a discrepancy between the graph of the anode characteristics and the circuit diagram. The graphical estimation assumes a bypassed cathode resistor of the driver stage while the circuit diagram shows it un-bypassed. That decision is ultimately up to you.

With un-bypassed R_K the gain will be reduced from 49 down to 33, so the local NFB, in this case, is $20\log(49/33.7) = 20\log 1.452 = 3.24$ dB!

If voltage gain of 33 is sufficient for your application, as it should be, assuming an input signal of between 1 and 2 V_{RMS}, by all means, leave R_K un-bypassed, as per the diagram.

LEFT: The initial design of the VT-25A SET amplifier (audio section, one channel only)

LOCAL NEGATIVE FEEDBACK DUE TO THE UN-BYPASSED CATHODE RESISTOR

When R_K is bypassed to ground (for AC signal) by a capacitor, the voltage gain (without feedback) is $A = -\mu R_A/(r_I + R_L)$, and with local negative feedback through the un-bypassed R_K the gain is $A_F = -\mu R_A/[r_I + R_A + (1+\mu)R_K]$. By mathematical manipulation it can be shown that $A_F = A/[1+(R_K/R_A)A]$, so the feedback ratio is $\beta = R_K/R_A$ and the feedback factor is $FF = 1 + \beta A$!

Practical implementation

The available power transformer had a 380V secondary (unloaded voltage), which was a bit low, and it didn't have a CT (center tap), so a hybrid rectification scheme was chosen. Two 800V 3A diodes were added to the 5V4G rectifier. The indirectly heated rectifier provides a time delay in applying the high anode voltage to directly heated output tubes (and preamp tubes as well, of course), prolonging tube life. Directly heated tubes heat up quicker, so by the time the anode voltage ramps up, their filaments are ready to supply electrons through thermal emission.

The transformer didn't have two 7.5V 2A windings for VT-25A heaters either, but it had two 6.8V (unloaded) 4A windings, so they were used instead. Each one also powered the heaters for EC8010 lovelies.

Since the heater load in each winding was less than half of the rated amperage, the voltage only dropped from 6.8 to 6.6V; with a full 4A load, it would have been precisely 6.3V.

The power triodes are underheated by 12% (6.6/7.5=0.88 or 88%), which reduces hum significantly.

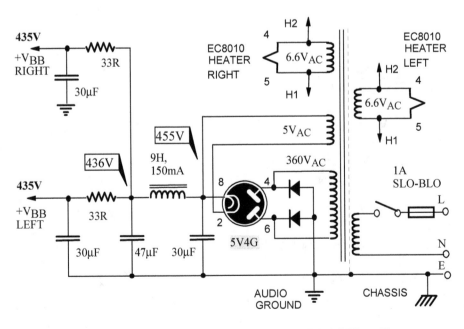

The as-built power supply and audio section (next page) of the VT-25A SET amplifier

The hum was just detectable on 87dB/W speakers.

Adding an auxiliary transformer to provide the nominal heating voltages and using DC heating would have increased the complexity and negatively affected the sonics; directly-heated triodes sound better when AC heated!

Filtering capacitors are polypropylene film motor-start type, rated at 280 V_{AC}, good for up to $2.8*280 = 780\ V_{DC}$.

The anode supply voltage for the first stage was higher than in the initial design, 380V instead of 300V, which increased the anode current from 13mA to almost 20mA. As a result, the bias voltage also rose to 2.35V.

The idle current of the output stage is 33.5V/1.7 = 19.7V, which is about 20% lower than the planned 24mA. Likewise, the bias is proportionally lower, -33.5V instead of the planned -40V.

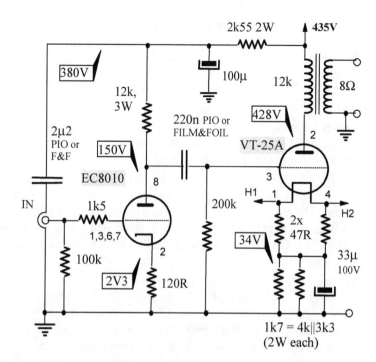

The decoupling capacitor (100µF) should be bypassed by a high-quality paper-in-oil or film & foil capacitor of 1-10µF (2.2mF here), soldered to the ground lug of the input RCA socket. This improves its effectiveness and reduces hum and hiss.

Obtaining output transformers with 12kΩ primary impedance is not easy or cheap (unless you wind your own, as we've done here). Alternatively, parallel two or even three VT-25A (in 2016 priced at US$150-200 per pair or even higher) and use more common and thus cheaper 5k transformers for two or 3k5 transformers for three in parallel.

Due to the much lower anode supply voltage, lower bias, and underheating of power tubes, the maximum output power was limited to 2 Watts; otherwise, 3 Watts would most likely be possible!

Topology and construction issues

The chassis was a black powder-coated cash drawer. The centered hole at the back (interface cable entry to the cash register) was enlarged and used to mount the rectifier tube's (1) octal socket.

The two output transformers are flanking it (2), with a 30µF motor start capacitor in front of each (3).

The power transformer is in the front enclosure (4) and the power switch (5) is on the side towards the back, very close to the mains inlet at the rear.

There is no volume control, so with RCA inputs less than an inch from input tubes (6) a minimal signal path length (MSPL) is achieved, obviating the need for shielded audio cables.

The front opening was covered by a jarrah (Australian hardwood of dark red color) fascia (7). Luckily, one of the standard widths sold in hardware shops is 80mm, exactly the height of the chassis, so only a single cut was required.

It was possible to attach the timber fascia with self-tapping screws from the inside; that way, nothing would be visible from the front, but we opted for thin bolts from the outside and nuts on the inside.

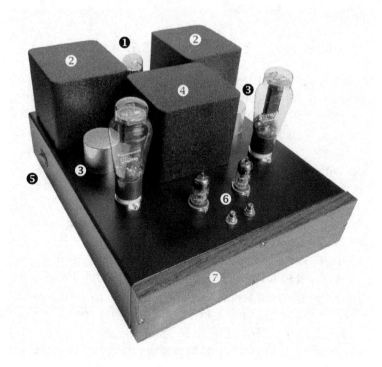

Measurements and sonic impressions

The damping factor was measured as $DF = V_L/(V_O-V_L) = 3V/(4V-3V) = 3.0$. Since $DF = R_L/Z_{OUT}$, the output impedance is very low: $Z_{OUT} = DF/R_L = 3/8 = 0.375\Omega$, a great result for a SET amplifier with no overall NFB.

The square wave reproduction isn't great; the 20kHz square wave is warbled beyond recognition, another example of an amp that sounds much better than it measures.

Pairing a 2 Watt amplifier with 87 dB/W speakers is usually considered insane, but in this case, it was a marriage made in heaven. The amplifier had so much driving power that you'd think you were listening to a 25-watt SET beast using one of the big bottles (211, GM70, 845). Indeed, that's how it sounded, plenty of slam, very dynamic and "confident".

Compared to these large transmitting tubes, which lack the "tenderness" and "microdetail" of smaller DHT brethren such as 2A3 or 300B, this VT25-A amplifier had better microdynamics and resolution than 300B amplifiers and, again, a better driving power. 300B is a much more powerful tube than VT-25A (more than twice the anode power rating), but it sounds meek in comparison.

The bass was very prominent, fast and tight for a SET amplifier. The seductive midrange and incredibly detailed and sparkling treble completed a well-rounded sonic presentation. In conclusion, this is one of the best sounding amps featured in this Volume.

MEASURED RESULTS:
- BW: 17Hz - 32 kHz (-3dB, 1W out)
- Output power for 1V input: P_{OUT} = 2W ($4V_{RMS}$ into 8Ω load))
- Z_{OUT} = 0.375 Ω (DF=3.0)

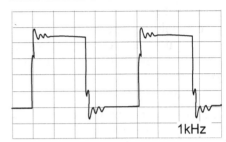

1kHz

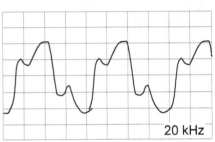

20 kHz

1626 (VT-137) triode

Introduced in 1941, VT-137 is a transmitting triode designed to work as an RF oscillator. It also comes under the names of CV-1755 and 1626. It has a mju of 5, and at low anode currents an internal resistance of 2.5 kΩ, meaning the transconductance is 5/2.5 = 2 mA/V. The grid can take some current to work in class A2; that is why the anode graph shows IA-VA curves for positive bias voltages (next page).

TUBE PROFILE: 1626 (VT-137)
- Indirectly-heated triode
- Octal socket
- Heater: 12.6V/0.25A
- V_{AMAX} = 250V_{DC}
- P_{AMAX}=6W, μ=5
- I_{GMAX}=8 mA, I_{AMAX}=25 mA

"Darling" and "Double Darling" SET amplifiers

Bob Danielak and Jeremy Epstein are two designers and constructors behind the family of flea-powered designs known as Darling amplifiers, ranging from a single, capacitive-coupled 1626 tube to a pair of DC-coupled 1626 as output tubes. Bob favored a 7-pin miniature 8532 triode as a driver tube, while Jeremy used 6SN7.

Most designs have 1626 cathode-biased at -27 or -28 V with 28mA of cathode current, while the anode-cathode voltage is around 220-230V.

As with many designs published in magazines such as "Sound Practices" and "Glass Audio" in the late 1990s, no performance parameters of any kind were given, no distortion levels, no damping factor, no frequency range.

Tiny output transformers were used, even those for guitar amps (that don't reproduce much above 10 kHz or below 100 Hz), so my prediction is that their frequency range is severely curtailed and that the bass is woefully inadequate.

Verdict? The distortion levels seem OK, but the operating point is positioned above the curve of the maximum plate dissipation, so these triodes will have a short and stressful life if operated that way.

Also, the 10k load for one or 5k load for two triodes in parallel seems too high - 3k and 6k would be better, so 0.7 Watts output power levels (1.5 Watts for two in parallel) should be possible.

In his design published in issue 15 of "Sound Practices," Danielak connects the cathodes of 8532 driver triodes together (both channels!), so they share the 500W cathode resistor and 470mF bypass capacitors. He does the same with cathodes of output tubes. The anode resistors are 46k, and the grid resistors of the output stage are 220k.

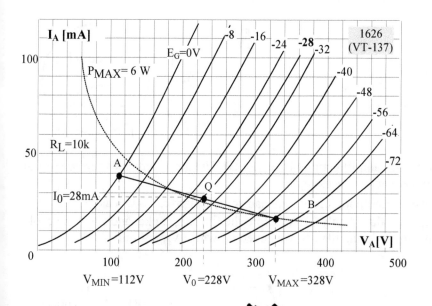

"DARLING" AMPLIFIER -
GRAPHIC ESTIMATION

$R_L=10k$, $V_0= 228V$
$I_0=28mA$, $V_{BIAS}= -28V$
$P_{IN} = 228V*28mA = 6.4 W$
$V_{MIN}=112V$, $V_{MAX}= 328V$
$\Delta V_A= 216V$
$I_{MIN}= 38mA$, $I_{MAX}=17mA$
$\Delta I_A= 21mA$
$P_{OUT}=(\Delta V \Delta I_A)/8= 0.6W$
$D_2= (V_{AQ}- V_{QB})/2\Delta V_A$
$= (116-100)/2*216 = 1.4\%$

GRAPHIC ESTIMATION

$V_{BB}= 265V$, $V_0= 130V$
$V_{MIN}= 35V$, $V_{MAX}= 194 V$
$\Delta V_A= 159V$, $\Delta V_G=4V$
$A= -\Delta V_A/\Delta V_G = -159/4 = -40$
$D_2= (V_+ - V_-)/2\Delta V_A= (95-64)/2*159 = 9.8\%$

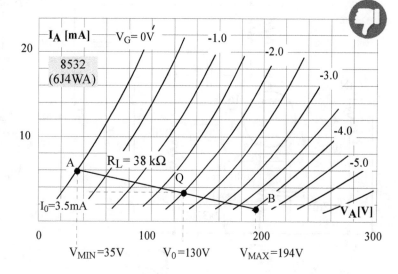

As far as the 8532 tube is concerned, the one Mr. Danielak used as a driver, we didn't have to calculate this horrendously high distortion of 10%; one look at its anode curves tells you everything. The curves are far from equidistant, meaning the distortion will be very high. This tube is highly nonlinear and thus unsuitable for audio use!

1626 line stage

Let's see how 1626 would perform in a line stage. With -8V bias and 22mA of idle DC current, the anode-cathode voltage is only 112V, which is a good sign!

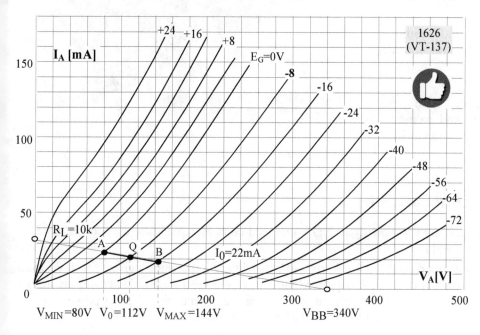

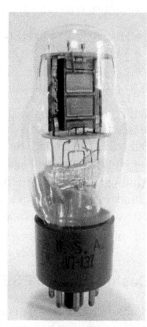

ABOVE: How cute can they get? 1626 seem large on the photo, but are tiny in real life!

GRAPHIC ESTIMATION

V_{BB}= 340V, R_L=10k, V_0= 112V, I_0=22mA
V_{MIN}= 80V, V_{MAX} = 144 V, ΔV_A= 64V, ΔV_G=16V, A= -64/16 = -4.0
D_2= $(V_{AQ}-V_{QB})/2\Delta V_A$= (32-32)/2*64 = 0/128 = 0%

The verdict? As a preamplifier, just enough voltage amplification (4x or 12 dB). The graph indicates zero distortion under these conditions, so no local or global feedback is necessary.

Our tubes, a NOS quad made by Sylvania, were microphonic, and one had some heater-cathode leakage (around 3 MΩ on Triplett 3444 tester). Our power supply produced only 315V before the filtering choke, and since the choke had a high DC resistance, its voltage drop was quite high, 40V, so we ended up with +V_{BB} of only 275V.

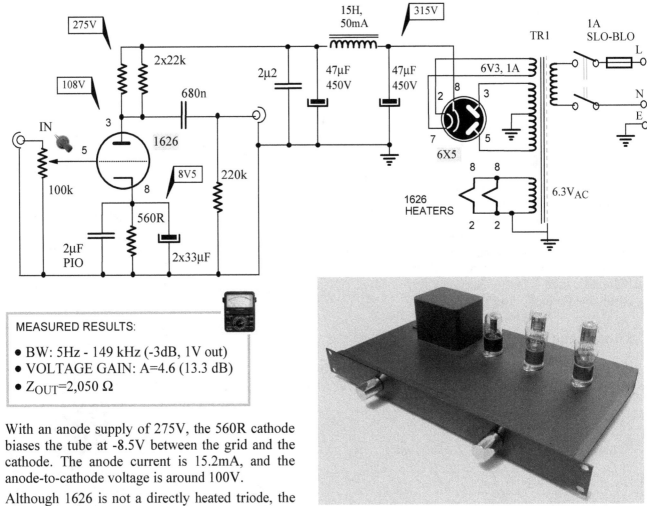

MEASURED RESULTS:

- BW: 5Hz - 149 kHz (-3dB, 1V out)
- VOLTAGE GAIN: A=4.6 (13.3 dB)
- Z_{OUT}=2,050 Ω

With an anode supply of 275V, the 560R cathode biases the tube at -8.5V between the grid and the cathode. The anode current is 15.2mA, and the anode-to-cathode voltage is around 100V.

Although 1626 is not a directly heated triode, the sound is very DHT-like, engaging, and palpable, yet very clean and transparent, a rare combination.

Recommended!

Design project: PSET 1626 power amplifier

Since 1626 triodes aren't expensive and there are plenty of NOS, matched pairs and quads are easy to get, a PSET power amplifier becomes a viable proposition.

We had a pair of toroidal output transformers, rated at 10 Watts and 2k5 primary impedance, perfect for three 1626 in parallel. Three paralleled 1626 have a higher anode dissipation than one VT-25A, so we should get slightly higher audio power output but at a much lower anode supply voltage, which means a simpler and cheaper power supply. The heating voltage is standard 12.6V; six tubes would draw 6x0.25A = 1.5A, too easy.

The published anode curves for 1626 go up to 120mA, which is weird; nothing over 50mA can be used, so we redrew the curves up to 50mA only, or, in this case, up to 150mA, for three tubes in parallel. Since the anode-to-cathode voltage isn't high, we'll use cathode biasing, so the anode supply voltage needs to be 235+32 = 267V_{DC}.

The cathode resistor needs to be 32V / 0.07A = 457Ω; a standard value of 470Ω can be used. The power wasted into heat on that boy will be P=V*I = 32*0.07 = 2.2 Watts, so a 5 or 7 Watter is needed. Alternatively, we could parallel two 920Ω, 2 Watt resistors, which will give us 460Ω, 4 Watts.

With a 47μF bypass capacitor, the lower -3dB frequency of the output stage will be 7.2 Hz, perfect! Alternatively, use two 33μF caps in parallel.

From the graph, we can estimate the voltage gain of the output stage as A_V= -($\Delta V_A/\Delta V_G$)= -(330-115)/64 = -3.4, so the anode-to-grid capacitance will be increased due to Miller's effect to 4.4*3.4 = 15 pF.

For one tube, the total input capacitance is around 18.2pF, or 55pF for three in parallel, lower than the input capacitance of one 300B triode!

With a 1V_{RMS} signal at the input, the 1st stage needs a gain of 32/1.41 = 22.7 times. Many single and duo triodes can achieve a voltage gain of 20-25 times.

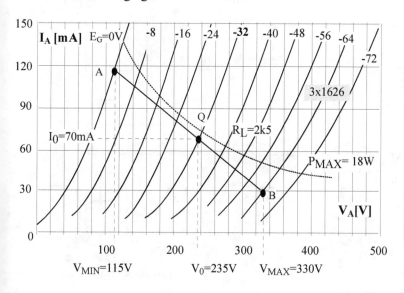

GRAPHIC ESTIMATION

V_0= 230V, I_0=70mA, V_{G0} = -32V
P_{IN} = 235V*0.07 = 16.4 W
V_{MIN}=115V, V_{MAX}=330V, ΔV_A=215V
I_{MIN}= 30mA, I_{MAX}=115mA, ΔI_A= 85mA
P=($\Delta V_A \Delta I_A$)/8= 2.3W
D_2= (V_{AQ}- V_{QB})/2ΔV_A
D_2= (120-95)/2*215 = 5.8%

Let's use our favorite, ECC40 duo-triode. With a lowish anode voltage of 165V, two paralleled ECC40 triodes achieve a voltage gain of 27 times, with 2.2% 2nd harmonic distortion, so let's proceed on that basis.

Just in case we opt for DC-coupling, we should not go wild and choose a very high anode current and high bias, which would move the quiescent point Q too far to the right, requiring a very high anode voltage. The cathode resistor of the 1st stage must be R_K=4/0.005 = 800Ω, so use 820Ω, a standard value.

We have already determined that the cathode voltage needed was 32V, but that was for capacitive coupling. Now we need 32+164 = 198V!

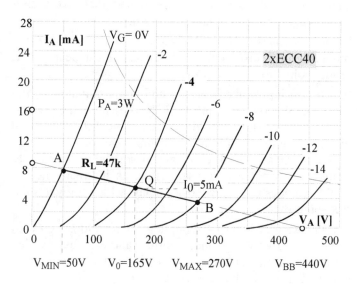

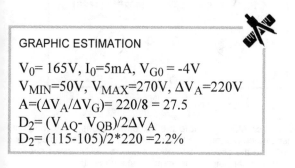

GRAPHIC ESTIMATION

V_0= 165V, I_0=5mA, V_{G0} = -4V
V_{MIN}=50V, V_{MAX}=270V, ΔV_A=220V
A=($\Delta V_A/\Delta V_G$)= 220/8 = 27.5
D_2= (V_{AQ}- V_{QB})/2ΔV_A
D_2= (115-105)/2*220 =2.2%

With DC coupling, the anode DC voltage on the driver stage must be added to the cathode voltage of the output stage, which in turn increases its anode voltage and the required anode supply voltage by the same amount, so V_{BB} now needs to be 235+198 = 433V_{DC}.

The new value for the output stage's cathode resistor is 198V/0.07A = 2,828Ω, so a standard value of 2k7 will be fine.

The power wasted into heat will now be P=V*I =198*0.07 = 13.9 Watts, so a 20 Watt resistor is required.

Alternatively, use three 8k2 resistors in parallel, rated at a minimum of 7 Watts each.

TUBE PROFILE: ECC40

- Medium μ dual triode
- RimLock socket, 6.3V, 0.6 A heater
- V_{AMAX}= 300V, V_{HKMAX}= 175V_{DC}
- P_{AMAX}= 1.5 W
- TYPICAL OPERATION:
- V_A=250V, V_G=-5.5V, I_0=6 mA
- gm=2.9 mA/V, μ= 32, r_I = 11 kΩ

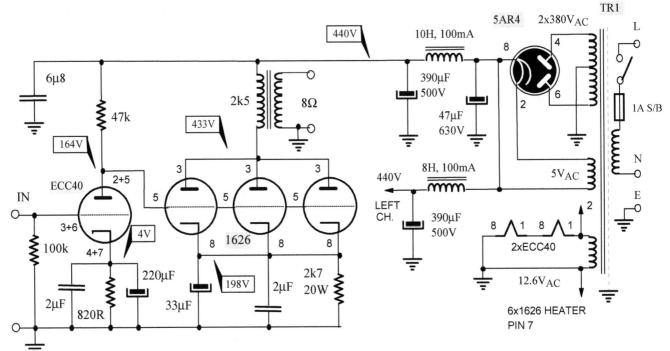

Since there are only two stages and we are not using any NFB, no anode decoupling is needed - oscillations cannot happen. It could even be argued that using the same anode supply voltage for both stages, especially when they are directly coupled, as they are here, makes them work as a single stage where all voltages swing in unison.

If your chassis is tall enough (above say 60-70mm), the two small filtering chokes, one for each channel, can be mounted inside, bolted to the side walls.

Before the input tube heats up and starts conducting, the anode supply voltage of 440V will be present on the grids of output tubes, which could shorten their life and cause premature failures. Therefore, some kind of time delay should be incorporated, especially if solid-state or directly heated rectifier tubes are used. Here, an indirectly heated 5AR4 provides a few seconds of delay, which mitigates the risk somewhat.

Just in case we opt for DC-coupling, we should not go wild and choose a very high anode current and high bias, which would move the quiescent point Q too far to the right, requiring a very high anode voltage.

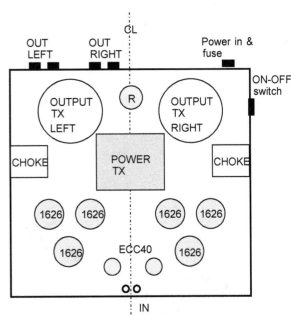

ABOVE: Although there are six power tubes, their footprint isn't much bigger than that of other small octal tubes such as 6V6, so the chassis can be relatively small. Filtering chokes and toroidal output transformers are under the chassis; the power transformer is mounted in its own enclosure on top.

THE STARS OF THE AUDION ERA: ANCIENT TUBES IN MODERN AMPS

The output power was higher than anticipated, with anode supply voltages of 450-470V, we measured up to 11 V_{RMS} on an 8Ω load! Most likely, the output stage crossed in Class A_2, since it should not be possible to get such high power levels in class A_1.

MEASURED RESULTS:
- V_{MAX} = 7.2V, P_{MAX}=6.5W
- BW: 21Hz-48kHz (-3dB@1W)

Design project: Line stage with #46 or #47 tubes

Type 46 dual grid triode

This obscure tube was initially included in this volume because it seemed an interesting educational case study. However, once we actually built a line stage with these cuties, we loved their sonics!

Although it has four electrodes, #46 is not a tetrode but a dual-grid directly-heated triode. When G2 is connected to the control grid G1, the amplification factor is high, as is the internal resistance, so the anode characteristics are reminiscent of a pentode (anode curves below).

Notice that biasing is positive, so a significant grid current flows. This is one service #46 was designed for, to work as a class B audio power amplifier in a push-pull output stage.

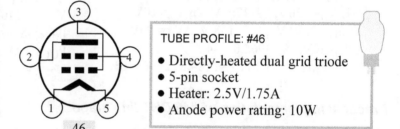

TUBE PROFILE: #46
- Directly-heated dual grid triode
- 5-pin socket
- Heater: 2.5V/1.75A
- Anode power rating: 10W

If you like McIntosh sound and wish to build its circuit, but with directly-heated triodes instead of the highly nonlinear 6L6 tubes that McIntosh chose for most of its designs, then #46 is your prime candidate. Sure, its anode power rating is much lower than 6L6, but connect two or three in parallel, and you are way in front!

You may not need the multiple (double, triple, and even quadruple) negative feedback that McIntosh had to use, simplifying the design and winding of the output transformers.

With a relatively low voltage of only 400 volts on anodes and 12mA of zero-signal anode current, two in push-pull can produce 20 watts of output power.

The average power input to the grids is 650 mW, so a powerful and low internal impedance driver tube is needed.

This brings us to the second task #46 triode was designed for, and that is if you haven't guessed it already, to drive itself! Or, more precisely, to drive two of its kind in the push-pull output stage. A very clever idea by the ever prudent and practical Yanks.

When G2 is strapped to the anode, we recognize typical DHT curves, equidistant and linear (next page).

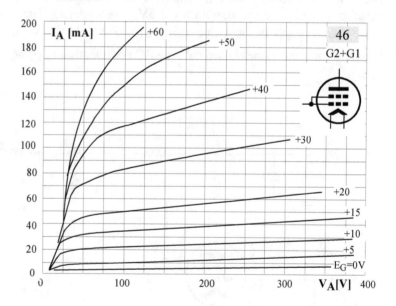

The maximum anode power dissipation is 10 Watts, and the static triode parameters are μ=5.6, gm=2.4 mA/V, r_I= 2k3. Compared to 300B, 2A3 and other DHTs with 700-800Ω internal resistance, a relatively high internal resistance of 2.3kΩ is #46's only drawback.

Verdict? As a preamplifier, a reasonable voltage gain at very low distortion levels (the graph indicates zero distortion under these conditions).

As a Class A SET power amplifier, almost 2 Watts output power at only 2.2% distortion, not bad at all. Two of these lovelies in parallel would use an output transformer designed for 300B or 2A3 service, 3k5 primary impedance, and deliver output power levels in the 2A3 class!

However, the output resistance will be higher and the damping factor lower, and to achieve the same frequency range as the 2A3 output stage, output transformers with higher primary inductance are needed.

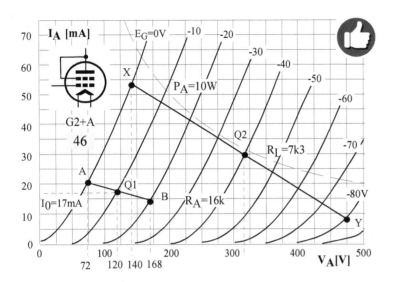

The -40V bias means the triode is relatively easy to drive, requiring up to 40/1.41 = 28V_{RMS} signal at its grid, so a single driver stage with voltage amplification of around 20 would be enough.

However, we have exceeded the maximum anode voltage of 250V, and the tube is operating close to its maximum dissipation, so 1.5W power output is a much more reasonable goal.

GRAPHIC ESTIMATION
CLASS A_1 DRIVER SERVICE (A-B):
Q1: V_0= 120V, I_0= 17mA, V_{BIAS}=-10V
V_{MIN}= 72V, V_{MAX}= 168 V, ΔV_A= 96V
ΔV_G=20V, A = -96/20 = -4.8
D_2= $(V_{AQ}-V_{QB})/2\Delta V_A$= (48-48)/2*96 = 0%

CLASS A_1 POWER SERVICE (X-Y):
Q2: V_{BIAS}=-40V, V_0= 315V, I_0=30mA
V_{MIN}= 140V, V_{MAX}= 475 V, ΔV_A= 335V
I_{MIN}= 8 mA, I_{MAX}= 53 mA, ΔI_A= 45 mA
P_{OUT}= $\Delta V_A * \Delta I_A / 8$ = 335*0.045/8 = 1.9 Watts
D_2= $(V_{XQ}-V_{QY})/2\Delta V_A$= (175-160)/2*335 = 2.2%

Practical implementation of #46 DHT line stage

The power transformer had only one 6.3V secondary and no 2.5V supply for #46 heaters, so an auxiliary heater transformer had to be added, with two 2.5V_{AC} secondaries. There was no 5V secondary for the rectifier tube either, so we used a couple of low-value resistors (0.33Ω each) to drop the voltage from 6.3V down to 5V.

The time constant of the chosen cathode RC combination is τ= 56ms, so the lower -3dB frequency is 2.8Hz.

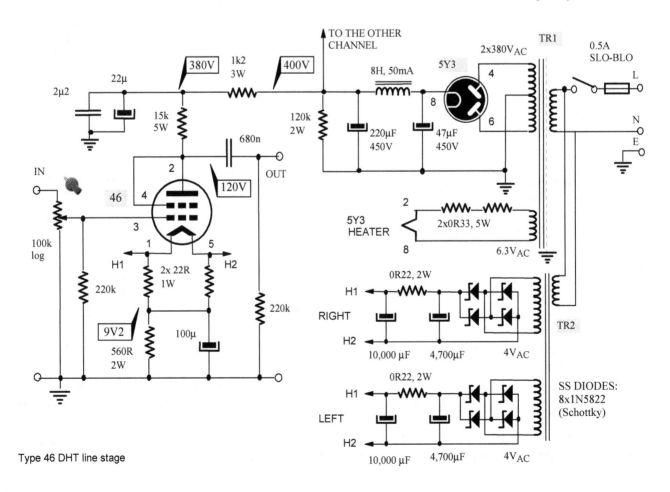

Type 46 DHT line stage

With 680nF coupling capacitor, the time constant of the output RC filter is $\tau = 220k*680n = 150ms$, corresponding to the -3dB frequency of around 1 Hz. With tube power amplifiers as a load, assuming their most common 100kΩ input impedance, that would increase to 3.4 Hz, which is still fine.

However, if this preamp is to drive a solid state amplifier with say 10kΩ input impedance, the -3dB frequency would rise to 23.5 Hz, which would be unacceptable, so in that case the coupling capacitor's value must be increased to at least 3μ3!

As in all minimalist designs, use the best quality parts possible.

Topology and construction details

The depth of the 19" rack-type chassis was close to optimal, but it was a bit too wide, so no matter which of the two topologies illustrated you choose, one channel will always have a longer input connection while the other will have a longer output connection.

Initially, we used shielded cables but then changed to the ordinary silver wire. I should say" unshielded" silver wire since there is nothing ordinary about those beautiful looking and even better sounding silver wires in various colors, salvaged from a couple of vintage HP microvoltmeters.

ABOVE: Top view of #46 triode, showing its VV filament, two grids and a boxed plate.

There was no hum introduced by the unshielded wiring, all magnetic components that radiate AC waves were enclosed and above the steel chassis, and all heater wiring was tightly twisted and run along the sides and corners of the chassis.

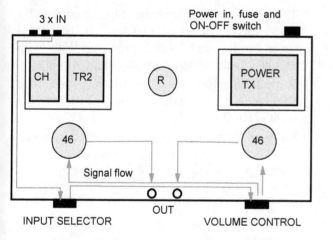

ABOVE: Preamplifier's topology, including major components and signal flow (top view), symmetrical front control knobs.

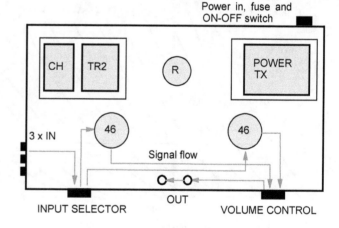

ABOVE: Alternative topology, shorter signal path and volume control at the output, still using symmetrical front control knobs.

AC hum in directly-heated tubes

AC heating was tried first, but there was too much hum, so another auxiliary transformer was substituted, with two $4V_{AC}$ secondaries. Once rectified, the value of the series resistors in the CRC filters was adjusted, so the heater voltages were exactly $2.5V_{DC}$.

Another alternative is to reduce the heating voltage from $2.5V_{AC}$ down to around $2.0V_{AC}$. The tube's life will be shortened, but #46 triodes aren't expensive, and the availability is relatively good for a tube that wasn't produced in large quantities even in its prime, probably because no commercial audio manufacturer has used this tube in their designs in the past 50 or 60 years.

Measurements and listening impressions

Of all the line-level preamplifiers featured in this book, this one sounded the best. A 3D soundstage with holographic imaging, palpable presence, smooth and chesty vocals, shimmering treble, fast and tight bass, it had it all. Even when paired with cheap & cheerful CD players such as Marantz CD52, CD63, CD67, and alike, it made them sound like $4,000+ sources!

Tube rolling option: triode-connected type 47 pentode

First introduced in 1932, type 47 pentode shares the same socket and has an identical heater (2.5V, 1.75A) as type 46 dual-grid triode. It comes under many other aliases, such as P2, 247, UY-47, UY-247, A247, VT-47, PH247, CV1772, F247, Z-47, 38047, G-47, 247S, EY647, NY-247, 447, and UY247.

As a pentode in a single-ended power stage, it can deliver 2.7 Watts to a 7kΩ load, with only -16.5V bias (see short-form specs) - it is a very sensitive output tube. If you study its internal construction, it is almost indistinguishable from the type 46 tube, apart from its additional grid, the suppressor, which is internally tied to the cathode.

Luckily, the pinout is identical to type 46, so it can be plugged into our 46 DHT line stage without any modifications.

The substitution of triode-connected #47 pentodes didn't change anything - the preamp retained its musicality and DHT character. No sonic differences between #46 and #47 tubes could be detected.

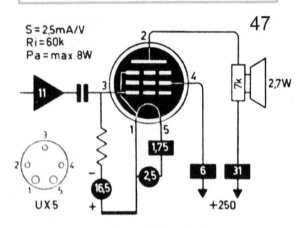

MEASURED RESULTS:
- BW: 2Hz - 169 kHz (-3dB, 1V output)
- VOLTAGE GAIN: A=3.9 (14 dB)
- Z_{OUT}=2.6 kΩ

ABOVE: Short-form specs of #47 pentode

Design project: #46 or #47 PSET power amplifier

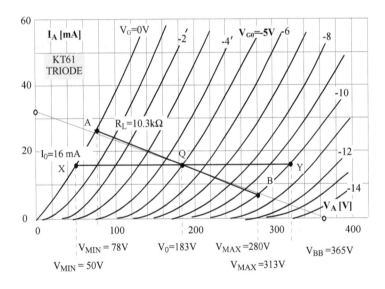

The previous graphic analysis indicated that 1.9 Watts was possible from one #46 in a SET output stage; two in parallel should produce 4 Watts with around 2% distortion, a promising possibility.

With 10k3 anode load, the preamp stage with EL33 will have a gain of around 20 times, A= -(280-78)/10 = -20.2.

Since the maximum peak signal on #46 tubes' grids is 40 V (or 28 V_{RMS}), the input sensitivity for full power will be 28/20.2 = 1.4 V_{RMS}.

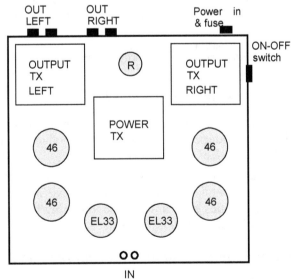

Using a 1:1 interstage transformer instead of capacitive coupling (the X-Y load line) would increase the gain of the driver stage to A= -(313-50)/10 = -26.3 times, requiring only 28/26.3 = 1.06 V_{RMS} at the input for full power.

An output transformer with a 3- 3k5 primary impedance would be needed, with only 60mA idle DC current, all at only 315V anode voltage, an easy task.

The DC bias is 40V, and allowing a $10V_{DC}$ voltage drop on the output transformer's primary winding, a high voltage power supply is needed with a 315+40+10 = 365V output.

ABOVE: The chassis for this project can be relatively small. A pair of octal sockets will accommodate tubes such as 6SN7, 6EM7 and many others, just in case you decide not to use KT33. Filtering chokes are under the chassis.

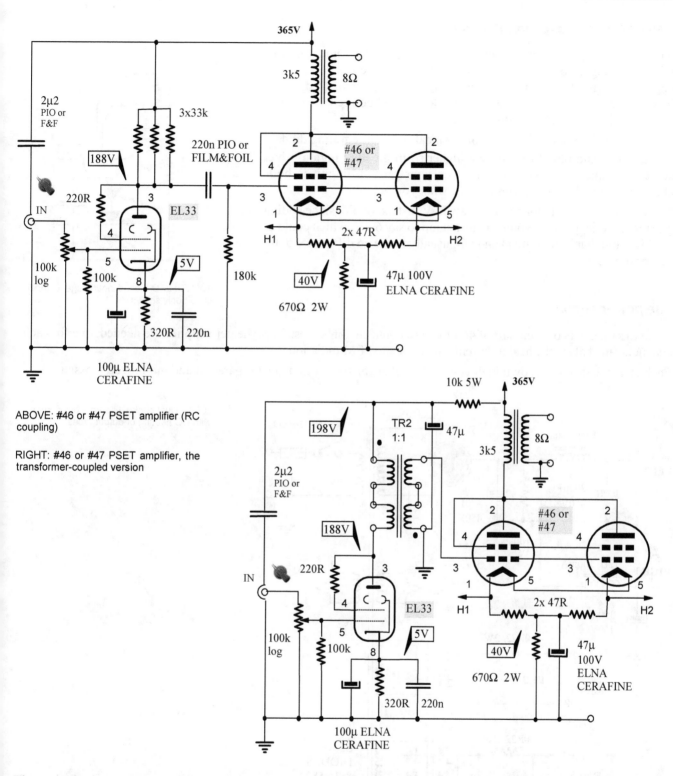

ABOVE: #46 or #47 PSET amplifier (RC coupling)

RIGHT: #46 or #47 PSET amplifier, the transformer-coupled version

The cathode time constant of the 670Ω resistance and 47μF bypass capacitor sets the -3dB of the output stage at around 5Hz, while the coupling capacitor and the 180k grid resistor form a filter with a -3dB corner frequency of around 4 Hz.

With an interstage transformer coupling, only $200V_{DC}$ is needed as the 1st stage anode supply. With 16mA current flow, the series resistor R_S must have a resistance of $R_S=(365-198)/16 = 10.3$ kΩ, so a standard 10k resistor will be perfect here. The heat it must dissipate is $P = I^2R = 0.016^2*10,000 = 2.6$ Watts, so use a 5 or 7 Watt resistor.

Over the years, we noticed that two or three paralleled anode resistors sounded better than a single one, so 3x33k were used in the final version.

The same anode supply voltage can be used for both the input and output stages. A decoupling circuit is not needed since there are only two stages; there is no possibility of the signal on that power line ever appearing in phase and thus forming a positive feedback loop and starting oscillations, as can happen in three-stage designs.

OPTION: LED-biased input stage

At relatively high currents (16mA in this case), LED (Light Emitting Diode) biasing becomes an option. The voltage drop across an LED will depend on many factors, the current level, its type or color (red, green, blue, etc.), and even its manufacturer. Not all red-colored LEDs will exhibit an equal DC voltage drop.

The dynamic internal resistance r_I of a typical red light-emitting diode is 4-5Ω. Green and blue LEDs have a higher r_I, in the 30Ω region. Thus, some experimentation is required. Once your preamp circuit is built, connect various LEDs in series and see what you get.

You may have to settle for a bias of say 4.6V instead of 5V, since adding one more diode in series would create a jump to say 6.2V, which could be too high, but that is not critical and will usually make no difference to the rest of the circuit.

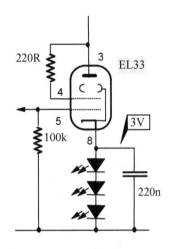

ABOVE: LED-biased input stage, cathode circuit only shown

The power supply

Since the filaments of each pair of 46 tubes are connected in series, $5V_{DC}$ heater supplies are needed, one for each channel. This halves the heater current and associated EM radiation.

Each channel also has its own high voltage LC filter for superior channel separation and minimized crosstalk.

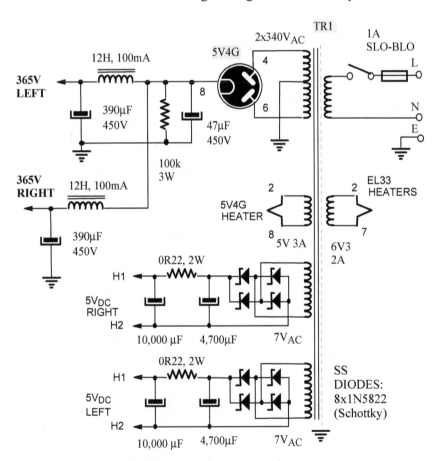

LEFT: Duel-choke filtered power supply with DC heating of output tubes

CHEAP & CHEERFUL: PREAMP & DRIVER TUBES FOR AUDIO EXPLORERS

- 6U10
- E90CC
- 12AY7
- 12B4A
- 6CK4 and 6AH4-GT
- ECC34
- E81L
- EC8010

"De gustibus non est disputandum."
("There's no disputing about taste." or "There is no point arguing over preferences.")
Latin proverb

6U10

6U10 is Compactron triple triode, one high μ triode (very similar to 12AX7), and two identical medium μ triodes of higher anode dissipation. The high μ triode #2 can be used in a high gain input stage, and the two medium μ triodes can be used in a two-tube phase splitter/driver, a paralleled driver stage, cathode follower, or SRPP stage; the possibilities are almost endless.

TUBE PROFILE: 6U10

- Indirectly-heated triode
- 12-pin Compactron socket, heater: 6.3V/600 mA
- TRIODES #1 & #3: P_{AMAX}=2W, I_{AMAX}=20mA, V_{AMAX}=330V
- gm = 2.3 mA/V, r_I = 7.7 kΩ, μ = 17.5
- TRIODE #2: P_{AMAX}=1W, V_{AMAX}=330V
- gm = 1.6 mA/V, r_I = 61 kΩ, μ = 98

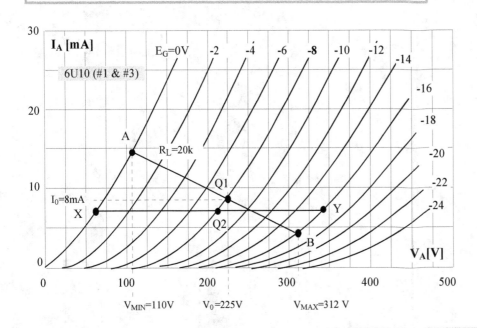

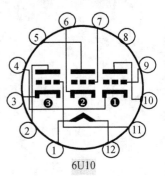

GRAPHIC ESTIMATION

X-Q2-Y operation, CCS:
V_G= -8V, V_0=210V, I_0 =7mA
VOLTAGE SWING: $\Delta V=V_Y-V_X$ = 343-63 = 280V
P_{IN}= I_0V_0 = 0.007*210 = 1.5W
A_V=-$\Delta V_A/V_G$ = -280/16 = - 17.5
DISTORTION (2nd harmonic):
ΔV_+=150V, ΔV_-=130V, $D_2=(\Delta V_+ -\Delta V_-)/2\Delta V$ *100 [%] =14/(2*280)*100 = 2.5 %

A-Q1-B operation, R_L=20k:
V_G= -8V, V_0=225V, I_0=8mA
VOLTAGE SWING: $\Delta V=V_B-V_A$ = 312-110 = 202V
P_{IN}= I_0V_0 = 0.008*225 = 1.8W
A_V=-$\Delta V_A/V_G$ = -202/16 = - 12.6
DISTORTION (2nd harmonic):
ΔV_+=115V, ΔV_-=87V, $D_2=(\Delta V_+ -\Delta V_-)/2\Delta V$*100 [%] = 28/(2*202)*100 = 6.9 %

The two medium μ triodes have an internal resistance of around 7-8kΩ. The mutual conductance isn't high, but if the two sections are paralleled, we get a respectful triode of doubled transconductance gm=4.6mA/V and halved internal resistance r_I=3.85kΩ, capable of anode currents of above 40mA (4 Watts anode dissipation)!

These are better specs than one 6SN7 triode, so that paralleled combination would make a promising driver tube. The distortion is high, so these triodes would benefit from choke-loading or a CCS (constant current source or sink), which would keep the AC load line horizontal and avoid operation in the low anode current-high distortion region.

With the constant current operation, distortion drops below 3%. With a 120k anode resistor, the amplification of a common cathode stage with #2 triode is around 67 times, which provides enough voltage swing for output triodes such as 300B and even 211, through a low impedance driver stage, of course. Zero distortion is indicated, incredible for such a large voltage swing!

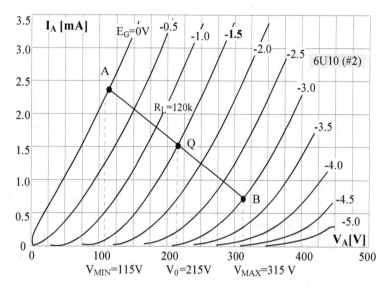

GRAPHIC ESTIMATION

R_L=120k, V_{bias}=-1.5V
V_0=215V, I_0=1.5mA
VOLTAGE SWING: $\Delta V = V_B - V_A$ =
315-115 = 200V
$P_{IN} = I_0 V_0 = 0.0015*215 = 0.32W$
$A_V = -\Delta V_A / V_G = -200/3 = -67$
$\Delta V_+ = 100V, \Delta V_- = 100V$
$D_2 = (\Delta V_+ - \Delta V_-)/2\Delta V *100 \, [\%] = 0\,\%$

E90CC

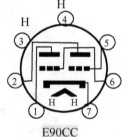

TUBE PROFILE: E90CC

- Indirectly-heated duo-diode
- 7-pin mini socket, heater: 6.3V/400 mA
- P_{AMAX}=2W, I_{AMAX}=15mA, V_{AMAX}=300V
- gm=6.0mA/V, r_I=4.5kΩ, μ=27

E90CC was developed for use in early digital computers. The inverse order of letters and numbers, as always, indicates an SQ (Special Quality) tube of high reliability and guaranteed long life of 10,000 hours. Since it wasn't developed for audio applications and its parameters are close to 12AU7 and 6SN7, commercial manufacturers did not bother to consider it, so cheap stocks of NOS tubes by Telefunken, Philips, and other reputable European brands are available.

However, linearity wasn't a factor when these triodes were designed. Hence, as with all "non-audio" tubes, do a provisional check of the anode characteristics to see what amplification factor and distortion you could expect.

These are not independent triodes, but share a cathode, so you can only parallel the two or use them in a differential amplifier or phase inverter, where two independent cathodes would be connected together externally anyway.

The -3V bias is perfect for an input stage. If the two triodes are paralleled, the anode current would be 16mA with CCS and 19mA with a load of around 13k. However, a 2nd harmonic distortion of more than 5% is not acceptable, so CCS operation is the only way to go. Even then, the distortion hovers around 3%.

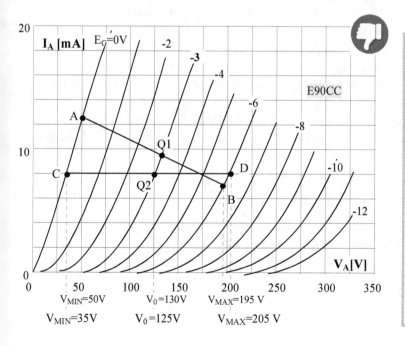

GRAPHIC ESTIMATION

<u>A-Q1-B:</u>
R_L=26k, V_G=-3V, V_0=130V, I_0=9.5mA
$\Delta V = V_B - V_A = 195-50 = 145V$
$A_V = -\Delta V_A / V_G = -145/6 = -24.2$
$\Delta V_+ = 80V, \Delta V_- = 65V$
$D_2 = (\Delta V_+ - \Delta V_-)/2\Delta V *100 \, [\%] =$
$15/(2*145)*100 = 5.2\%$

<u>C-Q2-D, CCS:</u>
V_G=-3V, V_0=125V, I_0=8mA
$\Delta V = V_D - V_C = 205-35 = 170V$
$A_V = -\Delta V_A / V_G = -170/6 = -28$
$\Delta V_+ = 90V, \Delta V_- = 80V$
$D_2 = (\Delta V_+ - \Delta V_-)/2\Delta V *100 \, [\%] =$
$10/(2*170)*100 = 2.9\%$

12AY7

12AY7 duo-triode never made the top 3 list (12AX7, 12AT7 and 12AU7) amongst Noval tubes. It has no European equivalent. By its gain (μ=40) it sits between the 12AU7 (μ=20), and 12AT7 (μ=60). It is sometimes used in guitar amps instead of the high gain 12AX7.

However, notice its highish internal resistance. With 3mA anode current, -4V on its grid and +250V on its anode, its r_I is almost 23kΩ! Remember, the lower the internal resistance of a tube, the better. High resistance means its mutual conductance is low, under 2 mA/V.

TUBE PROFILE: 12AY7

- Medium μ duo-triode
- Noval socket
- Heater: 6.3V/0.3A or 12.6V/0.15A
- V_{AMAX}=300V, P_{AMAX}= 1.5 W
- I_{AMAX}= 10mA
- gm=1.75 mA/V, μ=40, r_I = 22.8 kΩ

With a typical bias between -2 and -3V, and a power rating of 1.5 Watts, 12AY7 is only suitable for use as an input tube, not as a driver. Let's look at its anode curves and estimate the 2nd harmonic distortion.

Final verdict? With medium μ lovelies such as E86C, 6DJ8 and ECC40 around, why would one use 12AY7?

GRAPHIC ESTIMATION

<u>A-Q-B:</u>
R_L=47k, V_G=-3V
V_0=192V, I_0 =2.3mA
$\Delta V=V_B-V_A$=265-102 = 163V
$A_V=-\Delta V_A/V_G$ = -163/6 = - 27.2
ΔV_+=90V, ΔV_-=73V
$D_2 = (\Delta V_+ - \Delta V_-)/2\Delta V$ *100 [%]
= 17/(2*163)*100 = 5.2%

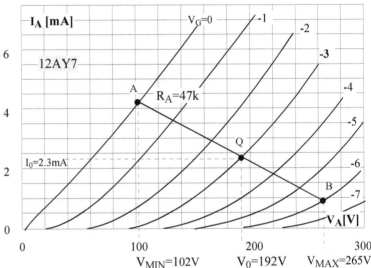

12B4A

12B4A is a miniature 9-pin low m triode, developed initially as a vertical deflection amplifier for small B&W TV receivers. Since it operates at high anode currents and low anode voltages, it seems sonically promising. It is suitable both as a single-stage preamplifier tube and in the first and second stages of power amplifiers, where gains of 4-5 times are achievable. With a total voltage gain of around 20, two triodes in a cascade could drive many power tubes to full power.

In a typical medium-current operating point (as illustrated below), its internal resistance is very low, around 1.2 kΩ, which compares favorably with high internal impedance triodes such as 6SN7 and 12AT7. The anode characteristics are fairly linear, indicating low harmonic distortion.

The tube was also used as a series-pass triode in HV regulated power supplies, most notably in HP and Tektronix instrumentation (electronic voltmeters and oscilloscopes).

TUBE PROFILE: 12B4 & 12B4A

- Indirectly-heated triode, Noval socket
- Heater: 6.3V/0.6A or 12.6V/0.3A
- 12B4A has a controlled heater warm-up time of 11 seconds
- V_{HKMAX}: 100V_{DC}, P_{AMAX}=5.5 W, V_{AMAX}=550V
- μ = 6, r_I= 1.25 kΩ, gm=4.8 mA/V

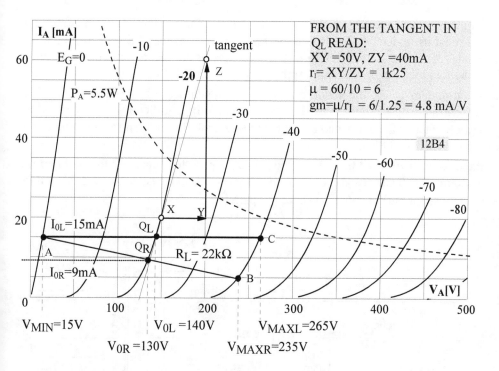

With a bias of -20V and 22k resistive anode load, the voltage gain is 5.5 times, which increases to 6.25 times with anode choke, an insignificant raise. However, the distortion drops from 2.3% to zero!

Triode parameters are not given in data sheets but can be read from the curves. The procedure is outlined in the shaded box on the left. Read r_I from the XYZ triangle and μ, and then calculate gm from the Barkhousens's equation.

GRAPHIC ESTIMATION

ANODE RESISTOR R_L =22k - loadline AB:
V_G=-20V, V_0=130V, I_0 = 9mA
VOLTAGE SWING: $\Delta V = V_B - V_A$ = 235-15 = 220V
CURRENT SWING: ΔI = 15-5 = 10 mA
Voltage gain: A = - V/V_G = 220/40 = 5.5
HARMONIC DISTORTION: ΔV_+=130-15 = 115V, ΔV_-=235-130 = 105V, $D_2 = (\Delta V_+ - \Delta V_-)/2\Delta V *100$ [%] = 10/(2*220)*100 = 2.3 %

ANODE CHOKE 100H-200H - loadline AC:
V_G=-20V, V_0=140V, I_0 = 15mA
VOLTAGE SWING: $\Delta V = V_C - V_A$ = 265-15 = 250V
CURRENT SWING: ΔI = 0 mA
Voltage gain: A = - V/V_G = 250/40 = 6.25
HARMONIC DISTORTION: ΔV_+=140-15 = 125V, ΔV_-=265-140=125V, $D_2=(\Delta V_+ - \Delta V_-)/2\Delta V*100$ = 0%

The 12B4 line stage from recyled parts

We had a pair of thick & deep timber photo frames, bought on a clearance sale for a few dollars. One was used as the power supply chassis, the other as the audio chassis. A few Neotech interconnect cables were bought at online clearance sales. These use three identical pure silver multi-stranded conductors of high current capacity (no shield), so we removed the RCA connectors from one and used them here as an umbilical cord between the two chassis.

From the HP 425A microvoltmeter, we reused a 120/240V mains power transformer, sanded down with a super fine sanding paper, and resprayed with sparkling silver paint to match the top covers cut out of a silver Perspex sheet.

The same HP voltmeter was completely wired up using pure silver hookup wire (in various colors!), which was all salvaged and reused in this and other projects.

We found an 8 Henry 50 mA filtering choke in our stash of salvaged transformers and chokes, too low a current rating for a power amp, but precisely what was needed here.

A 3-pin XLR plug and socket were used to terminate the umbilical cable on the power supply side while the audio side was permanently wired in.

For the accountants in our midst, you'd be pleased to hear that all the components for this line stage (excluding the pair of quality output coupling caps) cost us under US$60. If you are happy with "ordinary" film & foil coupling caps, that's where the total cost remains. Otherwise, feel free to line the pockets of companies making and selling "audiophile quality" or "hi-end" capacitors that cost more than a small LCD TV! Each!

As with any line stage with a low component count, the sound quality depends significantly on the quality of the components used. We used ELNA Cerafine cathode bypass capacitors; alternatively, parallel a few polypropylene film & foil (PPFF) beauties or MS (motor start) capacitors and eliminate the electrolytic caps. The 5µ6 or 6µ8 anode bypass capacitors should be high-quality PIO or PPFF.

The output coupling capacitor is the most critical of all here. We used Jensen copper foil; since all the parts for this baby are so cheap, use those savings to buy the best caps you can afford. The triode is biased at around -20V, and the anode current is 9 mA. The 22k anode resistor is three 66kΩ resistors in parallel.

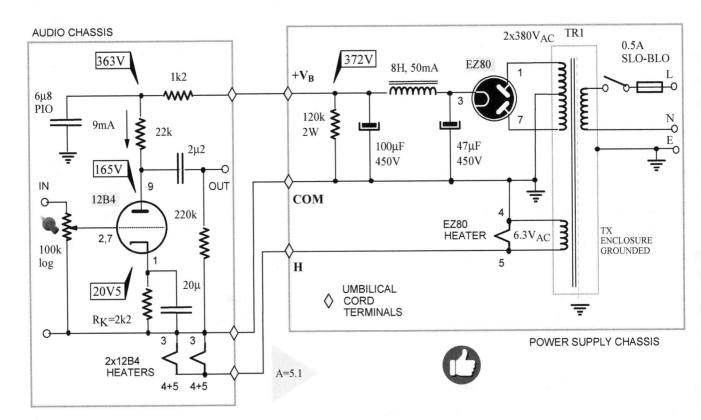

The value of the cathode bypass capacitor

The good news about using a relatively high resistance in the cathode circuit is that the bypass capacitor does not have to have high capacitance. Low capacitance means that we can avoid using the dreaded elcos! For instance, low voltage 100 µF film & foil capacitors cost a few dollars each and aren't physically large at all.

The time constant of the cathode circuit is approximately $\tau = R_K C$, and the low -3dB frequency is $f_L = 1/(2\pi R_K C)$. We get $f_L = 7.2$ Hz using a 10µF film capacitor, the smallest value to use here; 20 µF would be better, resulting in $f_L = 7.2/2 = 3.6$ Hz!

The value of the output coupling capacitor

If the preamp is to successfully drive low input impedance amplifiers, such as many solid-state amps and some tube amps with 22-47k input impedance, the output capacitor has to be sized for such a time constant. The 220k resistor to ground is fine by itself, but any input resistance of the following amp would be in parallel to it, so the total resistance would be $R = R_G \| R_{IN}$.

Again, $\tau = RC$ and the low -3dB frequency is $f_L = 1/(2\pi RC)$ If we assume a 10k input impedance, then R=220k∥10k = 9k56. We can see that in the first approximation, Rg does not matter much and that the input impedance of the following amp (the load) will determine the lower cutoff frequency and bass response.

Assuming a 5µ6 coupling capacitor and 10k load impedance, we will get $f_L = 1/(2\pi RC) = 2.8$ Hz. However, large values of the coupling capacitor tend to slow down the preamp response, so don't go overboard - consider lowering its value.

With 0.56μ we would get f_L=28 Hz, which is too high. 1μ of capacitance would result in f_L of 16 Hz, so the optimum is most likely between 1.5 and 3.3 μF.

Finding the optimal value is easy but time-consuming. Get six 1 or 1.5 mF capacitors (they don't have to be of equal value, this is just an example). During listening tests, keep soldering (or use crocodile clips) more and more caps in parallel. For instance, 1mF to start with, then another mF in parallel (2mF total), then another 1mF in parallel (3mF total) and so on.

Use a test record or CD with clean, tight, and prominent bass (but not boomy). Once you are happy with the speed/bass prominence balance, stop adding additional caps or return to the previous value by removing the last capacitor you added. Of course, you must change them in pairs for both channels.

Measurements and listening impressions

Everything worked like a charm from the start. Measurements confirmed a wide frequency range and lowish output impedance of around 1,400 Ω. As always, the sound was initially a bit strident and crisp. After a few hours, the edginess disappeared, the sound opened up and bloomed.

A very relaxed, natural, and effortless sound, well-balanced bass, midrange, and top end, in short - a flawless performance! It puts to shame many commercial line stages in the $3,000-$5,000 range.

We haven't tried different 12B4 tubes, only a NOS quad made by General Electric in the USA, but, in any case, a highly recommended design and a fantastic audio tube.

MEASURED RESULTS:
- BW: 9 Hz - 136 kHz (-3dB)
- Voltage gain: 5.1 (14.1 dB)
- Z_{OUT}=1.4 kΩ

6CK4 and 6AH4-GT octal triodes

TUBE PROFILE: 6CK4
- Indirectly-heated triode, octal socket
- Heater: 6.3V/1.25A
- V_{HKMAX}: 100V_{DC}
- P_{AMAX}: 12W, V_{AMAX}: 550V, I_{KMAX}: 100mA
- Typical operation:
- V_A=250V, V_G=-28V, I_A=40mA
- gm=5.5mA/V, r_I=1,200Ω, μ=6.6

TUBE PROFILE: 6AH4-GT
- Indirectly-heated triode, octal socket
- Heater: 6.3V/0.75A
- V_{HKMAX}: 100V_{DC}
- P_{AMAX}:7.5W, V_{AMAX}: 500V, I_{KMAX}: 60mA
- Typical operation:
- V_A=250V, V_G=-23V, I_A=30mA
- gm=4.3mA/V, r_I=1,780Ω, μ=8

If you prefer octal tubes and need an even higher anode power rating for your driver stage, these two beauties could be just what you've been longing for. Designed as vertical deflection amplifiers in smaller early TV sets, these obscure triodes are not widely used in audio applications. As a result, their NOS prices are quite low. If low output power levels are acceptable, you could also use them in SE and PP output stages, especially 6CK4, which has a respectable anode power dissipation of 12 Watts!

Very similar parameters to 12B4, internal resistance between one and two kΩ and amplification factor between 6 and 8. However, only a single heater voltage (6.3V) and higher input and Miller's capacitances due to their octal bases and larger dimensions, so not as wide upper-frequency extension as with the Noval alternative.

LEFT: 6CK4 is an impressive looking triode!
RIGHT: A comparison of inter-electrode capacitances of 12B4 and 6CK4 triodes

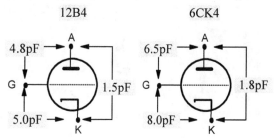

ECC34

Introduced in 1946, ECC34 is an octal duo-triode used in early TV sets, vintage tube oscilloscopes, and other instrumentation. Each of the two triodes has a relatively high anode power rating of 3.2 Watts, a medium internal impedance of around 5 kΩ, and a maximum cathode current of 50 mA. Its mju is low, only 11.5, limiting its use as a preamp tube in power amplifiers, but it would be ideal for a high current low gain line preamplifier.

The stage gain is nine times with a relatively low anode load of 18kΩ (graph below), and the 2nd harmonic distortion is only 1%. However, ECC34 is a rare and thus expensive tube, and as such, it does not represent value-for-money and cannot be recommended for new designs. Luckily, it is pin-compatible with ECC32 and 6SN7, so if you don't like the sound of ECC34, assuming you have correctly and conservatively designed its circuit, you can always replace them with 6SN7/ECC32 and sell them back to recoup your investment.

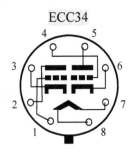

TUBE PROFILE: ECC34

- Indirectly-heated duo-triode
- Octal socket, heater: 6.3V/950mA
- V_{AMAX}= 300V, P_{AMAX}= 3.2 W, I_{KMAX}= 50mA, V_{HKMAX}= 50V
- TYPICAL OPERATION:
- V_A=250V, V_G= -16 V, I_0=10mA, gm= 2.2 mA/V, µ=11.5, r_I = 5.2 kΩ

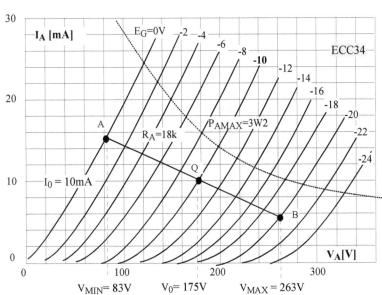

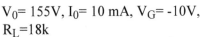

GRAPHIC ESTIMATION

V_0= 155V, I_0= 10 mA, V_G= -10V, R_L=18k
ΔV = 263-83 = 180V
A_V= -ΔV_A/ΔV_G= -180/20 = -9
ΔV_+ = 175-83= 92V
ΔV_- = 263-175= 88V
D_2= (ΔV_+- ΔV_-)/2ΔV_A = 4/360 = 1.1%

 HARD-TO-FIND & TOO EXPENSIVE

E81L

E81L is a special quality (low noise, high reliability, long life - 10,000 hours guaranteed) European pentode developed for telephony use. It has no American equivalent. Triode-connected it has a relatively high amplification factor and high transconductance. Considering its high (anode and screen together) power dissipation of around 5.7 Watts and low bias voltages (2-4V), it could be used both as a preamp and driver tube. This tube should also sound good triode-connected since high idle currents of 10-20 mA are involved.

GREAT SOUNDING & LOW DISTORTION WHEN TRIODE CONNECTED

TUBE PROFILE: E81L

- Indirectly-heated SQ pentode
- Noval socket, heater:6.3V/375mA
- V_{AMAX}= 210V, P_{AMAX}= 4.5 W, P_{SMAX}= 1.2 W
- C_{AG}=0.05 pF
- TYPICAL OPERATION (PENTODE):
- V_A=V_S=210V, V_G=-3 V, I_0=20mA, gm=11 mA/V
- AS A TRIODE: µ=36, r_I = 3 kΩ

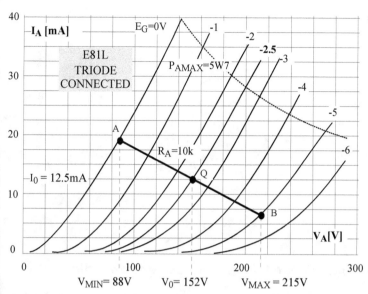

GRAPHIC ESTIMATION

E81L triode connected:
V_0= 152V, I_0= 12.5 mA
V_G=-2.5V, R_L=10k, ΔV = 215-88 = 127V
A_V=-$\Delta V_A/\Delta V_G$= -127/5 = -25.4
ΔV_+ = 152-88=64V, ΔV_- = 215-152=63V
D_2= (ΔV_+- ΔV_-)/$2\Delta V_A$= =1/254 = 0.4%

With a low anode load of 10kΩ and high anode current of 12.5 mA, the amplification is 26 times, and the 2nd harmonic distortion is only 0.4%, a very promising result.

EC8010

EC8010 is an SQ tube developed in the late 1960s for service as an amplifier and oscillator in UHF equipment up to 1GHz. Its parameters (μ, gm, r_I) fall within close tolerances. It's vibration and shockproof, not microphonic at all, and with a guaranteed life of 10,000 hours. Moreover, its cathode establishes no interface even when operated without anode current over prolonged periods. Sounds like a super triode? Let's look at its parameters.

A healthy anode power rating of more than 4 Watts, a high amplification factor of around 60 (in the 12AT7 class), and very high mutual conductance of 28 mA/V, meaning its internal resistance is very low, around 2 kΩ!

This is a high current triode, so it should sound better than low-current low power triodes of similar gain, namely 12AT7 and its close relatives. However, this tube will oscillate at every opportunity due to its extremely high gm, so grid and anode stoppers are mandatory!

SIEMENS AND VALVO QUALITY, BUT PRICES ARE INCREASING AND STOCKS ARE DWINDLING, SO BUY THEM ASAP!

TUBE PROFILE: EC8010

- Indirectly-heated UHF power triode
- Noval socket, hater: 6.3V/280mA
- V_{AMAX}= 200V, I_{KMAX}= 35 mA, P_{AMAX}= 4.2W
- Typical operation: V_{BB}=200V, V_A=140V
- R_K=47Ω, R_A=2k4, I_A=25mA
- STATIC PARAMETERS: gm=28 mA/V, μ=60, r_I=2.1 kΩ

GRAPHIC ESTIMATION

A-Q1-B:
V_{bias}= -1.75V, V_0=140V, I_0 =14mA
ΔV=V_B-V_A = 210-60 = 150V
A_V=-$\Delta V_A/V_G$ = -150/3.5 = - 42.8
DISTORTION (2nd harmonic):
ΔV_+=80V, ΔV_-= 70V
D_2 = (ΔV_+ - ΔV_-)/$2\Delta V$ *100 [%] = 10/(2*150)*100 = 3.3 %

X-Q2-Y:
V_{G0}= -1.75V, V_0=150V, I_0 =18mA
ΔV=V_Y-V_X = 228-70 = 158V
A_V=-$\Delta V_A/V_G$ = -158/3.5 = - 45
DISTORTION (2nd harmonic):
ΔV_+=80V, ΔV_-= 78V
D_2 = (ΔV_+ - ΔV_-)/$2\Delta V$ *100 [%] = 2/(2*158)*100 = 0.6 %

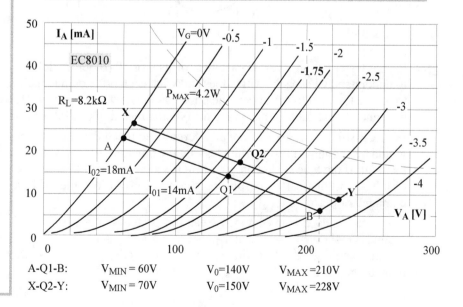

A-Q1-B: V_{MIN} = 60V V_0=140V V_{MAX}=210V
X-Q2-Y: V_{MIN} = 70V V_0=150V V_{MAX}=228V

The distortion under the initially-chosen conditions (A-Q1-B) was slightly high, over 3%, so further options should be investigated. A constant current operation should reduce the distortion and increase gain.

Alternatively, VBB could be raised to 300V. With X-Q2-Y swing, the situation is much improved; D2 distortion drops down to only 0.6%. It is fascinating how a slight change in operating conditions (Q-point moved towards higher anode currents, still the same load) reduced the distortion by 72%!

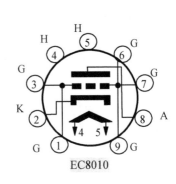

FAR RIGHT: The internal construction of EC8010 reminds one of ECC88 and E86C, two other lovelies!

How to test an unlisted tube on your mutual conductance tube tester - EC8010 example

Since EC8010 is a rare European tube, it is not listed on our Triplett 3444 tube analyzer's chart. Even the updated supplement makes no mention of it or its close American relative, the 8556. So, a bit of detective work was required to test a couple of Siemens tubes we got cheaply from an eBay seller in Slovenia. The exact procedure for determining the settings for an unlisted tube will, of course, depend heavily on a particular tube tester. Still, the methodology behind it and the overall approach stay the same.

Here are the major steps:

❶ TUBE PINOUT: You need a tube pinout and typical operating condition, meaning you need its data sheet.

❷ YOU NEED A SUITABLE SOCKET ON YOUR TESTER: If your tester uses prewired sockets (no switches for tube pins), such as B&K Dyna-Quick series or Precise 116, you need to find a tube that is listed on your tester's chart (means that it can be tested) that has an identical pinout. That sounds easy but can be a long and frustrating exercise, and, in the end, you may conclude that there is no prewired socket for such a tube.

In that case, you need to choose one socket (for EC8010 we need a Noval or 9-pin miniature socket) that you don't need - usually, that is a socket for some obsolete TV or RF tube not used in audio. Then, rewire it to suit the tube you want to test. Make sure you mark such a change in the tester's manual and on its faceplate.

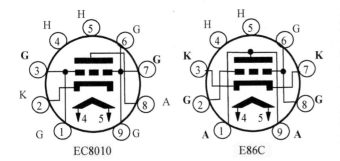

By the way, EC8010 is quite similar to E86C, another lovely tube, and with a bit of rewiring, it can be used as its substitute in audio amplifiers.

Let's say you have a tester that has a socket hard-wired to test E86C or 6CM4. Three pins are OK, 4 and 5 heater pins and 6, the grid. The others need to be rewired; the exact mod will depend on how the tester was wired in the first place. Alternatively, you can make a socket adapter.

❸ If your tester has an array of switches for its 9 or 10 (older models) or 12 pins (more modern ones), you need to understand what each switch does and what each position means. Triplett 3444 has nine switches, marked 1 to 9, one for each tube pin (see the photo on the next page). Easy. However, the 12 positions are marked 0 to 9, X and Y, so a table is needed to decode the meaning of each setting.

For EC8010, pins 1,3,6,7&9 are all internally connected to the control grid CG1. Connecting only one of those would be enough for testing (and leaving the other pin switches in position 0 - off), but we may as well connect them all. If the tube socket on your tester is a bit loose, as they all get with frequent plugging and unplugging of tubes, and that particular pin loses contact, you'd lose the grid bias, and the tube could overheat and get damaged or destroyed!

The heater ends are pins 4 & 5; we can put switch 4 in position 1 and switch 5 in position 2 or the other way around. On Triplett 3444 heater pins are marked + and - for some reason, but AC heating is used, so both settings would work. The cathode is pin 2, so switch 2 goes into position 3, and the anode is pin 8, so switch #8 must be in position 4! Thus, one of the few possible variations of the setting is (SW1-9): 5351-2-5545

CHEAP & CHEERFUL: PREAMP & DRIVER TUBES FOR AUDIO EXPLORERS

Switch position	Connection
0	OFF (pin not connected)
1	HEATER (filament) negative (-)
2	HEATER (filament) positive (+)
3	K1 (cathode tube 1)
4	P1 (plate tube 1)
5	CG1 (control grid tube 1)
6	G2 (screen grid)
7	G3 (suppressor grid)
8	K2 (cathode tube 2)
9	P2 (plate tube 2)
X	CG2 (control grid tube 2)
Y	Anode (plate) for "magic eye" tubes

Triplett 3444: The meaning behind "ELEMENTS" switches' positions.

Triplett 3444: A-heater voltage selector, B- bias adjustment, C - plate & screen voltage switch, and D-gm range (B and D are not shown on the photo). 1 to 9 are switches marked "ELEMENTS".

HEATER VOLTAGE SETTING: This is usually obvious (except for dual heater tubes), which can operate on 6.3 and 12.6V. In this case, the setting "A" for EC8010 is 6.3V.

❹ GRID BIAS, PLATE, AND SCREEN VOLTAGES; Consulting the anode curves below, we notice that at $100V_{DC}$ test voltage, depending on the setting of the bias potentiometer, we can test this tube at any point along the marked range. The anode current will vary from 2mA to 42mA, which is perfect for Triplett 3444, one of its two current measuring ranges is 50mA! So, we can match tubes not just in terms of gm (mutual conductance) but also anode current!

Unfortunately, the anode (and screen) test voltage on Triplett 3444 cannot be varied continuously as the bias voltage can, so there are only two switchable choices, $100V_{DC}$ and $250V_{DC}$! That is the most serious shortcoming of this otherwise great tester.

For instance, with 250V anode test voltage, the -4V bias setting would test the tubes in Q3. However, the allowable bias range would be very narrow, so we could easily under bias the tube, the anode current would shoot up, and the anode power rating would be exceeded. Plus, notice that the maximum allowed anode voltage for EC8010 is only $V_{AMAX}= 200V$, so we should not use the 250V setting.

The grid bias setting should be between 0 and -4V, so we need to select one of the 0-5V bias ranges on the tester. From the decoding table above, it is evident that positions 5, 6, 7, and 8 are the candidates. The screen grid voltage settings are irrelevant in this case; EC8010 is a triode and has no screen grid. Positions 5 and 8 use the 0-5V bias range, so either of those can be used with EC8010.

Switch C ("Plate Voltage") position	Plate V_{DC}	Screen V_{DC}	Bias range V_{DC}
1	250	250	0-5
2	250	250	0-50
3	250	100	0-50
4	250	100	0-5
5	100	100	0-5
6	100	100	0-50
7	100	45	0-50
8	100	45	0-5
9	30	30	0-5
10	30	12	0-5
11	12	12	0-5

The 11 possible combinations of the 3 factors, the bias range, the anode and the screen voltage, selectable by switch "C" on Triplett 3444 tube tester.

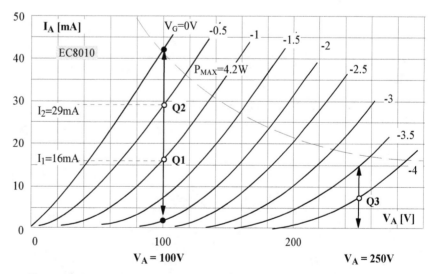

There re only two anode test voltages on Triplett 3444, $100V_{DC}$ and $250V_{DC}$, so EC8010 can be tested along the two ranges marked.

❺ THE GM RANGE: The data sheet for EC8010 lists nominal mutual conductance at around 28 mA/V (28,000 micromhos), which is very high. Older testers, such as most Hickoks, cannot go that high and cannot test high gm tubes.

Luckily the highest gm range on Triplett 3444 is 30mA/V, but even then, some very strong EC8010 tubes may go over the scale, especially at higher anode currents where gm increases (test point Q2).

To prevent that from happening, we should test these tubes at lower anode currents, in test point Q1, for instance, or even lower, below 10 mA of anode current. In that region, the slope of the anode curves (mutual conductance) is lower, from 20-25 mA/V (draw a tangent in a particular point and see for yourself); that will ensure that even the strongest tubes will not go off the 30mA/V top end of the tester's scale.

With a bias set at 8 (-0.8V), the results for our pair of tubes were gm_1=23.5mA/V and I_{A1}=15mA, and gm_2=22.5mA/V and I_{A2}=17mA, so the tubes were quite well matched.

SLEEPING GIANTS: OUTPUT TUBES FOR THOSE WHO WANT TO BE DIFFERENT

- 6AR6 (6098) or 6AV5GA PSE amplifier
- EL50 and EL51
- 6360 dual power tetrode
- EL41 RimLock power pentode
- Design project: PL83 UL PP amplifier
- Line output pentodes in audio applications

7

"It is the mark of an educated mind to be able to entertain a thought without accepting it."
Aristotle

6AR6 (6098) or 6AV5GA PSE amplifier

These octal beam power tubes were developed for service as horizontal deflection amplifiers in smaller black & white TV sets - they drove the electron beam in cathode-ray tubes in the left-right direction.

Notice that despite their heaters consuming the same power (P_H=1.2A* 6.3V = 7.6W), the anode power rating of 6AR6 is 19 Watts, compared to only 11 Watts for 6AV5. Thus, 6AR6 is a higher efficiency tube.

However, looking at the comparison photo on the right, the physical sizes of the anodes of the two tubes are identical, so by some logic, their power (heat) dissipation levels should be identical too.

As a triode, 6AV5 has a lower amplification factor (it needs a slightly larger grid drive signal), but it has a 25% lower internal impedance, meaning it gets closer to an ideal voltage source than 6AR6.

Bogen DB130, a mono push-pull amplifier from 1956, was one of the few vintage designs that featured the 6AV5GT tube. The anode voltage of 680V and screen grid voltage of 160V were used, with -45 fixed bias voltage, resulting in 35 Watts of output power at only 0.3% THD! 5U4G tube rectifier (1), large output transformer (2), and variable damping factor controls (3) are some of its distinguishing features.

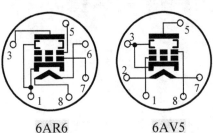

6AR6   6AV5

ABOVE: Rear detail of the vintage Bogen DB130 amplifier

SIDE-BY-SIDE	6AR6	6AV5GA
Type	Beam tetrode	Beam tetrode
Heater	6.3 V, 1.2 A	6.3 V, 1.2 A
Socket	Octal	Octal
P_A max.	19 W	11 W
V_A & V_{G2} max.	565/300 V	550/200 V
P_{G2} max.	3.2 W	2.5 W
I_A max.	115 mA	100 mA
r_I (pentode)	21 kΩ	25 kΩ
gm	6.0 mA/V	5.9 mA/V
μ / r_I as a triode	6, 1kΩ	4.3/740Ω

Triode operation of 6AV5

GRAPHIC ESTIMATION

Q-POINT:
R_L=3k3, V_{bias}=-45V
V_0=250V, I_0=50mA
VOLTAGE SWING:
V_B=375V, V_A=65V, $\Delta V=V_B-V_A$=310V
CURRENT SWING:
ΔI = 100-8 = 92 mA
$P_{IN}= I_0 V_0$ = 0.05*250 = 12.5W
OUTPUT POWER:
$P_{OUT}=\Delta V \Delta I/8$ = 310*0.092/8 = 28.5/8 = 3.5W

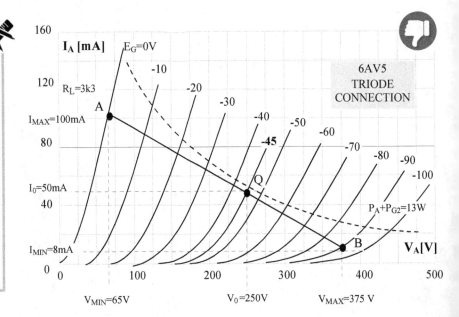

Two issues are immediately obvious. Firstly, the maximum anode + screen power dissipation of 6AV5 is only 13 Watts. Secondly, its maximum screen grid voltage is only 200V! With 250V in the quiescent point Q, we could be stressing the tube too much. Despite these two limitations, with a slightly higher anode load of 3.3kΩ, an output power of 3.5 Watts should be achievable.

The maximum anode current (point A) is lower than for 6AR6, only 100 mA, so if two tubes are paralleled, that is 200 mA, compared to 240 mA for the 6AR6 situation.

The 200V maximum screen voltage of 5AVG is a serious limitation, so we will proceed with the design and construction of a 6AR6 PSE amplifier. All that trouble for only 8 Watts from two paralleled triode connected 6AR6 lovelies does not seem justified. Let's see what we can get in the ultralinear SE connection.

Triode operation of 6AR6

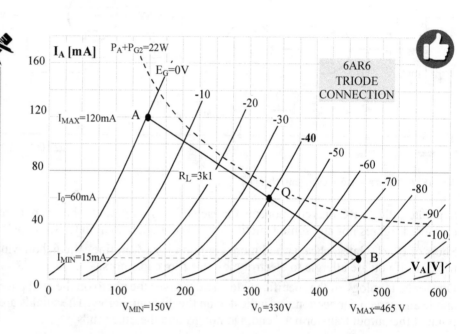

GRAPHIC ESTIMATION

Q-POINT:
R_L=3k1, V_{bias}=-40V
V_0=330V, I_0=70mA
VOLTAGE SWING:
V_B= 465V, V_A=150V
$\Delta V = V_B - V_A$ =315V
CURRENT SWING:
ΔI = 120-15 = 105 mA
INPUT POWER:
P_{IN}= $I_0 V_0$ = 0.06*330 = 19.8W
OUTPUT POWER:
P_{OUT}=$\Delta V \Delta I/8$ =
= 315*0.105/8= 4.1 W

Let's see how 6AR6 would behave as a triode. While these operating conditions have not been optimized but simply chosen using intuition and previous experience, the results look promising. Although the maximum screen grid voltage is 300V, it can safely be exceeded by 10% for almost all power tubes, so we have chosen a quiescent anode/screen voltage of 330V.

The output power is higher than that of the 2A3 triode, and since the maximum anode + screen power dissipation of 6AR6 is approx. half that of the standard 300B, that was to be expected.

6AR6 PSE ultralinear amplifier - the audio section

A modified SRPP input/driver stage popularly known as the "Bel Canto" circuit was used to drive a pair of 6AR6WA beauties in ultra-linear mode. The transformer's primary impedance was around 2.7 kΩ, but you can use anything between two and four kΩ!

A simple cathode bias is used in the output stage, 195Ω resistance (two 390Ω resistors in parallel), with 134 mA total anode current or 67mA per tube. This is a very conservative operating regime, so a long tube life is anticipated. Some claim that these Tungsol beauties can take even higher screen voltages, up to 450V, but we didn't want to risk frying them.

You can experiment if you want more output power or if you opt for a triode connection and a higher anode/screen voltage of around 400V!

KEY FEATURES:

- Single driver stage
- Choice of 12AU7, 12AT7 or 12AX7 driver tubes
- Self or cathode biased output stage, no adjustments of any kind needed
- Very simple power supply

MEASURED RESULTS:

- BW: 20 Hz- 40 kHz (-3dB, 10W into 8Ω)
- V_{OUTMAX}=16V_{RMS}
- P_{MAX} = 32W

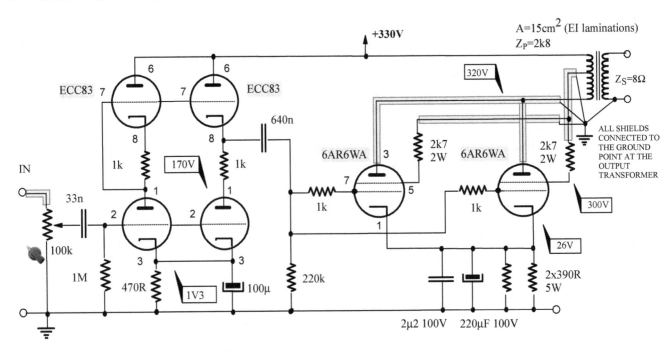

ABOVE: Amplifier's audio section (one channel)

Separate cathode RC bias components for each of the four output tubes were tried, but there was no improvement in measured figures nor any sonic benefit. Use the best quality cathode elcos you can find.

Short of Black Gates, cheaper and more common Elna Cerafine, Nichichon Muse and similar lovelies will do the trick, or, if you are allergic to electrolytics, use film capacitors.

Due to the topology of the specific enclosure used (see the photo on the next page) we had to use shielded cables for anode and screen connections, as indicated on the diagram below. The shields are all grounded (earthed) at the same point at the output transformers' end. Do not ground the other ends.

The power supply

Since the cathodes of the upper triodes are at an elevated DC voltage (around 170V!), the 6.3V heater supply is connected to a resistive voltage divider (120k and 39k) that supplies around +81V to elevate the heaters with respect to ground. If ECC83 is used, that is unnecessary since its maximum heather-cathode voltage is 200V, but it is recommended to improve the signal-to-noise ratio. This must be done if you want to use 12AT7 and 12AU7, whose V_{HKMAX}=100V.

The upper cathodes are at 170-81=89V, and the bottom cathodes are at 1-81= -80V with respect to the heaters, all within the 100V limit of 12AT7 or 12AU7.

A 35µF motor start capacitor is the last in the filtering chain, significantly improving the sonics.

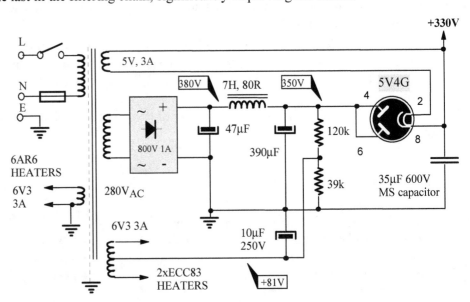

The chassis (enclosure)

This plastic instrument enclosure sat for years in storage as if it was waiting for this project. Drilling and hole cutting was fast & easy. The front and rear panels had slots, so no ventilation holes needed to be drilled on the top and bottom panels.

The sides are made of a thick extruded aluminium profile with equally-spaced grooves (channels), so transformers and chokes can be bolted onto it. The power transformer and the filtering choke are bolted onto one and the two output transformers onto the other side of the chassis, resulting in a very quiet operation and high S/N ratio.

Listening impressions

We started with 12AX7 in the input stage, which resulted in plenty of gain when directly driven by a CD player, but too much gain when driven by an active preamp. The sound was open and detailed, but for my liking, too crisp. I don't want to use the word "harsh"; that would not be fair. Solid-state is harsh; tubes can only be "crisp."

Changing the four input tubes to 12AT7 (without any changes to the circuitry) reduced the input sensitivity slightly (volume control setting from 9 o'clock to 10 o'clock for the same loudness) but removed the edge from the sound and made the sonics smoother, more "emotional" and involving.

Finally, 12AU7 triodes significantly reduced the input sensitivity. However, there was still plenty of drive with a direct CD connection and plenty of headroom with an active tube preamplifier with a gain of around ten. The sonics improved with 12AU7 - now there was true magic.

The bass remained tight and well-controlled, but the midrange and treble improved with each trial. The midrange became smoother, the voices "chestier." You get the feeling of singers' lungs and chests; you don't just hear their vocal cords. The higher frequencies become more refined, the microdynamics improved.

Again, this listening evaluation confirmed our preferences for lower m, higher anode current tubes such as 12AU7 over higher m duo-triodes such as 12AT7 and 12AX7.

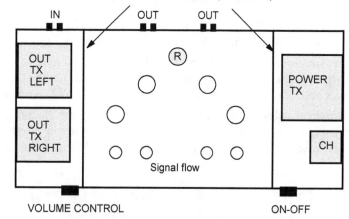

ABOVE: Amplifier's topology, including the major components
BELOW: Topside and rear-side view of the finished amplifier

Tube choices

6384 tubes (6AR6 equivalent) labeled Bendix or Cetron are recognizable by their white bases and white or red ceramic spacers. These tubes have an impressive internal construction and are considered a superior option, which prompted the greedy sellers to raise their prices incredibly high.

To save you some anguish so you don't spend hours wondering if you should "upgrade" and spend half of your monthly income on a quad of those tubes, we concluded that there is no sonic benefit at all. The sound with Bendix 6384 was slightly harsher. A great deal of the "musicality and magic" was gone. You'd be better off investing money in the remaining Tungsol 6AR6, buying all you can find; my crystal ball tells me they will get more and more expensive very soon!

Experiment: Testing a batch of 6AR6 power tubes on Triplett 3444 tube tester

We had fifteen 6AR6 lovelies, made by Tungsol in the USA, all of the identical construction. We tested them for gm and anode current on Triplett 3444 Tube Analyzer, one of the best vintage tube testers ever made.

The first four tubes are the perfectly matched quad used in the described parallel SE ultra-linear amplifier. If anode current is taken as the primary criterion (as it should be for output tubes), it is possible to find another perfectly matched quad (four tubes at 41 mA), marked with (1), and two well-matched pairs (at 34 and 36 mA).

However, all seven tubes in the group (2) are within 2mA of each other, which is considered a very close match! The precise vertical scale exaggerates tiny differences in anode current. Compared to the currently produced output tubes, whose parameters vary widely, all 15 of these tubes would be considered closely matched.

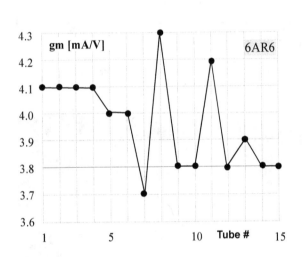
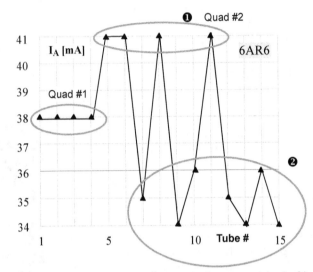

ABOVE: Measured results in a visually-friendly form. Notice the incredible uniformity (compared to most other power tubes) of both mutual conductance (from 3.6 to 4.3 mA/V) and anode currents (ranging from 34 to 41 mA)!

EL50 and EL51

These power pentodes were developed by Philips and copied by Tesla in then Czechoslovakia. The Philips versions are rare nowadays, but Tesla pairs are occasionally available for sale online. With only 18 watts of anode dissipation and a relatively low screen voltage of 400V, there are much better alternatives to EL50, so this tube is of no practical interest.

EL51, however, is a very large and impressive-looking tube with 45 Watts anode rating. With both anodes and screens at their maximum ratings of 750 Volts, two in push-pull can produce 140 Watts!

The mutual conductance is relatively high, 11 mA/V, as is the amplification factor, triode connected μ=16, so low driving voltages are needed. As a comparison, 300B has a very low μ of 3.85!

Notice that the anode is quite short compared to the height of the top-heavy tube. Let's see how EL51 would behave in SET conditions.

TUBE PROFILE: EL50

- Indirectly-heated power pentode
- European side contact socket (P8A)
- 6.3V, 1.35A heater, C_{AG1}=0.8pF
- V_{AMAX}=800V, V_{G2MAX}=400V, V_{HKMAX}=100V
- I_{K2MAX}=120mA, P_{AMAX}=18W, P_{G2MAX}=3W
- As a pentode with V_A=400V & V_{G2}=425V: gm=6mA/V, r_I = 30kΩ, bias -33V, I_A=45mA, I_{G2}=5.5mA

TUBE PROFILE: EL51

- Indirectly-heated pentode, P8A socket
- 6.3V, 1.9A heater, C_{AG1}=1.5pF
- V_{AMAX}=750V, V_{G2MAX}=750V, V_{HKMAX}=50V
- I_{K2MAX}=200mA, P_{AMAX}=45W, P_{G2MAX}=7W
- As a pentode with V_A=500V & V_{G2}=500V: gm=11mA/V, r_I = 33kΩ, bias -22V, I_A=95mA, I_{G2}=12mA
- As a triode: μ= 16, r_I = 1.3kΩ

Triode operation of EL51

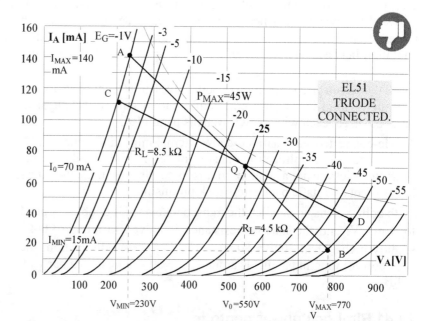

GRAPHIC ESTIMATION

A-Q-B (R_L=4k5):
V_G=-25V, V_0=550V, I_0=70mA
$\Delta V = V_B - V_A$ =540V
ΔI=140-15 = 125 mA
$P_{IN}= I_0 V_0$ = 0.07*550= 38.5W
$P_{OUT}=\Delta V \Delta I/8$= 540*0.125/8 = 67.5/8 = 8.4W
$D_2=(\Delta V_{OUT+} - \Delta V_{OUT-})/2\Delta V_{OUT}$
D_2=(320-220)/2*540 = 9.3 %

C-Q-D (R_L=8k5):
V_G=-25V, V_0=550V, I_0=70mA
ΔV=640V
ΔI=120-35 = 85 mA
$P_{IN}= I_0 V_0$ = 0.07*550 = 38.5W
$P_{OUT}=\Delta V \Delta I/8$ = 640*0.085/8 = 28.5/8 = 6.8 W
$D_2=(\Delta V_{OUT+} - \Delta V_{OUT-})/2\Delta V_{OUT}$
D_2=(350-290)/2*640 = 4.7 %

With R_L=4k5, a horrendous distortion of almost 10% with only 8 Watts is very disappointing. Plus, while 550V on the anode is fine, such a high voltage on the screen grids could reduce tubes' life and reliability. Let's increase the load impedance to R_L=8k5 and see if anything improves. We'll use the same quiescent point.

The distortion has halved but is still higher than 300B and similar triodes, with a lower power output of only 6-7 watts. Now we understand why Phillips only recommended this tube for push-pull operation and did not even specify its single-ended operating conditions, neither as a pentode nor as a triode.

Once factors such as the unfortunate top cap and limited availability are considered, the verdict for SE applications is a resounding no!

6360 dual power tetrode

6360 is a twin tetrode designed as a class C amplifier and oscillator and in push-pull circuits as an output tube, driver, or frequency tripler. Those applications include low-power transmitter stages operating at frequencies up to 200 MHz. Its anodes (plates) are Zirconium coated and have cooling fins, explaining the tube's relatively high plate dissipation relative to its diminutive physical size.

Two halves in class AB1 push-pull, with 300V on the plates and 200V on the screens, with a 6k5 PP load (plate-to-plate), can produce 12W in class AB_1 and 17.5 Watts of output power in class AB_2, at 2.5% distortion, which means one of these tubes can produce more audio power than a pair of EL84 or 6V6 and with a lower output transformer's primary impedance!

TUBE PROFILE: 6360

- Indirectly-heated dual tetrode (common cathode)
- Noval socket
- Heater: 6.3V/820mA or 12.6V/410mA
- V_{AMAX}= 300V, V_{G2MAX}= 200V
- P_{AMAX}= 2x15W, P_{G2MAX}= 2x2W
- I_{AMAX}= 2x55mA
- As a triode (each half):
- gm = 4 mA/V, μ= 7.5, r_I = 1.8 kΩ

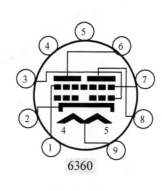

RIGHT: Two 6360 specimens from different angles, by Philips and RCA

The Amperex Application Bulletin explains the unusual construction of the tube: "The control grids and the screen grid are 'shadowed' which means that the screen grid wires are placed behind the control grid wires in the direction of the electron flow. This measure promotes the formation of a radial beam and ensures the correct space charge condition between the screen grid and the plate. This type of construction leads to a relatively low screen grid current."

Since the screen grid is common to both halves, 6360 cannot be used in a triode or ultra-linear push-pull configuration unless two physical tubes are used, each with its two systems paralleled.

Triode-connected, each half has a low internal impedance, under 2kΩ, and a decent amplification factor of 7.5. The triode anode characteristics are not available, so it is impossible to judge the tube's linearity, but its static parameters look promising.

With both systems in parallel, the transconductance is doubled, from 4 to 8 mA/V, and the internal resistance is halved to around 900Ω, slightly higher than 300B's 800Ω. Including screen dissipation, the total power rating in parallel is around 34 Watts, almost on a par with 300B.

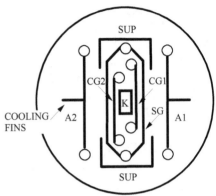

ABOVE: Horizontal cross-section of 6360 tube, showing its interesting construction, with common cathode and common screen grid.

EL41 RimLock power pentode

EL41 is a miniature output pentode developed by the mighty Phillips laboratories in Holland in the late 1940s. When triode connected, these lovelies have a very high amplification factor (μ=18) and thus low bias voltage (around -8V) and high input sensitivity.

Just like ECC40, the preamp duo-triode of the same vintage and using the same RimLock socket, EL41 is vibration-resistant, not microphonic at all, a very sturdy mighty midget indeed.

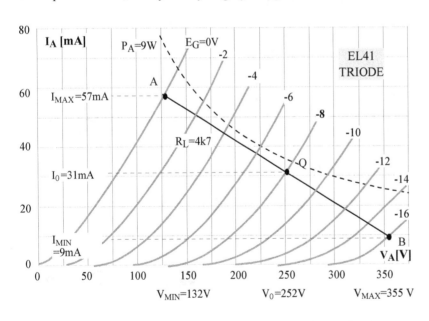

GRAPHIC ESTIMATION

R_L=4k7, V_G=-8V, V_0=252V, I_0=31mA
V_{MIN}=132V, V_{MAX}=355V, ΔV = 223V
ΔI = 57 - 9 = 48 mA
P_{IN} = $I_0 V_0$ = 0.031*252 = 8W
P_{OUT} = $\Delta V \Delta I /8$ = 223*0.05/8 = 1.35 W
ΔV_{OUT+}= 252-132=120V, ΔV_{OUT-} =355-252=103V
D_2=(ΔV_{OUT+}-ΔV_{OUT-})/2ΔV_{OUT} =17/2*223 = 3.8 %

ABOVE: Two different internal constructions of nominally the same tube, by Philips French subsidiary Miniwatt Dario (left) and German Telefunken (right).

Except for the modest output power, relatively promising results, less than 4% distortion at maximum output. The voltage gain of the output stage is $A_V = \Delta V_A/\Delta V_G = -223/16 = -14$!

Such a high voltage gain could be predicted from the very low bias voltage of -8V and the triode amplification factor $\mu=22$! This means we could even make this amplifier as output stage only and drive it directly with a line-level preamp, providing its gain is higher than, say, $8V/1.41 = 5.7$ (for 1V input sensitivity). This is easily achievable.

> **TUBE PROFILE: EL41**
>
> - Indirectly-heated power pentode
> - RimLock socket, heater: 6.3V, 0.7A
> - $V_{AMAX}=300V$, $V_{G2MAX}=300V$
> - $I_{KMAX}=55mA$
> - $P_{AMAX}=9W$, $P_{G2MAX}=3W$
> - As a triode: $\mu=22$, $r_I=1k8$

With two triodes in parallel and 2k5 output transformer's primary impedance, we are into the 2A3 power territory. The idle anode current would only be 62mA. Five watts is achievable with three tubes in parallel. Use output transformers capable of passing 100mA of primary idle current, which, again, all but the cheapest and tiniest output transformers can do. 1k5-2kΩ primary impedance would be optimal.

To get 8V cathode voltage at 31mA, each cathode resistor must be $8/0.031= 258\Omega$. Its power dissipation is $8*0.031 = 0.25$ W, so an ordinary 270Ω (1 Watt) will do. To get the lower -3dB (half power) frequency of the stage at around 3Hz, the bypass capacitance needs to be $C_K=1/(2*\pi*f_L*R_K) = 196$ μF, so use a 180μF or 220μF electrolytic capacitor. For the audio extremists in our midst, you could use two 100μF film capacitors in parallel.

We didn't have six well-matched EL41 pentodes to proceed with this project, but we had some lovely Telefunken PL83 video output tubes, and that's how the next project was conceived.

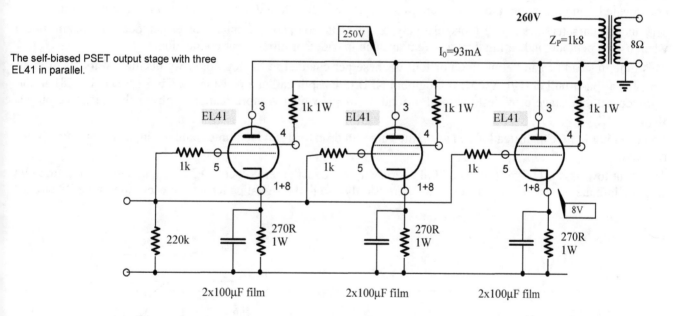

The self-biased PSET output stage with three EL41 in parallel.

Design project: PL83 UL PP amplifier

PL83 power pentode

This little cutie is also a very sensitive pentode; its anode characteristics (next page) go only up to -6V grid bias. Like its bigger brother, PL519, a horizontal deflection tube, PL83 was designed to operate on a 300mA series heater string in TV sets. Compared to most other power pentodes, its curves are also reasonably equidistant, indicating low distortion.

If you look closely at PL83's static parameters in its Tube Profile Box, you will notice that it is almost identical to EL41! 11mA/V versus 10 mA/V transconductance, triode μ of 24, compared to 22 for EL41, and an identical anode dissipation of 9 Watts.

> **TUBE PROFILE: PL83**
>
> - Indirectly-heated power pentode
> - Noval socket, heater: 15V, 0.3A
> - $V_{AMAX}=250V$, $V_{G2MAX}=250V$
> - $I_{K2MAX}=70mA$, $P_{AMAX}:9W$, $P_{G2MAX}=2W$
> - As a triode: $\mu=24$, $gm=11mA/V$, $r_I=2k2$

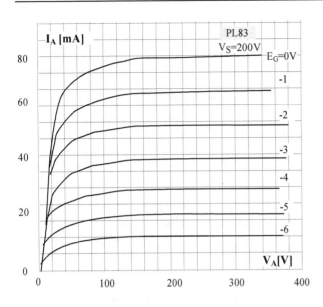
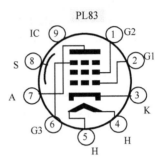

RIGHT: While the supply of desirable NOS Mullard, WE and Telefunken audio tubes has been practically exhausted by now, tubes not traditionally used in audio such as PL83, PL508 and many others, are still available. Prices range from very low to reasonable.

LEFT: Compared to other power pentodes, the anode characteristics of PL83 are relatively equidistant, forecasting lower distortion.

The socket and heater specs are different (the RimLock socket of EL41 was not popular and was abandoned soon after), but it is quite plausible that this tube is actually a repackaged EL41 with a different heater!

Neither triode nor ultra-linear curves are available for PL83, but that does not matter; curves are just a starting point for an initial assessment of the tube's potential. So, even if a tube you are considering using is not "well documented," and most non-audio tubes are in that category, that should not stop you from trying it in practice.

The anode characteristics above look relatively equidistant (compared to other power pentodes and beam tubes) when used a pentode, indicating they will remain such in triode or ultra-linear connection.

This time, instead of paralleling three or four of these per channel in SE topology, we decided to try them in an ultra-linear push-pull design. With a mild global NFB to its split cathode resistor (2k7 and 130Ω), ECC81 in the common cathode stage provides a gain of around 27 times, which is more than enough for this sensitive output stage.

You can plug in ECC82 instead if you wish; the gain will drop to around ten times, and the input sensitivity will be reduced.

The input low pass RC filter (10k and 30pF) serves to reduce RF interference. DC coupling between the first two stages eliminates one coupling capacitor, so there is only one in the signal path (one for each side of the PP circuit, that is).

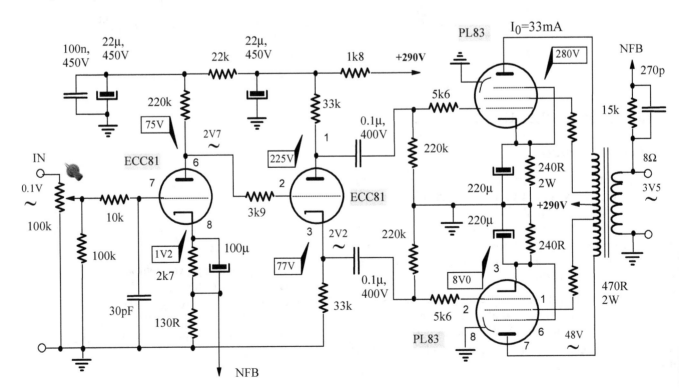

The output stage is self-biased with separate cathode resistors, so precisely matched tubes aren't necessary. Should you have tightly matched pairs, use a shared RC pair in the cathode, 120Ω and 330 μF.

The electrostatic screen inside the PL83 should be connected either to the ground or to the cathode. The power supply will not be shown here; all you need is a single anode supply voltage of +290 V (270 to 320 V will work). The idle current is 4 x 33 mA= 132 mA for the output stages plus 10 mA for the preamp stages. However, at full power, that will increase up to 250 mA, so use a rectifier that can cope with such current levels.

Feel free to use a separate transformer to supply 15 V for EL83 heaters. Universal 2x7.5 V or single 15 V secondary transformers are cheap and plentiful. Even if you don't have 6.3V for the heaters of the two ECC81 tubes, use the same 15 V supply with suitable-sized series resistors that will drop the voltage down from 15 V to 12.6 V! A balanced option with a humbucking rheostat is illustrated here.

Optimizing the performance of the suppressor grid

Most power pentodes have their suppressor grids internally tied to the cathode, so amplifier designers have no choice regarding its connection. Some, most notably EL34, have the suppressor grid wired out to its own pin. PL83 is one such pentode (pin 6).

Standard practice is to connect the suppressor grid to the cathode (pin 3). However, connecting the suppressor to a negative voltage source improves the efficiency and linearity of the output stage and the reliability of the power tubes.

The most convenient negative voltage source would be the negative bias circuit. In this case, we used cathode bias, but if you opt for a fixed bias, all you need is a source of around -15VDC.

Connect the four suppressor grids to that voltage and install two or four trimmer pots to adjust the actual bias on the control grids to between -8 and -10 volts.

Topology and construction details

The chassis was a standard 1U high (50mm), 19" wide rack chassis, made of powder-coated steel, with natural aluminium fascia.

The rotary on-off switch is on the left, together with a red neon indicator, RCA inputs are in the middle, and volume control is on the right.

The power transformer and two filtering chokes are under the left steel cover; two output transformers are under the right cover.

Short signal paths (RCA inputs very close to the volume control and input tubes) and power supply section separation and shielding contributed to a very quiet operation.

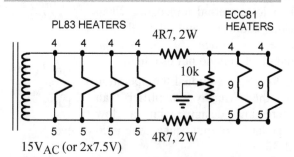

MEASURED RESULTS:
- BW: 8 Hz - 60 kHz (-3dB, 1W into 8Ω)
- BW: 14 Hz - 52 kHz (-3dB, 10W into 8Ω)
- $V_{OUTMAX}=11 V_{RMS}$
- $P_{MAX}= 15W$

ABOVE: Connecting the 15V heaters of output tubes and 12.6V heaters of preamp tubes to one 15V_{AC} secondary winding.

Line output pentodes in audio applications

There are a dozen line output pentodes, designed initially to drive cathode ray tubes in vintage TV sets in the vertical direction. The main criterion was the ability to withstand high voltage pulses; that is why almost all have their anodes taken to the top cap, physically as far away from other pins (in the base) as possible.

When used as audio tubes, the first problem is their low maximum screen grid voltage rating, typ. 200-250V. When used as triodes, the anode voltage cannot exceed 250V, or 300V tops, which reduces the output power available.

The second problem flows from the first, which is the presence of high anode DC voltage on top caps, so tube cages are needed for safety reasons.

Most importantly, sweep tubes are much less linear than "real" triodes, even when triode-connected. In Vol. 1 of this book, we had a quick look at the triode operation of EL36 and concluded that the distortion would be horrendous, around 12.1%.

The graph and calculations are repeated here as a reminder. Notice how the anode voltage swing QA is much larger than the swing QB.

The main advantages of horizontal and vertical deflection TV tubes are their low internal impedance when triode connected and relatively low price (since commercial manufacturers do not use them).

In Vol. 2, we studied the operation of the Murray amplifier, an interesting design with a low impedance output stage that used EL36.

However, Murray's design used a very strong negative feedback to linearize his very complex (and fiddly to adjust) design, which is not recommended to anyone except advanced constructors.

L-R: 6DQ6A by AWA Radiotron (Australia) two different views, EL36 by Philips, and PL36 by AWA Radiotron (Australia)

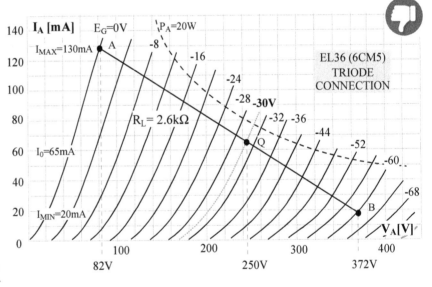

GRAPHIC ESTIMATION

$R_L = 2.6 k\Omega$, $V_G = -30V$
$V_0 = 250V$, $I_0 = 65mA$
$V_{MIN} = 82V$, $V_{MAX} = 372V$
$\Delta V = 190V$
$\Delta I = 130 - 20 = 90 mA$
$P_{IN} = I_0 V_0 = 16.25W$
$P_{OUT} = \Delta V \Delta I / 8 =$
$190 * 0.09 / 8 = 2.2W$
$D_2 = (\Delta V_{OUT+} - \Delta V_{OUT-}) / 2 \Delta V_{OUT}$
$D_2 = (168 - 122) / 2 * 190 = 12.1 \%$

SIDE-BY-SIDE	EL500	PL504	EL36 (6CDA)	6DQ6A
Heater	6.3V/1.4A	27V, 0.3A	6.3V/ 1.25A	6.3V/1.2A
P_A max.	12W	16W	12W	15W
V_A max.	250 V	250V	250 V	700V
P_{G2} max.	5 W	6W	5 W	3 W
V_{G2} max.	250 V	250V	250 V	200V
I_K max.	250 mA	250 mA	200 mA	140 mA
V_{HK} max.	220V	250V	100V	200V
gm [mA/V]	11.5		14.0	6.6
µ as triode		6.0	5.6	4.1

ABOVE: A comparison of a few line output pentodes, less commonly used in audio.

THE QUEEN OF HEARTS: SINGLE-ENDED AMPLIFIERS WITH 300B TRIODES

- D&Y A-300 SET amplifier
- Sun Audio SV-300BE SET amplifier
- Audio Note Kit 1
- Design project: PSET 300B monoblocks with interstage transformer coupling

"There is a big rip-off going on. Companies are selling extra high-priced equipment that has no benefit except a high profit to the company that sells it. I don't think that many of these fads that come along are true advances."
David Hafler, of Dynaco fame, in an interview for Vacuum Tube Valley magazine, issue 15, 2001

The "Lemon of the Decade": D&Y A-300 SET amplifier

Apart from the mainstream amp makers such as Yaquin, Yarland, Bez, Psvane, Music Angel, and many others, there seems to be a whole cottage industry in China, most likely individuals building kits. The names are often just screen-printed words on the chassis, with various people building amplifiers in them, while the "factory" with such a name does not exist at all. So, consistency and quality control is an issue here; you are never sure what you are getting for your money.

In Oct 2015, this stereo 300B amplifier sold for US$529.00 +US$145 shipping (US$674 total). The circuit is so simple, and the workmanship is so neat that you could reverse-engineer the amp's topology from the internal photograph alone (page 104). There is also a circuit diagram floating on the web.

A PayPal promotion of 15% discount (US$80 saving) sealed the deal, so we bought one on eBay.

The specifications

The specs are suspiciously similar, almost identical to many other Chinese-made 300B amplifiers being sold on eBay, so I would not place too much trust in it: rated output power of 8 Watts per channel, frequency range 20Hz-35 kHz, SNR 88 dB.

Notice that frequency range is not specified in terms of attenuation. Are these -1dB, -2dB, or -3dB points? The output power level when those "measurements" were taken is also missing, which renders the specification meaningless! The same applies to the SN ratio - is it weighted or unweighted, referenced to 1Watt or to the nominal power output of 8 Watts? So, another useless piece of information.

However, one figure looked promising (assuming it was truthful), which is the weight of this amplifier, 18 kg! That would mean that output and power transformers are of decent size or that its chassis is heavy. The seller also specified the transformers as:

- Power transformer: 0.35 / Z11 / 105x60
- Output transformer: 0.35 / Z11 / dual 86x40
- Choke 0.35 / Z11 / 66x36

How to decipher these figures?

The first number is the thickness of the EI laminations used, 0.35mm in all three cases, which is good. Z11 is the type or grade of the magnetic material EI laminations were made of. Z11 is the old name used by Nippon Steel, equivalent to the current M6 AISI or 35Z145 specification. It is the lowest grade of GOSS (grain-oriented stainless steel), 0.35 mm thick, with core losses specified at 1.47 Watts/kg at 1.7 Tesla induction levels and 50 Hertz signal frequency.

The third figure seems to be the core dimensions where (choke's example) 66 would mean EI66 laminations (66mm being the largest dimension), and 36 would mean stack thickness of 36mm.

The output transformers use EI86 laminations with a 40mm stack, so the cross sectional area is $A = a*S = 3.8*4.0 = 15.2$ cm^2, with a power rating of $P \approx A^2 = 15.2^2 = 231$ VA.

The mains transformer's cross-sectional area of the center leg is $3.5*6 = 21$ cm^2, so its power rating is $P \approx A^2 = 21^2 = 440$ VA! This was very good news.

The financial analysis

The first test any candidate for upgrading must pass is the CCC test, the "Cumulative Cost of Components." A quick search on eBay, Ali Express, and other websites would give you approximate figures (all including separate transport costs from various sellers): chassis US$120, power transformer US$100, output transformers US$150 (a pair), choke US$50, a pair of 300B triodes (Psvane), plus rectifier and preamp tubes (Chinese) US$180. Allow $70 for RCA sockets, speaker binding posts, tube sockets, connectors, switches, feet, potentiometers, resistors, and capacitors, which brings the total to US$670, which is exactly what this amp sold for (including air courier transport).

Thus, you are getting free labor, which, considering that all the drilling, wiring, testing, and fine-tuning would take an average DIY constructor 3-4 days, makes this a seemingly good deal. Let's see if it is ...

Audio circuit analysis

This classic topology has been used in many commercial 300B designs. Direct coupling between the first and second voltage amplification stages is used, so only one coupling capacitor is in the signal path. The capacitively-driven output stage can only operate in Class A_1.

The circuit diagram found online had all resistance and DC voltage values, except the power supply voltages V1, V2, and V3. The anode and cathode currents for triodes are identical, so $I_{A1} = V_{K1}/R_{K1} = 1.95/750 = 2.6$ mA Since the voltage drop on the 30k anode resistor is $30*2.6 = 78$V, adding 78V to the anode voltage of 74V gives us V_1, so $V_1 = 78+74=152$V!

Repeating the same process for the 2nd stage, $I_{A2} = V_{K2}/R_{K2} = 81/15 = 5.4$mA.

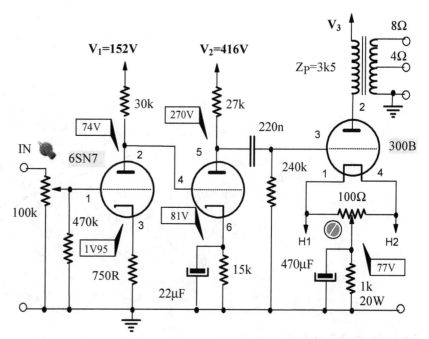

ABOVE: The published schematics of D&Y A-300 amplifier or something similar

Since the voltage drop on the 27k anode resistor is $27*5.4 = 146$ V, adding 146 V to the anode voltage of 270 V gives us V_2, so $V_2 = 416$ V!

The quiescent current through the output tube is $I_{A3} = V_{K3}/R_{K3} = 77/1k = 77$mA, which is on the high side; normally, we would use 65-70 mA. We don't know that anode voltage, so we cannot comment if the maximum anode power dissipation of 35-40 W has been exceeded or not.

1st stage DC analysis

To analyze the first stage (next page), we need to refresh our knowledge of load-lines. The DC load line can be drawn by determining two significant points, marked with "o." First, when anode current IA is zero, there is no voltage drop across the 750R cathode resistor, and there is no voltage drop across the 30k anode resistors. Thus, the voltage between the anode and the tube's cathode is the full anode supply voltage, in this case, $152V_{DC}$.

When the anode-cathode voltage (V_{AK} or V_A for short) is zero, the maximum current through the circuit is determined purely by the supply voltage and the sum of the cathode and anode resistances, so $I_{MAX} = 152/(30k+0.75k) = 4.95$ mA. Now we can draw the DC load line. Since the cathode resistor in the first stage is not bypassed (for AC signals) by a capacitor and there is no grid resistor from the 2nd stage's grid to ground (which for AC signal would be in parallel with the anode resistor), the AC or signal load line is identical to the DC load line.

We know that the cathode is at around 2V with reference to the ground, meaning that the grid is at -2V with respect to the cathode, so the quiescent point (no signal present) Q_1 must lie at the intersection of the AC & DC load line with VG=-2V curve!

The vertical projection to the V_A axis (horizontal axis) gives us anode voltage of 74 V, while the horizontal line to the I_A (vertical) axis indicates an anode current of 2.6 mA, both precisely as per the schematic.

1st stage AC analysis

Although the graph is not very precise, we can estimate the voltage gain of the 1st stage. The grid (input) voltage swing between points A1 and B1 is 4V, while the anode (output) voltage swings between 40V and 105V. The voltage gain with a bypassed cathode resistor would be $A_1 = -\Delta V_A/\Delta V_G = -(105-40)/4 = -65/4 = -16.25$, but due to local NFB it is now $A_F = A/[1+(R_K/R_A)A] = -16.25/[1+0.025*16.25] = -16.25/1.4 = -11.6$

1st stage distortion estimation: $V_0 = 74$V, $V_{MIN} = 40$V, $V_{MAX} = 100$ V, $\Delta V = 60$V

$\Delta V_+ = 34$V, $\Delta V_- = 26$V, $D_2 = (\Delta V_+ - \Delta V_-)/2\Delta V *100$ [%] $= 8/(2*60)*100 = 6.7$ %

This high distortion level is unacceptable; negative feedback should be used to correct it, or the stage should be redesigned. We could have anticipated such high distortion since the load line enters the region of low anode currents where the characteristics are curved (nonlinear), and "nonlinear" means high distortion!

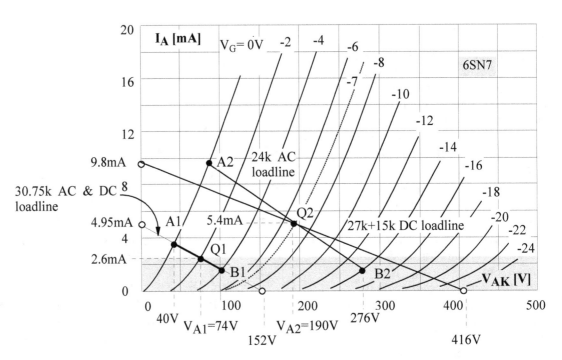

2nd stage DC and AC analysis

The cathode resistor of the 2nd stage is bypassed for AC signals, the DC load line is drawn using a total anode and cathode resistance of 27+15 = 42kΩ. We know that the anode current is 5.4mA and the anode-to-cathode voltage is 190V, so we can position Q2 immediately and read that grid bias is -7V!

The AC load line must pass through Q_2, but its slope is different, determined only by the anode resistance of 27k and the grid resistor of the output stage in parallel, for a total of 24kΩ, so it is steeper than the DC line.

The grid (input) voltage swing between points A2 and B2 is 14V, while the anode (output) voltage swings between 82V and 276V. The voltage gain is $A_1 = -\Delta V_A / \Delta V_G = -(276-82)/14 = -194/14 = -13.9$!

2nd stage distortion estimation:

$V_0 = 190V$, $V_{MIN} = 82V$, $V_{MAX} = 276 V$, $\Delta V = 194V$

$\Delta V_+ = 108V$, $\Delta V_- = 86V$, $D_2 = (\Delta V_+ - \Delta V_-)/2\Delta V *100 [\%] = 22/(2*194)*100 = 5.7 \%$

Again, this is way too high, B2 is in the "forbidden zone" of high distortion, highlighted in gray.

The power supply

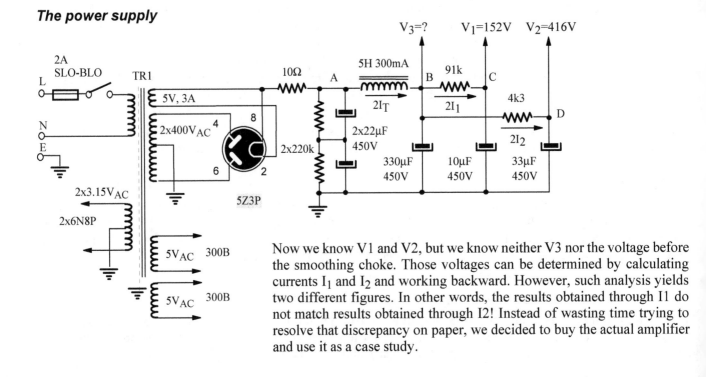

Now we know V1 and V2, but we know neither V3 nor the voltage before the smoothing choke. Those voltages can be determined by calculating currents I_1 and I_2 and working backward. However, such analysis yields two different figures. In other words, the results obtained through I1 do not match results obtained through I2! Instead of wasting time trying to resolve that discrepancy on paper, we decided to buy the actual amplifier and use it as a case study.

The first impressions and external tests

The amplifier arrived in just over a week and worked straight away. Well, kind of. There was a very strong hum from the speakers. The amp did not look new at all. It was dusty, and the RCA inputs were dull, meaning the thin gold plating had almost worn off due to constant plugging and unplugging of the interconnect cables, and then the underlayer started to oxidize. There were also a few minor dings and scratches!

Without opening the amp up, we plugged in one of our octal test jigs (pictured) and measured all AC and DC voltages on one channel's preamp tube, 6N8PJ (6SN7). The results are in the table below.

The first problem became apparent straight away - 5.5V_{AC} on 300B heaters and 7.1V_{AC} on 6N8P's heater pins. 10-15% higher heater voltages are a sure sign that the mains transformer was designed for 220V and not for 240V. We specifically requested a 240V version which the seller promised to ship, but clearly, that was a big fat lie!

300B triodes were AC heated, which was one possible cause of the very strong hum from both channels.

	6N8P V1	6N8P V2	300B	5Z3P
Anode	82 V_{DC}	230 V_{DC}	440 V_{DC}	430 V_{AC}
Cathode	1.8 V_{DC}	87 V_{DC}	74 V_{DC}	450 V_{DC}
Heater	7.1 V_{AC}	7.1 V_{AC}	5.5 V_{AC}	5.5 V_{AC}

The high cathode DC voltage of 300B tubes indicates a self-biased output stage, and the high cathode voltage of the driver stage (V2) indicates DC-coupling between the stages.

With a signal generator feeding the input of one channel, AC voltages were measured on the anodes of V1, V2, and V3 and on the speakers. An Analog multimeter with FET input was connected through a 0.1μF/630V film capacitor to remove the high DC voltages from the instrument's input; otherwise, the meter would get damaged!

With 0.1V_{AC} input (RMS voltage), there was 1.75VAC at the output of the 1st stage, 25V_{AC} at the output of the 2nd stage, 73V_{AC} on 300B anodes, and 3.5V_{AC} speaker output. From there, you can calculate the voltage gains of all stages and the voltage ratio of the output transformer.

Next, we measured the damping factor, which was very low (2.1), meaning the amplifier's output impedance was high, 2.1 times lower than the 8Ω load or 3.8Ω!

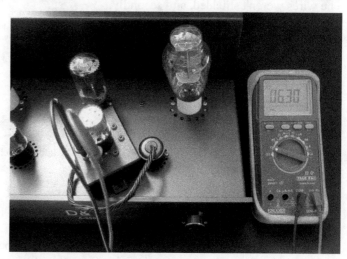

ABOVE: The octal test jig plugged into D&Y amp, heater voltage on pins 7 and 8 of 6SN7 being measured. Notice a spot-on value of 6.3V_{AC}, that was after the faulty Chinese power transformer was replaced by ours.

The frequency range of the two channels differed significantly; one went up to 23 kHz (a poor result), the other only up to 18 kHz (both -3dB points at 1 Watt output level). Since all left and right channel audio tubes were tested on Triplett 3444 tube tester and had very close gm and anode currents, we suspected that the output transformers were not identical.

MEASURED RESULTS (original amplifier):

- Left channel frequency range:
- 26 Hz - 18 kHz (-3dB, at 1W into 8Ω)
- Left channel max. power out: 7.5 W
- Right channel frequency range:
- 30 Hz - 23 kHz (-3dB, at 1W into 8Ω)
- Left channel max. power out: 8.2 W
- AC hum voltage on the 8Ω output: 98mV
- Damping factor: 2.1 (Z_{OUT}=3.8Ω)

ABOVE: These DIY test sockets are the most useful tool of all! Build one for all commonly used socket types (7-pin mini, Noval, Octal, Magnoval, RimLock, etc.) Pin 1 is marked with a plastic sticker on the Noval jig, the one on the Octal jig fell off.

So, the specified S/N ratio was a dream, as was the claimed frequency range, a pure fantasy! THD was measured as 3.9% @ 1 Watt and a whopping 9.8% @ 7W, so the amplifier did not even come close to the claimed THD of under 1%! We knew that from the start, this kind of design, with a high distortion input and driver stage and no global negative feedback, cannot have a THD figure below 3% under any circumstances, and that is on a good day!

With Bob's Black Box (a simple wattmeter described in Vol. 2 of this book), the power draw of the amplifier while idling (without any signal) was around 160 VA.

Under-the-chassis inspection and tests

Twisted wires run from the power inlet (1) to the on-off switch at the front, parallel with the chassis sidewall, which is proper practice. However, on the other side (cut off from the photo), the same type of wire was used between the RCA inputs (3) and the volume control potentiometer (4).

The wires weren't shielded, and secondly, twisting the signal wires for both channels together is a major no-no. Twisted wires are electromagnetically coupled; the signal from one will induce an identical signal in the other and vice versa, thus causing significant crosstalk and reducing channel separation. Whoever constructed this amplifier did not understand even the most fundamental principles of physics.

The under-the-chassis-view of the amplifier reveals almost a dozen bloopers!

The HV rail (6) was not insulated, so if you work on or test this amplifier while energized, there is a real danger of accidentally touching it. Notice how the legs of coupling capacitors (turned upwards) support the thick copper "bus bars," the most bizarre reversal of their roles!

The filtering choke (7) was placed in the worst possible location, right between the input tubes, as if the designer made sure that its radiated magnetic field affected both channels equally! If possible, a choke must be placed above the chassis; if not, it should be at the very back, as far away from the audio tubes as possible.

Look how much unused space the designer had on both sides of the mains transformer (8), under the back cover. The mains transformer did not have to be centrally positioned; it cannot be seen anyway. Even placing the choke in position (8) under the chassis would be fine.

And finally, notice how wires carrying heater currents were twisted and bundled together with the signal carrying wires just so the wiring looks neat & tidy, another fax-pass. No wonder this amp hummed and buzzed like crazy!

The power supply saga

The power supply was a true shocker, the worst we have ever seen. One-half of the high voltage secondary was left unused (1), so to be able to use the other half only, the builder had to add two silicon diodes, effectively turning the rectifier tube into a hybrid rectifying bridge.

Furthermore, the unfiltered voltage at the output of the bridge (before the filtering choke) was fed to both output tubes (2)!

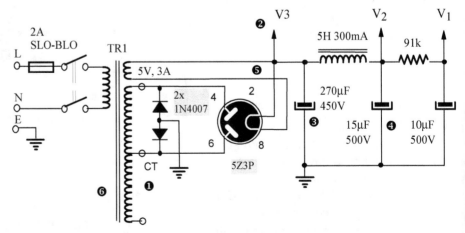

The actual power supply. Heater circuits are not shown to save space.

The capacitance of the first capacitor was too high for the rectifier tube; the limit is 33-47µF (3). The higher the capacitance, the higher the inrush current through the rectifier tube when the amp is powered on. Internal flashes caused by such high current pulses can damage and even destroy rectifier tubes.

After the choke (4), the capacitors were of very low capacitance, 15µF and 10µF, instead of the other way around, low capacitance before the choke and much higher capacitance after the choke. Again, it seems that the designer/constructor of this amplifier had no clue about transients and physical phenomena in power supplies!

Pin 2 of the rectifier tube was used as a DC output (5). Normal practice is to use pin 8, so other tube types with the cathode at pin eight can be plugged in instead, such as the indirectly-heated 5V4G, for instance.

Finally, the power transformer did not have an electrostatic shield (6), whose purpose is to prevent the capacitive coupling between the primary and secondary windings and the buzz thus transferred to the secondary.

Since this is an educational discussion, here is a challenge for you. Please go back to the photograph on the previous page and study it carefully to identify another error. Hint: look at the three transformers at the back.

Rewinding of the faulty power transformer

The power transformer is oriented the same way the output transformers are, which increases the magnetic coupling between them. The power transformer (or both output transformers) had to be rotated 90° either way!

The one half of the HV secondary that was in use could barely provide enough current, the winding was getting hot, and there was a significant voltage drop across it, so its failure was just a matter of time. There was something wrong with the other half, most likely a few shorted turns; that is why the amp's builder disconnected it and modified the rectification circuit.

So, we had no choice but to disassemble the transformer and rewind it properly, adding an E/S shield. Since we decided to use DC heating for 300B tubes, we couldn't just replace two 5VAC windings, the voltage after rectification would be too low, so two 6V windings were wound instead. If you ever decide to perform such a task, you need to count the number of turns in each winding during the process of unwinding.

The primary had 440 turns of 0.75 mm² wire (for 220V mains, 1A current), which means that the designer of the original transformer used the turns-per-volt formula of TPV=42/A, where A is the cross-sectional area of the center leg, in this case, 21cm². So, the turns-per-volt figure is TPV=42/21=2!

For 240V mains and 0.75A primary current (since the power draw of the whole amp is around 150 Watts), we wound 480 turns of 0.65 mm² wire.

The output transformers bungle

The output transformers' primary windings were wound using 0.25 mm wire; for the secondaries, 0.5mm diameter wire was used. The laminations were indeed EI86. The left channel output transformer had primary DC resistance of 220Ω and primary inductance of 12.6 H at 120Hz. Its primary impedance was 4.4kΩ with an 8Ω load.

The right channel output transformer was different. Its primary DC resistance was 200Ω, and the primary inductance was lower, only 9H at 120Hz. Its primary impedance was lower as well, 3.6kΩ. It seems as if the winder forgot to wind one whole primary section or at least a layer or two of the primary winding. A lower number of primary turns would explain lower DC resistance, lower impedance, and lower primary inductance.

All transformers were in galvanic contact with the chassis, metal on metal, which is a poor practice since the chassis made of mild steel is magnetic; it "collects" the leakage flux and magnetically couples the three transformers. So, we inserted cardboard cutouts between each transformer and the chassis.

MAINS TRANSFORMER:
- EI106 laminations
- a=36mm, stack thickness S=60mm
- Center leg cross section A=aS = 21.6cm²
- Primary inductance: L_P=4H (@120Hz)
- Primary DC resistance: R_P=6.1Ω
- Magnetizing current: 18mA

OUTPUT TRANSFORMERS:
- EI86 GOSS laminations
- a=28mm, stack thickness S=40mm
- Center leg cross section A=aS = 11.2cm²
- Primary: 3 sections, 0.25 mm² wire
- Secondary: 2 sections, 0.5 mm² wire

LEFT CHANNEL TRANSFORMER:
- L_P= 12.6 H @ 120Hz, 35 H @ 1kHz
- Primary impedance: Z_P=4.4kΩ
- Primary DC resistance: R_P=220Ω

RIGHT CHANNEL TRANSFORMER:
- L_P= 9 H @ 120Hz, 19 H @ 1kHz
- Primary impedance: Z_P=3.6kΩ
- Primary DC resistance: R_P=200Ω

The required fixes

1. New power transformer (rewound on the original core)
2. DC heater circuits added for 300B tubes, hum-balancing pots removed
3. Completely redesigned power supply section
4. Choke moved to the back of the chassis, away from input tubes
5. Twisted wires replaced by low capacitance (30pF/m) shielded cables (RCA inputs to the volume control pot)
6. Audio cables unbundled and separated from heater wires
7. Rewinding of one output transformer.

The final circuit after all fixes and mods

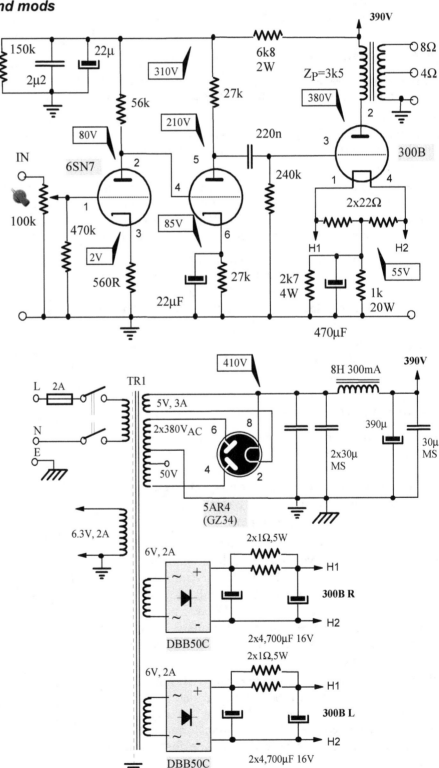

A 2k7 resistor was added in parallel with the main 1k cathode resistor, effectively making the cathode resistance 730W. The anode current is around 75mA and the anode dissipation is $P_A = V_{AK} * I_A = (380-55)*0.075 = 325*0.075 = 24.4$ Watts, which is very conservative.

A single choke shared between the channels was retained; adding another identical choke would have been better, so each channel would have its own CLC filter.

The maximum capacitance following the 5AR4 rectifier is 60μF; two 30μF motor start capacitors were used in parallel.

We didn't rewind one of the output transformers as initially planned, so one channel remained slightly different, which was noticeable during the listening evaluations.

The frequency range did not improve at all, but the hum was gone entirely (less than 1 mV on the speakers), and the sound improved immensely. It was cleaner, more detailed, and transparent.

The final verdict

Upon purchasing this abomination, an audiophile without technical knowledge or someone buying this amp with a view of upgrading it would be stuck with a substandard amplifier. The high hum and other issues would have made the amp unusable, either for listening or upgrading.

The overloaded power transformer's secondary would burn out soon, and due to heater voltages being way too high, the audio tubes would have a short life. Which of the two would happen first is hard to predict.

To return this amplifier to China or Hong Kong from Australia would cost around AU$550 (US$385 in Dec. 2015) due to brazenly high prices by Australia Post, FedEx and other couriers, so not a financially viable option.

Assuming that this or a similar amplifier (same topology) was decently constructed and suffered from no major problems, the input and driver stages could easily be modified into different topologies. Let's look at a few.

Conversion option #1: Paralleling input triodes into a single stage with LC coupling

By paralleling the two triodes inside one 6SN7 tube, we half the equivalent internal resistance. At 8mA anode current through each triode (a total of 16mA), the resistance is around 8kΩ each or 4kΩ total when paralleled.

The highest gain we can hope for is the mju of the tube, which is around 20. The load line must be horizontal, meaning we need to use a CCS (constant current source) or an anode choke of high inductance.

The only question now is if such a gain is sufficient. The output tubes were biased at 55V, meaning the RMS grid signal cannot be larger than 55/1.41 = 39 Volts. With a gain of 20, the input sensitivity would be 39/20 = 1.95 V.

With 1.95V_{RMS} at its input, the peak grid voltage would be 1.95*1.41 = 2.75 V. Since our bias is -6V, this is fine; the required grid swing would be from point X to point Y, leaving plenty of overload margins on both sides.

You may recall one similar project in Vol. 1 of this book, where we had a choke-loaded single input stage with two paralleled ECC40 triodes, providing a voltage gain of 30, which is 50% higher than in this case. However, that would necessitate a change of preamp tubes' sockets.

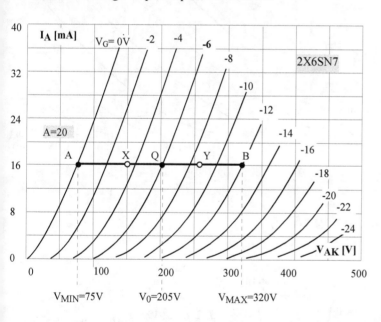

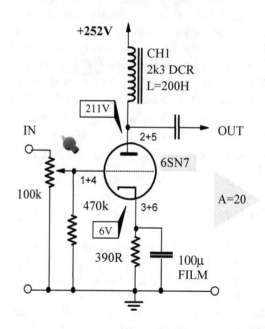

Conversion option #2: 6EM7 or 6DN7 instead of 6SN7

6DN7 and 6EM7 were designed for service in TV sets. The first triode was an oscillator working at the mains frequency of 60Hz, and the second triode was employed as a frame amplifier, driving the vertical deflection plates of cathode ray tubes (TV screens).

Both are pin-compatible but very different in terms of basic parameters apart from the anode power ratings. The triodes in 6DN7 have amplification factors of the same order of magnitude, 22 and 15, while one triode in 6EM7 has mju more than ten times higher than the other (μ_1=68 and μ_2=5.4).

TUBE PROFILE: 6EM7

- High μ triode + low μ power triode
- Octal socket, heater: 6.3V, 925 mA
- TRIODE #1: P_{AMAX}=1.5W
- V_{AMAX}=330V, V_{HKMAX}=200V_{DC}
- μ=68, r_i=40k, gm= 1.6 mA/V
- TRIODE #2: P_{AMAX}=10W
- V_{AMAX}=330V
- μ=5.4, r_i=750Ω, gm=7.2 mA/V

Fortunately, both tubes have the same pinout as 6SN7 triodes, so no wiring modifications are required. Let's see how would 6EM7 work in a two-stage amplifier driving a 300B power tube.

Obviously, we will use triode #1 (P_{AMAX}=1.5W) in the first gain stage and triode #2 (P_{AMAX}=10W) as the driver!

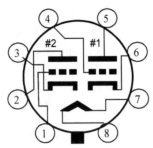

6EM7, 6DN7, 6SN7

TUBE PROFILE: 6DN7
- Dissimilar duo-triode
- Octal socket, heater: 6.3V, 900 mA
- TRIODE #1: P_{AMAX}=1.0W
- V_{AMAX}=350V, V_{HKMAX}=200V_{DC}
- μ=22, r_I=9k, gm= 2.5 mA/V
- TRIODE #2: P_{AMAX}=10W
- V_{AMAX}=550V, V_{HKMAX}=200V_{DC}
- μ=15, r_I=2k, gm=7.7 mA/V

1st stage analysis (using 6EM7)

T1 in 6EM7 is a low-bias tube, so to preserve the bias of around -2V, we need to position the Q-point at around 170V anode-to-cathode voltage. We also need to keep the load line as horizontal as possible so the voltage swing to the lowest point (B1) does not get into the region of very low anode currents, close to the cutoff point, where distortion is high. That means we must use a very high anode resistance of, say, 167 kΩ!

The cathode resistance must be $R_K=V_K/I_K$ = 2/0.0013 = 1,538 Ω and the voltage gain is A_1=-$\Delta V_A/\Delta V_G$ = -(275-65)/4 = -210/4 = -52.5.

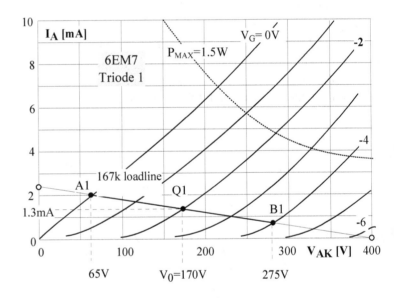

2nd stage in common cathode configuration (using 6EM7)

With DC coupling, we know that the grid of the 2nd triode is at +170V. Choosing -15V bias and 20mA current, the cathode of the 2nd triode must be at 170+15=185V. Since the graph shows the anode 90V above the cathode, the anode must be at 185+90 = 275V.

If we choose 3k9 for the anode resistor, the voltage drop on that resistor is ΔV_A=3.9*20 = 78 V, giving us the required anode supply voltage of 275+78 = 353 V. The cathode resistance must be $R_K=V_K/I_K$ = 185/20 = 9k25 (use 9k2). The DC load line is 9k2+3k9 = 13k1.

Even with a bypassed cathode resistor, the voltage gain of the 2nd stage is low: A_2=-(125-50)/20 = - 3.75! However, with A_1 over 50, even unity gain would be sufficient.

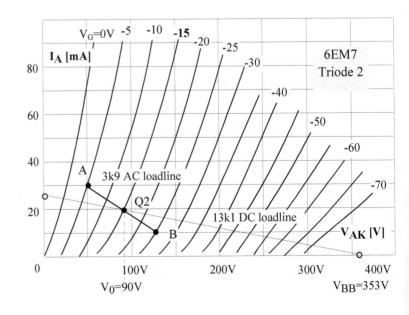

Higher heater current draws

Two 6SN7 triodes draw 1.2A heater current, while two 6EM7 tubes load the heater secondary winding with almost 2A, a 67% increase. So, before you just plug 6EM7 lovelies into your amplifier, make sure that the transformer's secondary winding can supply that much current. Before you carry out modification of this kind, load the heater secondary provisionally with 2A (use crocodile clips and two 6EM7 with just their heaters connected) and measure the nominally $6.3V_{AC}$ voltage.

If it drops significantly, say down to 5.9V, the secondary was wound with a thin wire and cannot support a 2A load! Of course, you could always add a small ancillary 15VA or larger (6.3V, 2A or 12.6V, 1A) transformer specifically for this purpose and connect two heaters in parallel or series.

Our rewound mains transformer had a 6.3V/2A winding, so the heater voltage remained at 6.24V even with two 6EM7 tubes.

Finally, notice that this tube substitution only applies if the first stage triode is the one with pins 5-4-6, because that is the high mju triode in 6EM7. Alas, that was not how the D&Y amplifier was wired. When 6SN7 is used, it does not matter how the amplifier is wired, which of its two triodes is used for the 1st and the 2nd stage.

Is it worth it?

Technically, the main benefit of this modification is the more powerful driver stage. Instead of 6SN7 with $P_A=5W$ and high internal impedance of 7-8kΩ, the 300B would now be driven by a 10 Watt triode with 2kΩ output impedance.

From the tube rolling perspective, the choice of quality NOS 6EM7 is much more limited than that of new production and NOS 6SN7.

Despite the mismatch in tube section between the amp and the 6EM7 tube, we bravely tried them together anyway. The 1st stage was triode #2, the 10W rated one, and the driver was triode #1. Everything worked fine!

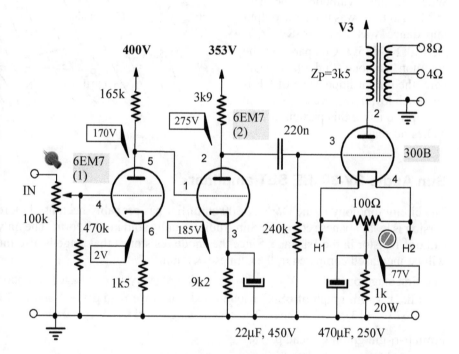

As for DC voltages, triode #2 was pulling a much higher anode current than 6SN7, so its anode voltage dropped to low levels. Incredibly, despite the reversed and thus mismatched input and driver triodes, the sonic benefits were significant. The sound with both 6EM7 and 6DN7 was noticeably superior to both the Chinese 6N8P and any brand of 6SN7 we tried (GE, RCA, Westinghouse). Another perplexing example of how designs that may not make technical or theoretical sense often make beautiful music!

Conversion option #3: The 6EM7 as cathode follower option

Since the first stage can provide a voltage gain of around 50 times, we could use the power triode (#2) of 6EM7 as a cathode follower. We have a high voltage of $500V_{DC}$ available, so let's choose 450V for the anode supply of the CF (point Y on the load line). The cathode resistor could be a parallel combination of four 47kΩ resistors, so the total resistance is 47/4=11.75 kΩ.

To get the point X, we divide $V_{BB}=450V$ with the load resistance and get $I_X=450/11.75= 38.3$mA. To determine Q2, we need to draw the grid bias line. We simply take two grid voltages, for instance -20V and -60V.

Due to DC coupling, a bias of -20V on the grid means the cathode voltage is 20V+170V = 190V. The anode and cathode current is now 190V/11.75Ω = 16.17mA, so we mark point Z.

Likewise, a bias of -60V on the grid means the cathode voltage is 60V+170V= 230V. The anode and cathode current is now 230V/11.75Ω = 19.6 mA, so we mark point W. The intersection of the grid bias line and the load line determines the quiescent point Q2. We read $V_{A2}=230V$ and $V_{G0}= -47V$, meaning the cathode is at 47V+170V (due to DC coupling) = 217V. Finally, we read the anode current in Q2 to be around 18.5 mA!

To double check if our figures are correct, with 18.5mA of anode current, the cathode is at 18.5*11.75 = 217V, the anode is 230V higher than the cathode, so the anode voltage is 217+230 = 447V. This should be 450V, but we are close. It means the anode current is slightly higher than 18.5mA, but this is an estimation only anyway. The tube we use could be slightly different from this typical tube used for the anode characteristics, so we will need to fine-tune the currents by adjusting the value of cathode resistor.

The power dissipated on this resistor is $P=R_K*I^2 = 11,750*0.0185^2 = 4$ Watts. Since we are using four 47k resistors, each must be rated at 2 Watts minimum for 8 Watts total power rating (double the heat dissipation).

The output impedance of this cathode follower is $Z_{OUT} = Z_K \| R_K = [r_I/(1+\mu)] \| R_K = [2,000/(1+5.4)] \| 11,750 = 304\Omega$. The output impedance of the 2nd stage using 6SN7 triodes was $Z_{OUT} = R_L \| r_I = 9k \| 27k = 6k75$.

When common cathode stage with 6EM7 triode #2 was used, the output impedance was $Z_{OUT} = R_L \| r_I = 2k \| 750\Omega = 545\Omega$. Compared to the common cathode 2nd stage with gain, the output impedance of CF is not much lower, and we lose gain, so this option using this particular tube makes no sense.

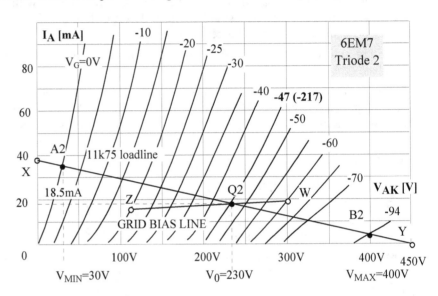

Sun Audio SV-300BE SET amplifier

The circuit topology of the D&Y A-300 amplifier has probably been used more than any other in audio history. This version is by a Japanese company Sun Audio, usually sold in a kit form. The only noticeable difference is the bypassed cathode resistor in the 1st stage. Since the anode resistor of that stage is also much higher (double the value), its gain will be increased compared to the Chinese version.

The first stage operates at $V_{AK} = 70-1.5 = 68.5$ V (very low voltage) and anode current of $I_1=1.5/470 = 3.2$mA, so from the μ-r_I-gm graph of 6SN7 triode (next page) we read $\mu=19.2$ and $r_I=13k\Omega$. Its gain is $A_1 = -\mu R_L/(r_I+R_L) = -19.2*62/(62+13) = -15.8$ times. The second stage works at $V_{AK}= 210-76=134$V and $I_2=76/27= 2.8$mA.

From μ-r_I-gm graph we read $\mu=18.7$ and $r_I=15k\Omega$, and calculate the gain of $A_2= -\mu R_L/(r_I+R_L)= -18.7*27/(27+15) = -12$ times.

Strictly speaking, these curves are only valid for anode-to-cathode voltage of $V_A=250V_{DC}$, but the parameter values obtained that way are perfectly adequate for estimation purposes.

The total voltage amplification before the output tube is thus -15.8* - 12 = 190 times, which is way too high for the output stage with a bias of 52 Volts!

The driving signal's peak cannot exceed 52V; its maximum effective value is thus 52/1.41 = 37V. With 37V after the 2nd stage, the input voltage into the amp needs to have an RMS value of 37V/190 = 0.2V! Thus, this is a high input sensitivity design.

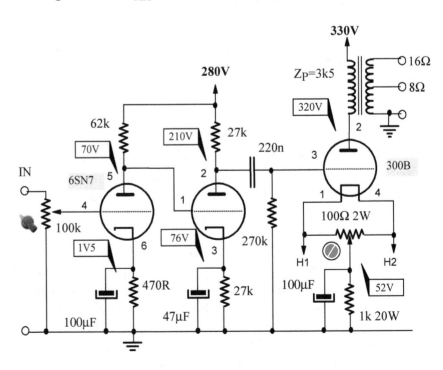

ABOVE: Sun Audio SV-300BE, audio section, one channel only shown

The anode current of the output stage is 52V/1k = 52mA, which is very low for 300B. The anode-cathode voltage is V_{AK} = 320-55 = 265V, which is also too low.

The quiescent power input is P_{IN} = 265V*0.052 =13.8W, so the power tubes are significantly underutilized. Positioning the load line on 300B's curves indicates less than 4 Watts output power, a 2A3 territory!

The load line enters the region of low anode current. The anode characteristics in that region are very nonlinear, resulting in very high distortion. With a load ellipse (an actual loudspeaker load), the output tube will go into cutoff very early, so not even 3.5 Watts will be possible.

For some reason, Japanese audio designers seem to be highly regarded, but in this case at least, a very substandard design.

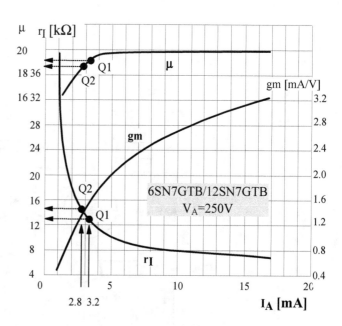

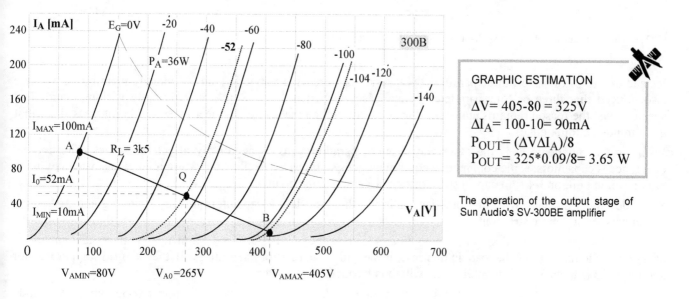

GRAPHIC ESTIMATION

ΔV = 405-80 = 325V

ΔI_A = 100-10 = 90mA

P_{OUT} = $(\Delta V \Delta I_A)/8$

P_{OUT} = 325*0.09/8 = 3.65 W

The operation of the output stage of Sun Audio's SV-300BE amplifier

CASE STUDY: Audio Note Kit 1

This kit started Audio Note's kit side of the business in 1995. In 2015 the cost for standard configuration with EI output transformers was US$2,250.- The first versions were butt-ugly, with open-frame transformers (not even end-bells) and a flimsy-looking frame on both sides, either to act as handles or as a support for a cage or top mesh, although I have never seen a photo of one with top cover installed.

However, the transformers and chokes were of decent size and quality and the design was basically sound, so the kit gained a good reputation.

1st and 2nd stage - DC and AC analysis

The first stage (diagram on the next page) operates at V_{AK}=170V and anode current of I_1=4.8/680 =7mA, so from the μ-gm-r_I graph for 6SN7 we read that for 7mA anode current the parameters are approx. μ=20 and r_I=9kΩ.

The voltage gain of the 1st stage is affected by local NFB through the un-bypassed R_K so the gain is A_F= $-\mu R_A/[r_I+R_A+(1+\mu)R_K]$ = -20*27/[9+27+21*0.68] = -20*0.54 = 10.7, approximately half of tube's mju.

Notice that the volume control potentiometer is positioned after the 1st stage, so its value needs to be added in parallel to the anode resistance of approx. 27k, which will reduce the anode load to 21k3 and lower the gain of the 1st stage to around 9.5.

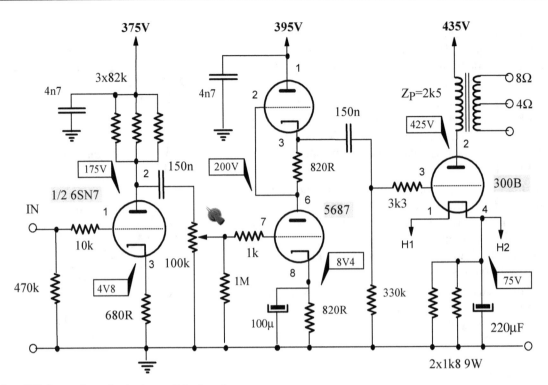

ABOVE: The audio section of Kit 1, one channel only shown, © Audio note

Instead of one 27k resistor, three paralleled 82k resistors were used as anode load. As already mentioned, we have also noticed that three or four paralleled resistors sound better than a single unit. Despite various claims and speculations, the reason remains unknown, the sonic difference makes no engineering sense, but it does show that great minds do think alike!

With most components, it is the other way around; paralleling tubes negatively affect the sonics. The notable difference is capacitors; some commercial equipment goes so far as to parallel 20 or so smaller elcos instead of using only a single unit, or 3-4 film caps instead of one larger capacitance unit.

Some designers disagree with that practice, claiming that paralleling capacitors increases ringing and makes the sound harsher and more stringent.

The second stage is an SRPP amplifier with an anode current of $I_2 = 8.4/0.82 = 10.2$ mA. Its voltage gain cannot be accurately calculated since the load, in this case, is the 330k grid resistor in parallel with the 3k3 grid stopper resistor and the input impedance of the 300B tube, which is frequency-dependent.

However, it can be estimated by assuming an infinite load, which would make the upper triode an active anode resistor for the lower one, whose value is $R_{A1} = r_{I2} + (\mu_2+1)R_2 = 2.7 + (16+1)*0.82 = 16.6$ kΩ.

The estimated voltage amplification is $A = -\mu R_{A1}/(r_I + R_{A1}) = -16*16.6/(2.7+16.6) = -16*0.86 = 13.8$! The voltage gain is approx. 86% of tube's mju (16 in this case).

So, Audio Note designers had discovered very early on that high current preamp stages sound superior to stages with tubes biased to pass low currents. Both 6SN7 and 5687 are high current triodes with highly regarded sonics, so their pairing with 300B output princesses was a clever choice!

The output stage and the power supply

The anode-cathode voltage of 300B tubes is $V_{AK} = 425-75 = 350$V, which is as per Western Electric's operating point, and their anode current is $75V/0.9k = 83$ mA. The quiescent power input is $P_{IN} = 350V*0.083 = 29$ W out of the maximum 40 Watts, which is very conservative.

The secondary winding supplying the heaters of the two 5687 tubes (next page) is elevated to $86V_{DC}$ through a voltage divider (82k and 330k). Since the cathodes of the upper triodes in both 5687 tubes are at $200V_{DC}$, the maximum heater-cathode voltage for 5687 of 100V is still exceeded by around 24V, but quality tubes should be able to tolerate such stress on their insulation!

The heater supplies for output tubes are regulated DC, provided by simple three-terminal solid-state regulators.

Finally, notice the RC decoupling between the audio ground and the chassis ground, the 100R resistor, and the 1nF film capacitor in parallel.

THE QUEEN OF HEARTS: SINGLE-ENDED AMPLIFIERS WITH 300B TRIODES

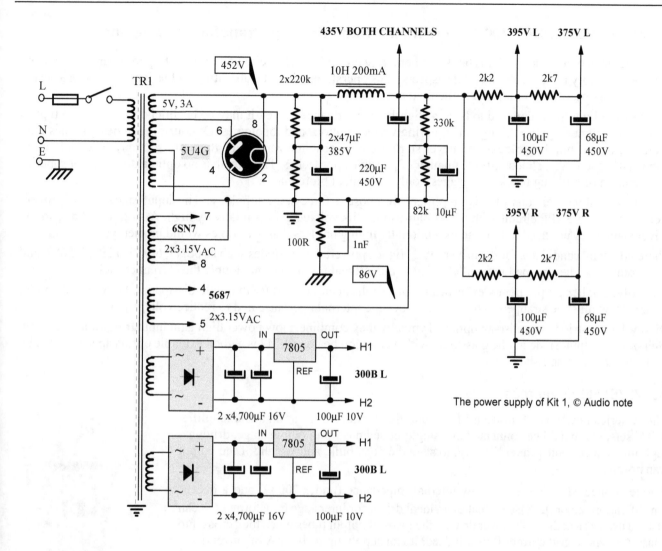

The power supply of Kit 1, © Audio note

The filament wear issue with DC-heated filamentary tubes

With direct heating, one end of the filament is at a higher DC potential. The other is at a lower potential, so there is a voltage gradient along the filament. The grid bias also changes from one end (pin 1) to the other (pin 4).

With reference to the filament end connected to the positive heater supply (pin 4 on the diagram), the grid is biased more negatively. Hence, the cathode emission of that end is lower. In contrast, the filament end connected to the zero end of the 5V heater supply (pin 1 on the diagram) emits more electrons and will wear out quicker, i.e., it will lose its emission faster.

One way to prolong the life of the DHT tubes is to install a switch that will reverse the polarity of the heater supply to the tube sockets and to manually operate that switch every time just before the amplifier is switched on.

Alternatively, tubes can be swapped between the channels (or monoblocks), but that requires the DC heater supply to be wired differently in each channel or monoblock. For instance, +5V would be brought to pin1 in the left channel but to pin 4 in the right channel. So, every 20, 50, or 100 operating hours simply swap the DHT tubes around.

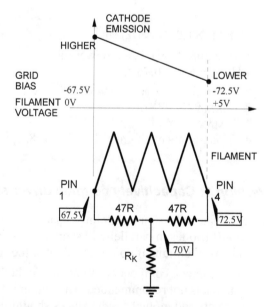

ABOVE: DC voltage distribution on 300B filament with 70V quiescent DC voltage on the cathode resistor R_K

Design project: PSET 300B monoblocks with interstage transformer coupling

In the previous two volumes of this book, we generally presented a final circuit diagram of a particular amplifier and a few explanatory remarks about certain aspects of its operation, without elaborating on the genesis of such a design or the process by which we arrived at the final circuit.

Since this volume was supposed to be at a slightly more advanced level, it may be beneficial to illustrate a typical process a designer may go through in exploring various options and topologies. Of course, each designer has their preferred circuit building blocks and never starts from an absolute zero; plus, there are so many tube and circuit options that there is an almost infinite number of possible approaches. Thus, there is no optimal topology or superior combination of building blocks and stages - only individual preferences.

For the first three versions of this amplifier, we chose a three-stage topology. The input stage is our proven, well-performing SRPP circuit with 6BQ7 duo-triodes. Its voltage gain is relatively high, A_1=-25 or 28 dB, so the driver stage does not need to have an exceptionally high gain; less than five times would be plenty.

There are many candidates for the driver tube. Higher-powered duo-triodes such as 6SN7, 6CG7, 12BH7, 5687, and many others can be strapped in parallel to halve the internal resistance and double the current capacity.

Secondly, smaller output pentodes or beam power tubes can be used (with anode dissipation between 10 and 20 watts). They must be strapped as triodes to reduce the internal resistance and to reduce voltage gain.

Dissimilar duo-triodes are also an option. Typically they combine a low power high gain input triode and a low mju high power output triode in one glass bulb with a common heater. 6EM7 is a good example of this approach, a tube we have already discussed.

6CW5 (EL86) driver tube

The American 6CW5 or European EL86 pentode was developed as an audio amplifier in TV sets; two would be connected in a single-ended push-pull output stage, driving a high impedance loudspeaker directly, to save the cost, bulk, and weight of the output transformer.

Triode connected, it has a very low internal impedance (under 700 Ω) and a decent amplification factor (μ=8), so it makes an ideal driver for interstage transformers. It can also be used as a cathode follower driving the grids of output tubes when they cross into Class A_2, where grid current flows. Indeed, it can supply up to 100mA of current!

Noval socket means the inter-electrode capacitances are much lower than for similar octal tubes, which means wider high-frequency extensions!

TUBE PROFILE: 6CW5 (EL86)
- Miniature Noval beam power tube
- Heater: 6.3V, 0.76 A
- V_{AMAX}= 270V_{DC}, V_{SMAX}= 200V_{DC}
- V_{HKMAX}=100V_{DC}
- P_{AMAX}=12W, P_{SMAX}=1.75W
- Triode-connected: gm=12mA/V, μ=8, r_I =670Ω

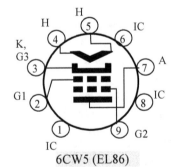

6CW5 (EL86)

Version 1: Capacitively coupled driver stage

To eliminate at least one coupling capacitor from the audio signal path, the first two stages will be DC-coupled. The two 300B triodes in parallel will share a self-biasing circuit, a resistor, and a capacitor. You can separate them into two identical circuits if you wish - double the resistor's value and halve the value of the bypass capacitor.

Separate cathode components allow unmatched tubes or tubes that are not perfectly matched; however, in a parallel design, that is not recommended. The output tubes should be as close to identical as possible, which means their idle DC currents and mutual conductances should be the same!

With 2.5mA idle DC current through the 1st and 14mA through the 2nd stage, the amplification factors were A_1=25 and A_2=4, so with 0.5V signal at the input, the signal at the grid of output tubes was 50V. The voltage gain of the output stage was 2.2 times.

The output power was a respectable 28 watts or 14 watts per 300B triode, which is pretty much the maximum that can be expected from cathode biased stage operating in class A_1 only.

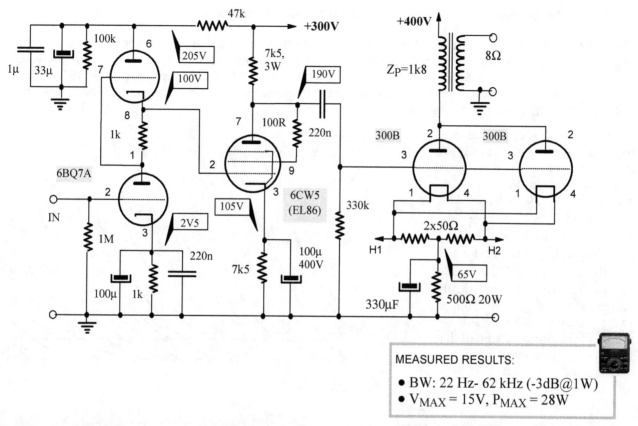

Version 2: LC-coupling

The 7k5 anode resistor of the 2nd stage was replaced by a 30H anode choke with 3,300 Ω DC resistance. The frequency range at 1 Watt remained approximately the same, 22Hz to 66 kHz, but the maximum output signal voltage dropped to 10V, meaning the output power was only 12.5 Watts, so this option was abandoned immediately.

Version 3: 6CW5 transformer-coupled driver stage

Since the output stage can now be driven into class A_2 by the low impedance driver stage, the maximum obtainable power with IS transformer coupling is a whopping 45 Watts. The -3dB frequency range at 1 Watt was a respectable 22 Hz - 36 kHz.

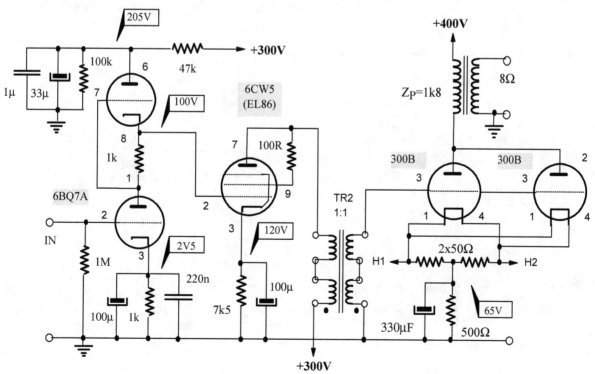

Choosing the quiescent current level through the driver stage

Since EL86 can pass up to 100mA of current, higher anode currents would make the tube operate in a more linear region (above 25-30 mA) and thus reduce harmonic distortion. However, higher anode DC currents would also require a larger magnetic core for the interstage transformer and a bigger air gap. That would reduce the core's permeability and thus primary inductance and narrow the frequency range of the transformer, especially at the lower end, thus attenuating the bass frequencies.

So, as always, a compromise is required. We wound these IS transformers on smallish magnetic cores, so had to keep the DC current below 20 mA for best performance.

At 1kHz, the primary impedance of the interstage transformer is $Z=2\pi fL=2*3.14*1,000*36 = 226$ kΩ, so the load line is practically horizontal. The anode voltage swing is approx. from 24V to 272V, so the estimated gain of that stage is $A_2= 246/40 = 6.1$, while the measured gain was $A_2=5.4$!

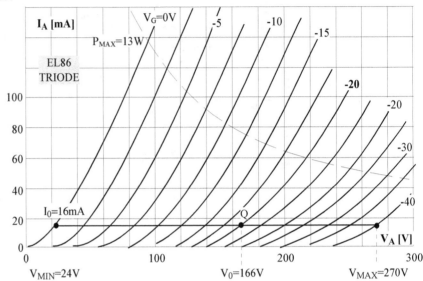

MEASURED RESULTS:

- BW: 22 Hz- 36 kHz (-3dB@1W)
- Class A_1: $P_{MAX} = 25W$
- Class A_2: $P_{MAX} = 45W$

KEY FEATURES:

- DC-coupling between the first and driver stage
- Transformer-coupling to the output tube
- No coupling capacitors in the signal path
- Self biasing, no adjustments of any kind are needed
- No global negative feedback
- Significantly increased output power (Class A_2)

Version 4: 6EM7 - 6DN7- 6SN7 transformer-coupled driver stage

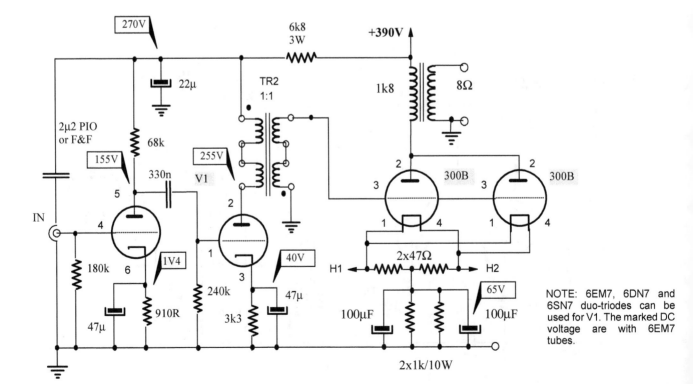

NOTE: 6EM7, 6DN7 and 6SN7 duo-triodes can be used for V1. The marked DC voltage are with 6EM7 tubes.

The first stage is biased at 1.55mA of current and has a voltage gain of 35. The anode current of the second stage is around 12mA, and the voltage gain is 3.7 times. Finally, the output stage at 130 mA (65mA per 300B triode, which is very conservative) amplifies 1.74 times. After the output transformer, the load voltage is 15V for 1V at the input, so the overall voltage gain 15 times or 23.5 dB.

Tube rollers will love this relatively simple self-biased design in which various rectifiers and input tubes can be used, both of which impact the amplifier's voicing. There is no need to change resistor values; simply plug & play!

KEY FEATURES:

- Class A_2 capable
- Various rectifier tubes can be used (6Y3, 5U4G, 5V4G, GZ34)
- Various input tubes can be used, 6SN7, 6EM7, 6DN7

MEASURED RESULTS:

- $V_{MAX} = 18V$, $P_{MAX}=40.5W$
- BW: 22Hz-36kHz (-3dB@1W)

The importance of interstage transformer phasing

Notice the connection of the IS transformer's windings. The ends with the same phase are marked with a dot. When connected as per the circuit diagram, the -3dB frequency range of the amplifier extends up to 36 kHz. When the secondary leads are reversed, so that the end with the dot is connected to the output tubes' grids, the range narrows to only 25 kHz. This phenomenon is caused by inter-winding capacitances which in the second case act as a low pass filter, shunting higher frequencies to ground.

TRADE TRICKS

Various octal rectifier tubes can be used, from 5Y3 to 5V4G. The $+V_{BB}$ voltage with 5V4G is about 10V higher than with 5Y3 (390V versus 380V) because 5Y3 exhibits a very high internal voltage drop.

Higher current capacity rectifiers such as GZ34 can be used for an even higher $+V_{BB}$ of almost 400V.

The filaments of both output tubes are connected together, in parallel, to a source of $5V_{DC}$, obtained by a center-tapped 2x6.3V secondary winding feeding a pair of solid-state rectifiers (actually two out of four diodes in a 35A diode bridge), followed by a CRC ripple filter.

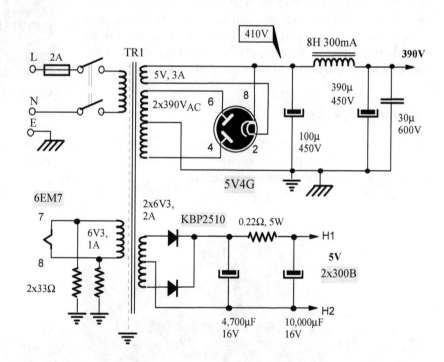

Construction details

All wires from the power transformer exit through a rectangular slot in the chassis (simply cut with an angle grinder and filed smooth), just below the filtering choke on the photo (9). The heater wiring for the input tube runs along one side of the chassis (1). The RCA input is so close to the input tube's socket that no shielded cables are necessary.

All audio components are mounted on one terminal board (2), except the cathode RC components for the power tubes, which are on an identical terminal board containing the power supply components (7). Two cathode resistors in parallel have copper heatsinks on their leads (8).

The 35A-rated rectifier bridge is mounted directly on the steel chassis (6), which acts as its heatsink. Remember to remove the powder coating from underneath it; otherwise, the heat flows onto the chassis will be severely impeded! The 35A rating isn't necessary, but since those bridges are pretty cheap, having a physically large unit helps the heat dissipation.

The first filtering elco is mounted directly on the rectifier tube's socket (5). Mains DC filter (4) is on its own small board, mounted close to the AC inlet, while its output feeds the primary of the power transformer.

The interstage transformer (3) is in its own enclosure, mounted in the front corner, close to the input tube, while the motor start capacitor is mounted close to the filtering choke.

The chassis were steel drawers (turned upside down) finished in metallic gray. All we had to do was manufacture the bottom covers. IKEA drawer fronts in cherry color were perfect for side trim.

Steel transformer covers in wrinkle black finish were purchased on eBay, as were the chrome-plated mounting plates for 300B sockets. The rectifier tube is behind the rear 300B tube.

The mirror image topology was used; if that is too complicated for you, make both monoblocks identical, as most commercial manufacturers do (because it's cheaper), but mirrored units look more expensive and "high end"!

Again, the MSPL topology, RCA input right next to the input/driver tube, no need for shielded cables at all.

The mains IEC inlet, fuse, and on-off switch are at the back, a bit inconvenient to reach to the back to turn the monoblocks on and off, but they aren't that deep, and it is a small price to pay for the quieter operation. You can run a twisted mains cable to the front-mounted on-off switch if you prefer (along the side opposite the RCA input), but that may increase the hum due to the proximity of the input and output tubes.

ABOVE: The internal view of the prototype with 6EM7 tubes.

Listening impressions

While some other 300B amplifiers sounded better with 6EM7 tubes instead of 6SN7, the opposite was the case in this design. Both 6SN7 and 6DN7 sounded slightly better than 6EM7, although the amp was designed specifically for 6EM7! In any case, the news was excellent.

The monoblocks had the slam of high voltage triodes such as GM70 and 211 but still retained the tenderness and microdynamics of 300B, a very rare combination, the refined magic we haven't been able to achieve with those high voltage big bottle triodes.

The bass was fast and prominent; the amps kept their rhythm & pace even with the most complex and demanding pieces of music, something those 8 Watt run-of-the-mill 300B amps cannot do.

These monsters could easily drive even inefficient and hard-to-drive speakers, those that could never be paired with 300B lovelies before!

TRIODES, PENTODES AND BEAM TUBES: MORE SINGLE-ENDED DESIGNS

- Shindo Cortese amplifier
- Design project: F2a SET amplifier
- Design project: EL12 SE monoblock amplifier
- Design project: EL156 SET amplifier
- Design project: PSE ultralinear EL84 amplifier with a separate power supply
- Design project: PL508 PSE "enhanced triode" amplifier
- Design project: Battery-powered PL508 SET amplifier
- Yarland FV II and FV35A amplifiers
- Audion Sterling EL34 amplifier

"There's plenty of breathless hype in PSA's literature as well - all of which I'm convinced that they, in all sincerity, believe."
Michael Fremer, Stereophile, Feb 2016

Shindo Cortese Amplifier

A 10W stereo amplifier with a brazenly high retail price of US$9,900, the original version of Cortese used one 6AW8A triode-pentode tube as a driver and one F2a beam tetrode as a power tube. The new version uses two 6AW8A tubes per channel, but the output power remains the same.

The circuit diagram was obtained from an unverified source on the web, so we cannot ascertain its accuracy. DC voltages were not marked, so we cannot analyze the circuit; it is shown here for discussion purposes only.

The design cannot be simpler, a quasi- SRPP stage using dissimilar tubes (triode and triode-strapped pentode), driving F2a in pentode mode using fixed biasing and capacitive coupling. Only Class A1 operation is possible.

The 10R cathode resistor enables measuring of the cathode current, but the bottom cover must be removed to do that! The bias pots are also mounted inside the chassis, without any access from the outside, so to adjust the bias, the user needs to turn the amplifier upside-down and remove the bottom cover. All mains terminals are exposed (not shrouded at all), making biasing or voltage checks a dangerous task!

Likewise, the mains fuse is of the open type, mounted inside the chassis, which isn't user-friendly, making a simple fuse replacement into a major screwdriver exercise.

Ordinary components (resistors, capacitors) are used, with no expensive "hi-end" brand names. Coupling capacitors seem glued to the top metal plate that carries all tubes. Generally, that is a bad practice. Any vibrations transferred to the metal chassis from the power transformer and chokes are then transferred to the coupling capacitors, resulting in smudged, incoherent sound.

The wiring is point-to-point, using terminal strips.

There is no need to bypass the 10R cathode resistor (low resistance), let alone use such a high capacitance (500µF). With 100mA of cathode current, the DC voltage drop on that resistor would be only 1V!

Judging by the distance between output transformers' mounting bolts, they are smallish and mounted side-by-side along the narrow side of the top cover. Three vertically-mounted capacitors flank the output transformers, followed by the power transformer. No chokes are used at all.

Two solid-state regulator circuits with discrete components make their power supply time-consuming to draw, so we won't reproduce it here, see the summary of its features in the "Power supply" frame.

Shindo laconically specifies the frequency range being from 20 Hz to 20kHz, which does not seem to be an actual measurement, and secondly, if it is an accurate measurement, it isn't great.

Distortion figures and damping factor were not specified, and the web search yielded no results; it seems nobody has tested this amp on the bench (yet).

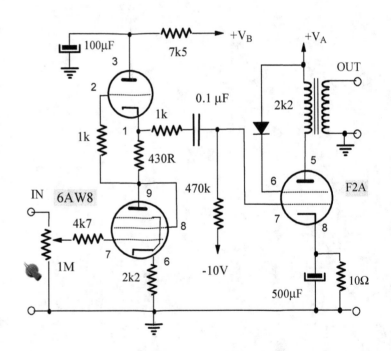

Shindo Cortese (old version) audio section (one channel), © Shindo Laboratories

Power supply:

- solid state rectifiers (bridge), EY88 vacuum tube diode in series in the anode voltage line
- Both HV and bias voltage supplies are regulated using LM317 solid state voltage regulator ICs
- Individual bias adjustment of output tubes
- AC heating of all tubes
- $6.3V_{AC}$ supply for preamp tubes is raised to half of the anode voltage due to low maximum heater-cathode voltage of 6AW8 tubes

Driver and output tubes

F2a11 and sonically slightly inferior F2a are special quality power beam tetrodes. The p-factor is 0.15%, meaning that 0.15 tubes out of every 100 will fail within each 1,000 operating hours period, or 15 tubes per 10,000! F2a has lower voltage ratings and uses a different socket.

While far from rare, eBay sellers keep prices high despite large stocks of NOS F2a Siemens tubes. The prices rose significantly in the last decade due to their use in Shindo amplifiers.

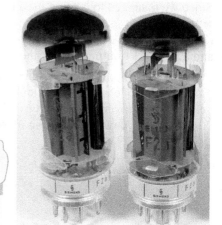

TUBE PROFILE: 6AW8A

- High μ triode + sharp cutoff pentode
- Noval socket, heater: 6.3V, 600 mA
- TRIODE: V_{AMAX}=330V, P_{AMAX}=1W, V_{HKMAX}=100V_{DC}
- μ=70, r_I=17k5, gm=4 mA/V
- PENTODE: V_{AMAX}=V_{G2MAX}=330V, P_{AMAX}=3.25W, P_{G2MAX}=1W

TUBE PROFILE: F2a & F2a11

- Indirectly-heated beam tetrode, 6.3V, 2.0A heater
- Max. anode/screen voltage 425V/425V (600V/425V for F2a11)
- Max. H-K voltage: 80V (120V for F2a11)
- Max. anode / G2 dissipation: 30W / 5W, max. cathode current: 140 mA
- STATIC DATA (PENTODE):
- gm=18 mA/V, r_I=23kΩ, μ=414, $μ_{G1G2}$=17.5, C_{GK}=18pF, C_{AK}=128pF
- STATIC DATA (TRIODE): gm=21 mA/V, r_I=800Ω, μ=17

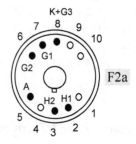

Typical pentode operation (not necessarily that of Shindo amplifier):

With a screen voltage of 250V_{DC}, the -9V bias in the Shindo amp would be too high, so it seems that the amplifier's G_2 voltage is higher than 250V. Consulting the curves for V_{G2}=330V reveals that for this load line, the bias would need to be around -11V, which means that V_{G2} used in the Shindo amp should be roughly halfway between 250V (-7V bias) and 330V (-11V bias), around 290V.

The pentode characteristics aren't even remotely equidistant, meaning the distortion will be unacceptably high. Using F2a in pentode mode without any negative feedback does not seem a good idea!

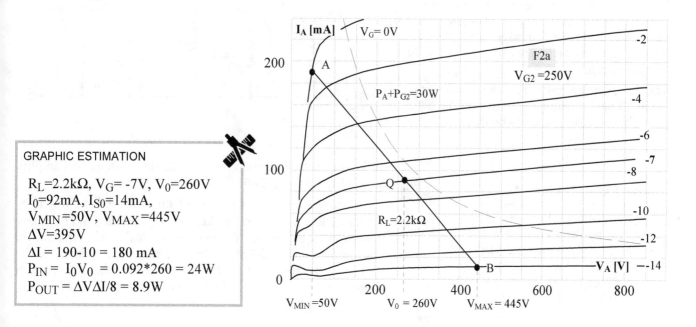

GRAPHIC ESTIMATION

R_L=2.2kΩ, V_G= -7V, V_0=260V
I_0=92mA, I_{S0}=14mA,
V_{MIN}=50V, V_{MAX}=445V
ΔV=395V
ΔI = 190-10 = 180 mA
P_{IN} = $I_0 V_0$ = 0.092*260 = 24W
P_{OUT} = ΔVΔI/8 = 8.9W

Design project: F2a SET amplifier

Since we managed to get two used F2a for under $70 each, we simply had to investigate this power tube and see (or rather "hear") if its sonic reputation is deserved or not. We also had a pair of in-house wound interstage transformers (leftover from the 300B PSET monoblocks, also featured in this volume), toroidal output transformers by Ertoco, and a pair of EI output transformers wound by us, so most major components were at hand.

In contrast to Shindo and its pentode amplifier, we decided to go the puritan way and use the output tubes in a triode mode. The triode results don't look promising, 4.5 Watts of output power at 5% distortion, but the reality is often very different from paper analysis. In other words, based on our experience, graphical analysis usually indicate 20-40% lower power levels than those obtainable in practice.

Triode operation of F2a

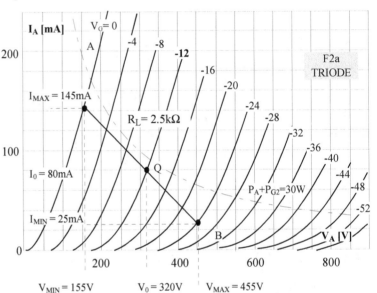

GRAPHIC ESTIMATION

$V_G = -12V$, $R_L = 2.5k\Omega$
$V_0 = 320V$, $I_0 = 80mA$
$V_{MIN} = 155V$, $V_{MAX} = 455V$, $\Delta V = 300V$
$I_{MIN} = 25mA$, $I_{MAX} = 145mA$, $\Delta I = 120mA$
$P_{IN} = I_0 V_0 = 0.08*320 = 25.6W$
$P_{OUT} = \Delta V \Delta I / 8 = 300*0.12/8 = 4.5W$
$\Delta V_+ = 165V$, $\Delta V_- = 135V$
$D_2 = (\Delta V_+ - \Delta V_-)/2\Delta V *100 [\%] =$
$= 30/(2*300)*100 = 5\%$

Version 1: 6SN7 driver stage

One of the aspects of a design is its tube-rolling potential. Since F2a tubes were made by only one manufacturer, such potential for the output stage is zero. Thus, a decision was made to use a sonically-proven driver tube widely available in various brands, price levels, and construction variations, namely 6SN7, so a DIY audiophile can roll driver tubes to his heart's content!

Since the triode-strapped F2a output stage is biased at around -16 Volts, it seemed that a single driver stage would be enough, so we paralleled two 6SN7 triodes (one physical tube) to halve its internal resistance and double its current capability.

The first stage current is 8.6mA (which is low for two 6SN7 in parallel), so to get -4V bias, the cathode resistor needs to be $R_{K1} = 4/0.0085 = 470\ \Omega$, a standard value.

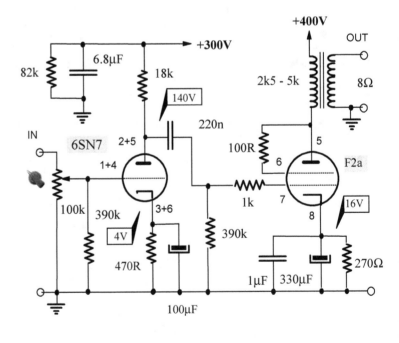

MEASURED RESULTS:

- Output voltage: $V_{MAX} = 7V$
- Output power: $P_{MAX} = 6W$
- BW (EI transformers): 20Hz - 23 kHz (-3dB@1W)
- BW (Toroidal transformers): 22 Hz - 124 kHz (-3dB@1W)

With a bypassed cathode resistor, the driver stage will amplify around 15 times or 24dB.

Two output transformers were tried, one designed and wound by us on EI laminations, with 5kΩ primary, and a 2.5kΩ primary toroidal unit by Ertoco. The maximum obtainable output power did not vary, around 6 Watts in both cases, and neither did the lower -3dB frequency, 22 Hz. However, the upper -3dB frequency limit was much higher with the toroidal transformer, 124 kHz!

TRIODES, PENTODES AND BEAM TUBES: MORE SINGLE-ENDED DESIGNS

GRAPHIC ESTIMATION

$V_0=140V$, $I_0=8.5mA$
$V_{MIN}=65V$, $V_{MAX}=190V$, $\Delta V = 125V$
$A= -\Delta V_A/\Delta V_G = -125/8 = -15.6$

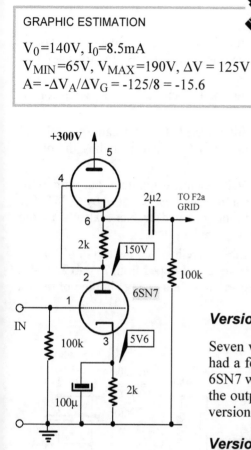

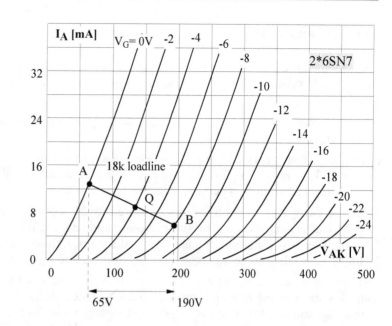

Version 2: SRPP driver stage

Seven volts maximum output voltage seemed in line with expectations, but we had a feeling F2a as a triode could deliver more, so an SRPP driver stage with 6SN7 was tried next (schematics on the left). Alas, both the frequency range and the output power remained the same as with a single paralleled triode driver in version 1, so the quest continued.

Version 3: Two-stage front end

We rationalized that if two 6SN7 triode stages could drive a 300B power stage, they must drive a much more sensitive F2a output stage. Sure, there could be too much voltage gain, but that would be easily fixed, so let's see what happens.

The voltage gain of the 1st stage is around 12, the gain of the 2nd stage is only 2, and the output stage amplifies voltage six times. Sure enough, a maximum of $12V_{RMS}$ at the output means the maximum power has increased to 18 Watts!

MEASURED RESULTS:

- $V_{MAX} = 12V$, $P_{MAX}=18W$
- Bandwidth: 21Hz-36kHz (-3dB@1W)

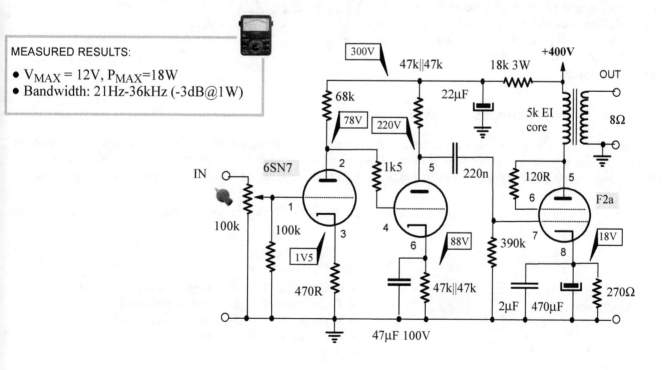

Listening impressions

With the 6SN7 preamp and driver tubes, the sound was impacted significantly by the quality of the 6SN7 triodes. Some sounded strident, others smoother but not emotionally engaging. It was hard to distinguish the sonics of F2a from that of the 6SN7.

Despite a relatively high lower -3dB frequency (f_L=21 Hz), the bass was prominent and tight, quite good for SE design without global NFB.

Once 6SN7s were replaced by 6EM7 lovelies, the sound changed significantly for the better. It became smooth and euphonic. The transients gained speed and impact, the kicks of the percussion instruments, including the piano, became lifelike; in short - everything improved.

Since the D&Y A-300 amplifier (also featured in this volume) had a virtually identical topology but with 300B output triodes, some meaningful comparisons could be made if its 6SN7 preamp tubes were replaced by 6EM7. The F2a amplifier had a more prominent bass and a more refined (cleaner and more detailed) treble. It also had more "slam." The 300B amp, on the other hand, sounded more seductive, sweeter, more "musical."

As a side note, for the measurement aficionados in our midst, when 6EM7 tubes were inserted instead of 6SN7, the cathode voltage of the 1st stage increased from 1.6 to 2.1 Volts, and its anode voltage dropped from 78V to 22 Volts. This is way too low for anode voltage, but there was no audible sign of distortion, let alone any clipping of the 1st stage. The idle current through the 1st stage increased from 3.2 to 4.5 mA.

Likewise, the cathode voltage of the 2nd stage dropped from 88 to 27 Volts, so the bias of the 2nd stage reduced from -10V to -5V. Its idle current dropped from 3.7mA down to only 1.1mA, which is very low for such a powerful triode (10 Watts dissipation).

In short, the voltages indicated unsuitable operating points for both stages with 6EM7 beauties, but the sound was divine. Another example of the discrepancy between good measurements and good sound!

The power supply used a 5AR4 tube rectifier and separate LC filters for each channel. Common for both channels, the first filtering capacitor is not an electrolytic but a film cap. This cleaned up the sound a bit, making it even more transparent and inviting.

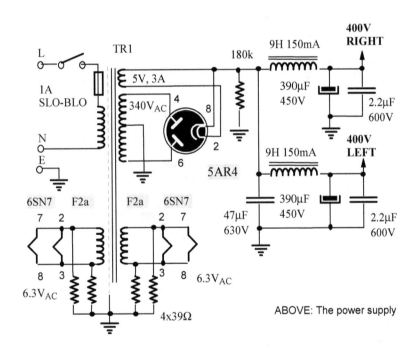

ABOVE: The power supply

Version 4: Transformer-coupled driver stage

Since the 300B PSET monoblocks with interstage coupling sounded great, we decided to use the same driver stage in this design. There were two 300B triodes in parallel per channel and different output transformers. That way, we could again compare the sonics of 300B and F2a amplifiers capable of Class A_2 operation and higher power levels.

This circuit (next page) has been described in detail in the 300B PSET project notes, so its analysis won't be repeated here.

MEASURED RESULTS:
- V_{MAX} = 14V, P_{MAX}=24W
- BW: 18Hz-68kHz (-3dB@1W)

KEY FEATURES:
- DC-coupling between the first and driver stage
- Transformer-coupling to the output tube
- No coupling capacitors in the signal path
- No global negative feedback
- Significantly increased output power (Class A_2)

TRIODES, PENTODES AND BEAM TUBES: MORE SINGLE-ENDED DESIGNS

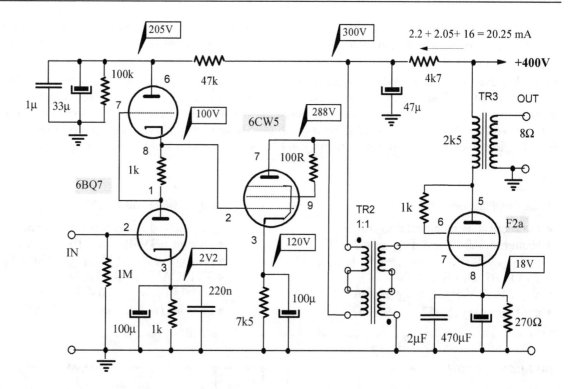

Compared to the previous Class A_1 F2a design, the maximum output power here increased from 18 Watts to 24 Watts due to the circuit's ability to cross over into Class A_2. Interestingly, the frequency range widened significantly using our EI output transformers, especially towards high frequencies.

Sound-wise, we again preferred the transformer-driven 300B triodes in parallel. F2a proved to be a solid performer but did not live up to its stellar reputation and even higher prices. LS50 (GU50) sounded similar; EL153 sounded more refined and detailed, not to mention VT-25A, which sounded superior to F2a in every sense!

Design project: EL12 SE monoblock amplifier

Now that we have seen what some of the best Japanese tube designer minds came up with, let's see if we can design, if not better, then an equally good amplifier for 10-20 % of the Shindo Cortese's cost! Of course, we designed and built a few of these amps many years ago, around the same time the late Shindo-San finalized his Cortese design.

However, instead of overpriced F2a, we used EL12N beauties. They may not last as long as F2a, but since they are ten times cheaper and will last at least 1,000 hours, we are in front of them again!

The similarities with Shindo Cortese are apparent; both designs use a single driver stage (a quasi-SRPP circuit) and a SE output stage with a German beam tetrode. However, this is where similarities end. Our design uses a grid choke and a switchable triode/ultra-linear output stage, while Cortese operates in the pentode mode.

Our power supply is much simpler, mainly due to the use of self- or cathode biasing for the output stage and the avoidance of solid-state voltage regulators, which aren't necessary for Class A designs.

The input stage can use 6BQ7A, with lower gain (lower input sensitivity), and 6AQ8 (ECC85), a higher μ tube, which provides a higher gain and a higher input sensitivity. The user can change the sensitivity and voicing of the amplifier by simply plugging in a different tube, making it easier to match these power amps with various preamps.

TUBE PROFILE: EL12N

- Indirectly-heated beam tetrode, 6.3V, 1.2A heater
- Max. anode/screen voltage 350V/350V, but takes 425V
- Max. anode / G2 dissipation: 18W / 3W, max. cathode current: 90 mA
- Static data (pentode): gm=15 mA/V, r_I=30kΩ, μ=450, $μ_{G1G2}$=18
- As a triode: gm=18 mA/V, r_I=1kΩ, μ=18

Shindo owners have no choice of either preamp or power tubes for substitutes and tube rolling. Finally, ours are monoblock amps, with zero crosstalk and no intermodulation between channels due to entirely separate power supplies.

We wish we could have an A-B comparison with the Cortese. We have only heard these babies with our output transformers. Still, should you wish to use expensive Audio Note, Bertolluci, or similar transformers in your project, you may achieve even higher pinnacles of musical bliss. After all, we are not professional transformer makers, and probably you aren't either. Since all parts used in this design are relatively cheap, you may want to buy the best chassis, grid choke, and output transformers you can afford, which would be money well spent!

TUBE PROFILE: 6AQ8 (ECC85)

- Indirectly-heated Noval duo-triode
- Heater: 6.3V/435mA, P_{AMAX}= 2.5 W
- V_{AMAX}=300V, V_{HKMAX}=90V, I_{KMAX}=15 mA
- TYPICAL OPERATION:
- V_A=250V, V_G=-2.3V, I_0=10 mA
- gm=5.9 mA/V, μ=57, r_I= 9.7 kΩ

TUBE PROFILE: 6BQ7 (6BZ7)

- Indirectly-heated Noval duo-triode
- Heater: 6.3V/400 mA, P_{AMAX}=2W
- V_{HKMAX}: 200V, V_{AMAX}= 250V, I_{KMAX}= 20mA
- TYPICAL OPERATION:
- V_A=150V, V_G=-2.0V, I_0=9 mA
- gm = 6.0 mA/V, μ= 35, r_I=5.8kΩ

The audio circuit

Grid chokes are the best financial investment an audiophile can make. They improve dynamics, increase output power and do pretty much everything else except serving croissants and cappuccino! You don't have to wind your own; if you value your time and your sanity, you shouldn't. Winding chokes and transformers is not for the impatient or dexterity-challenged! Affordable Chinese-made chokes are available for sale online.

With 6AQ8, the first stage amplifies 25-30 times or 18-20 times with 6BQ7A.

The output stage amplification factor is around 25 in ultra-linear mode, with a 43% U/L tap. The turns and voltage ratio of the output transformer is 20:1, so the total gain of the amplifier with 6AQ8 is A=30*25/20 = 37.5 or 31.5 dB. With 6BQ7 the overall gain is A=18*25/20 = 22.5 or 27 dB.

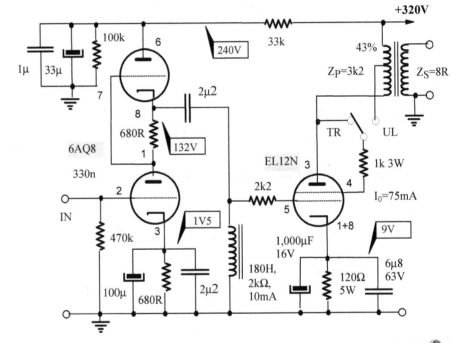

KEY FEATURES:

- Single driver stage, grid choke drive
- Regulated power supply for all stages
- Choice of 6BQ7A or 6AQ8 driver tubes
- Switchable triode or ultralinear operation of the output stage
- Self biasing, no adjustments of any kind are needed

The power supply - voltage doubler + tube regulator option

Although a voltage doubler is used here, the series regulator with 6080 duo-triode can be used with two diodes in a CT (center tap) secondary arrangement or with four silicon diodes in a bridge arrangement, should your power transformer lack the center tap on its secondary winding.

Some audiophiles like the sound of amplifiers with high voltage regulators; others don't. Voltage regulators do violate the Simplicity Rule. One could argue that they aren't required at all in Class A designs, where the power draw of the audio circuitry is more-or-less constant, compared to push-pull amps in Class AB or B, where it increases significantly with rising signal amplitude.

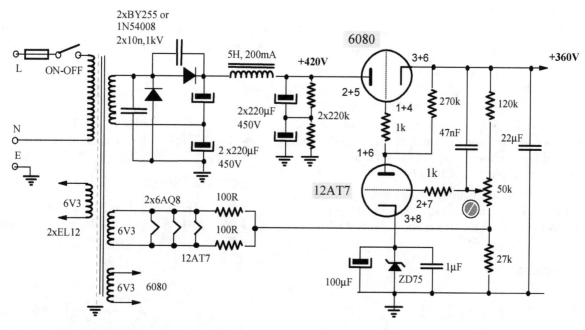

The power supply - hybrid rectifier option

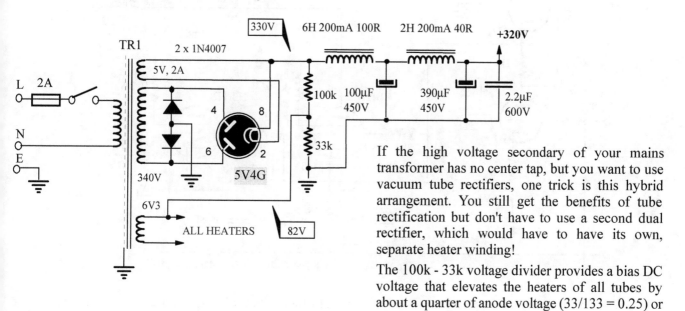

If the high voltage secondary of your mains transformer has no center tap, but you want to use vacuum tube rectifiers, one trick is this hybrid arrangement. You still get the benefits of tube rectification but don't have to use a second dual rectifier, which would have to have its own, separate heater winding!

The 100k - 33k voltage divider provides a bias DC voltage that elevates the heaters of all tubes by about a quarter of anode voltage (33/133 = 0.25) or 82V in this case.

The power supply - mercury tube rectifier option

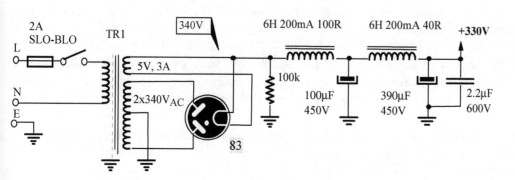

Mercury vapor rectifier is used here for its unmatched sonics and to minimize the voltage drop on the rectifier tube. MV rectifiers reduce the DC voltage only around 15V, compared to around 35V for 5V4 and 50V for 5U4 rectifiers. Since LC filtering is used, two chokes are necessary to reduce the ripple and hum below the audible level.

Listening impressions

The choice of the power supply significantly impacts the sound, especially the voltage regulator with 6080 triodes, which then also contributes to the sonics. Mercury vapor rectification is much simpler, and it sounds better.

Between 6BQ7A and 6AQ8 input triodes, the lower gain 6BQ7A sounds warmer and softer, 6AQ8 crispier, more like ECC81, which have never been my favorites.

Likewise, triode output sounds softer and warmer, although not as clean or dynamic as ultra-linear mode. However, the grid choke compensates for that, so even in triode mode, this amp is anything but lazy or laid-back.

Notice an extra-wide frequency range (test results on the right). Interestingly, there wasn't too much difference in output power between the triode and U/L modes!

MEASURED RESULTS (triode mode):

- BW: 14Hz - 70 kHz (-3dB, at 1W into 8Ω)
- BW: 17Hz - 42 kHz (-3dB, at 10W into 8Ω)
- $V_{MAX} = 11 V_{RMS}$, $P_{MAX} = 15W$

MEASURED RESULTS (ultralinear mode):

- BW: 16Hz - 90 kHz (-3dB, at 1W into 8Ω)
- BW: 20Hz - 42 kHz (-3dB, at 10W into 8Ω)
- $V_{MAX} = 12 V_{RMS}$, $P_{MAX} = 18W$

PSET version

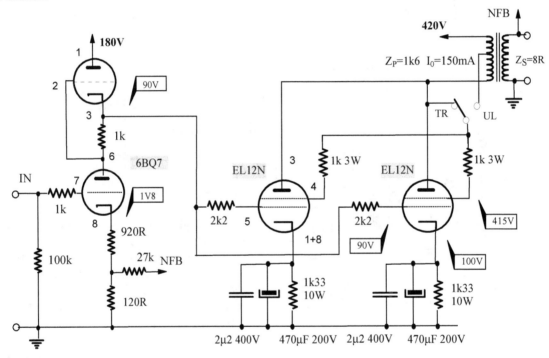

If 15 or 18 Watts isn't enough for you, how about doubling it to 30-36 Watts of single-ended bliss? And, since we are in the extreme Puissance Mode (puissance, a sexy French word for power, that sounds like, well, sexy), let's eliminate the only coupling capacitor and DC-couple the driver stage to the output grids!

We know that EL12N lovelies need to have their cathodes 10V above the control grids. From our previous experience, we also know that we need a minimum of around 180VDC on the SRPP input stage and that the cathode of the upper tube will be at one-half of that, or $90V_{DC}$. Since the control grids of the output tubes will then also be at $90V_{DC}$, to get -10V grid bias (DC voltage between CG and K), their cathodes must be 10V higher, or at 100V!

Assuming we keep the idle DC current per output tube at the same level of 75mA, the cathode resistance must be $R_K = V_K/I_K = 100/0.075 = 1,333Ω$! The power that will be dissipated (a nice Swedish technical word meaning "wasted") into heat is now $P_K = V_K^2/R_K = 100^2/1,333 = 7.5$ Watts! So, 10W resistors are mandatory; 20 Watt beasts would be better.

The DC voltage between the anodes and cathodes of the output tubes must remain the same as in the previous design, around 310 V_{DC}, meaning the power supply must now provide 310 V_{DC}, plus 100 V_{DC}, plus around 10V voltage drop on the output transformer's primary, so 420 V_{DC} in total!

This isn't that bad;, it is below the 450 V limit of most cheap elcos, but still too close to it for real comfort, so either buy the best quality elcos you can get or go for 500 V units! Of course, for those of you who are allergic to electrolytics, there are always large 47μ film caps rated at 630 V_{DC}. However, they are bulky, and to get the ripple and hum down to acceptable levels; you'll need lots of them at such large power supply currents.

Design project: EL156 SET amplifier

The battle of titans: EL156 against KT88 and KT120

EL156 was released in 1952 by Telefunken in Germany. As one of the most powerful audio tubes in its day, it found immediate popularity in hi-fi, PA, and professional amplifiers made by Klangfilm, Siemens, Telefunken, and many other reputable German manufacturers such as Neumann, which used it to drive its record cutting lathe LV60.

However, getting original Telefunken tubes 65 years later is not easy and not cheap. Luckily, the ever entrepreneurial Chinese spotted a gap in the market and released their version of EL156. The inner construction seems identical to Telefunken's, but the Chinese version uses the octal socket with the same pinout as EL34.

KT120 power tubes were released a few years ago, followed by the even more powerful KT150. However, these beasts aren't cheap, especially not the KT150. A side-by-side comparison reveals EL156's much higher m as a triode, 15 versus 9. The KT120 bias is at -40V, more than double the EL156 bias of -19V, so the demands on the driver stage are higher in the KT120 case. As a triode, EL156 is much easier to drive. However, KT120 has a lower internal resistance, so, for tennis lovers in our midst, "15 all"!

SIDE-BY-SIDE	KT120	EL156
	Beam tetrode	Power pentode
Heater	6.3 V/1.7-1.95 A	6.3 V, 1.9 A
P_A max.	60 W	50 W
V_A max.	850/650 V (TRIODE)	800 V
V_{G2} max.	650/600 V (TRIODE)	450V/500 V (TRIODE)
P_{G2} max.	8 W	12 W (8 W Chinese)
I_K max.	250/ 230 mA (TRIODE)	180 mA
r_I (PENTODE)	25 kΩ	25 kΩ
gm as a triode	12.5 mA/V	11 mA/V
μ / r_I as a triode	9/720Ω	15/1,300Ω

ABOVE RIGHT: Russian-made ElectroHarmonix KT88 (left) and Shuguang (Chinese-made) EL156 (right). Despite its 10 watt lower power rating, EL156 has a larger anode surface area!

Commercial benchmark: Audion Super Sterling 120

The promotional literature and website descriptions of this amplifier summarize its sonics as "The sweetness of the EL34 with twice the punch of the KT88." KT120 has an anode power rating of 60 Watts, while KT88's anode is rated at 42W; therefore, it is impossible to get twice the output power from KT120! Assuming, of course, that "punch" means output power (which can be objectively measured and verified) and not a vague expression of subjective "slam." Indeed, Audion's own similar model using JJ KT88 tubes produces 18 watts of output power per channel.

The promotional spiel continues, "The Audion Super Sterling 120 uses Tungsol KT120 pentode tubes the amp delivers approx. 24 watts into an 8 ohm load. Therefore, it has the power of competitors boasting 35-45W." A 24 Watt amplifier cannot have the power of amplifiers that measure 35-45 watts, so the second sentence is pure bull!

Super Sterling's output stage works in pentode mode. Strictly speaking, KT120 is a beam power tube, not a pentode; it does not have a suppressor grid but a pair of beam deflection plates, but a pentode mode has been used as an expression for both types of power tubes. Both have a screen grid which in pentode mode is connected either to its own source of high positive voltage, illustration a), or to the same HV source that supplies the anode via the primary of the output transformer, illustration b).

For many pentodes and beam tubes, the voltage rating of the screen grid is lower than the maximum voltage allowed on the anode, so the screen has its own lower voltage power supply.

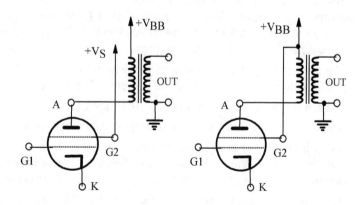

a) separate screen grid and anode supplies

b) common screen grid and anode supply

Estimating EL156 SET output power and distortion

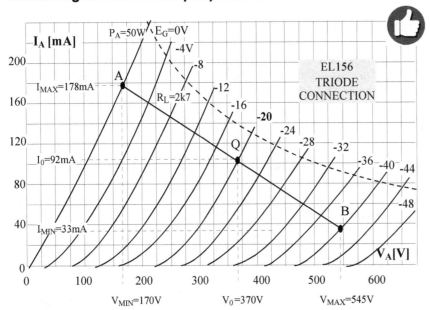

GRAPHIC ESTIMATION

$R_L=2k7$, $V_0=370V$, $I_0=92mA$
$V_{MIN}=170V$, $V_{MAX}=545V$
$\Delta V = 375V$
$\Delta I = 178-33 = 145$ mA
$P_{IN}=I_{A0}V_0=0.092*370=34$ W
$P_{OUT}=\Delta V \Delta I/8=375*0.145/8=6.8W$
DISTORTION (2nd harmonic):
$\Delta V_+=200V$, $\Delta V_-=175V$
$D_2=(\Delta V_+ - \Delta V_-)/2\Delta V *100$ [%]
$D_2= 25/(2*375)*100 = 3.3$ %

Pentode mode aside, how about KT120 strapped as a triode?

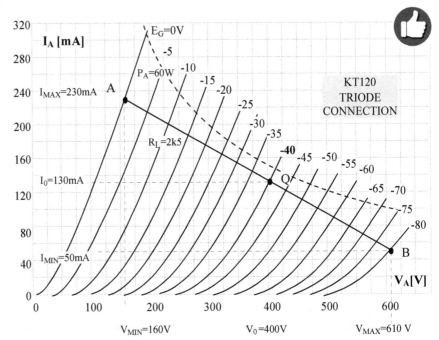

GRAPHIC ESTIMATION

Q-point: $V_0=400V$, $I_0=130mA$
Load: $R_L=2k5$
$V_{MIN}=160V$, $V_{MAX}=610V$
$\Delta V = 450V$
$\Delta I = 230-50 = 180$ mA
$P_{IN} = I_{A0}V_0=0.13*400 =52W$
$P_{OUT} =\Delta V \Delta I/8=450*0.18/8=10W$
DISTORTION (2nd harmonic):
$\Delta V_+=240V$, $\Delta V_-=210V$
$D_2 =(\Delta V_+ - \Delta V_-)/2\Delta V *100$ [%] = $30/(2*450)*100 = 3.3$ %

You may be tempted to think, "10 Watts versus 7 Watts, KT120 wins hands down". However, notice that we are not comparing apples with apples. First of all, EL156 was biased and run very conservatively; the idle dissipation is only 33 Watts for a 50 Watt-rated tube. The KT120 is run close to its maximum and will not last that long in an amp so designed! Prediction: 10 Watts from EL156 in triode mode could be achieved as well!

The parasitic input capacitance of the original EL156 is 19 pF (29 pF for EL120), while its G1-anode capacitance (the one that gets increased by the Miller's effect) is only 0.5pF, compared to 1.8pF for EL120.

Our tennis game continues, and it's 30:15 for EL156.

Interestingly, the 2nd harmonic distortion was identical to that of KT120.

As with their other amps, Audion claims that the distortion (not just of the output stage but the whole amplifier!) is "Distortion @ 1Watt: <0.1% No Feedback". That is impossible to achieve without any negative feedback. The damping factor and the amp's lower -3dB frequency aren't even mentioned in Audion's promotional blurbs.

Anyway, to continue our tennis analogy, EL156 leads 40:15! And finally, subjectively speaking, EL156 has a warmer sound that resembles 300B more closely than the dry sonics of the KT family, including KT120. Game EL156.

TRIODES, PENTODES AND BEAM TUBES: MORE SINGLE-ENDED DESIGNS

Circuit diagram

The audio circuit is fundamentally simple; both halves of ECC88 paralleled (to halve the already low internal resistance) and operated at a high current for best sonics and lower distortion. The global negative feedback can be eliminated, use a 150Ω cathode resistor, and omit the 47kΩ and 47Ω resistors from the 1st stage's cathode circuit.

Triode or U/L operation is switch selectable (selector mounted at the amplifier's back, behind the power tubes).

The power supply is also straightforward. There is no need to elevate any heater supply circuits or fixed-bias power supplies.

However, we did separate the LC filters for each channel to minimize crosstalk and for a better soundstage.

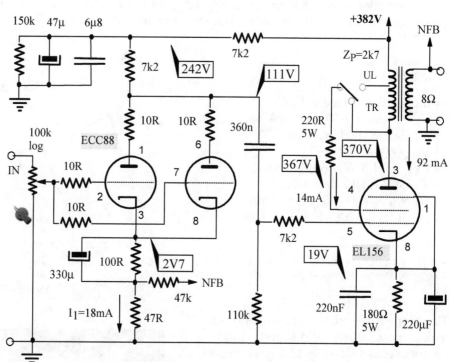

The total energy reserve stored in the charged HV power supply capacitors is impressive for such a low power SET amp, $E=0.5CV^2 = 85$ Joule, where C is the total capacitance (3x390μF).

The high resistance of the NTC resistor in a cold state limits the inrush current to some extent. If you have a spare 5V, 2A winding on your power transformer, insert an indirectly-heated rectifier tube there for a true soft start feature (connected in series with the load, instead of the NTC resistor).

Soft recovery ultrafast rectifier diodes were used in the HV section (UF5404).

UF5404 has a "maximum repetitive peak reverse voltage" of 400V, "maximum RMS voltage of 280V" and "maximum DC blocking voltage" of 400V.

The series continues with UF5405, UF5406, UF5407, and UF5408, each having 100V higher maximum repetitive peak reverse voltage than the previous one, up to 800V for UF5408. The maximum forward current is 3A.

If you cannot find such diodes in your local electronics shop or online, they can be salvaged from computer power supplies and DC-DC converters.

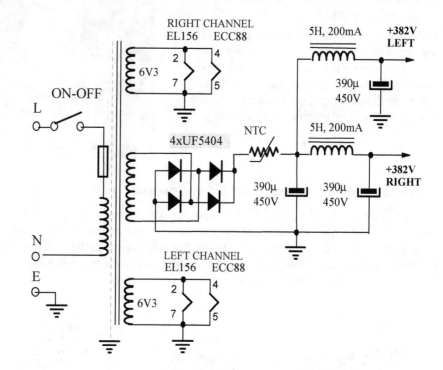

The input/driver stage

The voltage amplification or gain of the CC (common cathode) stage is $A = -\mu*R_L/(r_I + R_L)$ This assumes that the cathode resistor is fully bypassed, which in our case isn't strictly correct (100Ω resistor is bypassed but 47Ω resistor is un-bypassed). For a stage with two ECC88 triodes in parallel we have $R_L=7k2$, $r_I=1.3k\Omega$ and $\mu=33$, so $A = -33*7.2/(1.3+7.2) = -33*0.85 = -28$.

Notice that the internal resistance r_I is halved for two identical parallel tubes, but the amplification factor μ does not change. According to the Barkhousen Equation $\mu = gm * r_I$, the mutual conductance is doubled.

From the graph below we estimate the voltage gain as the quotient of the anode voltage swing and grid voltage swing: $A = -\Delta V_A/\Delta V_G = -(188-48)/5.4 = -140/5.4 = -26$.

Two factors cause this discrepancy. First of all, the curves are approximate, and our drawing isn't very precise. Secondly, the linear model is a simplification and, as such, assumes that I_A-V_A curves are straight lines, which they are not. In real life, depending on the particular tube used, you may get a gain of anywhere between 22 and 28. Since we also have local feedback through the partially un-bypassed cathode resistor and the mild global NFB from the speaker output coming into that point, that may drop down to 15 or so.

TUBE PROFILE: ECC88 (6DJ8)

- Indirectly-heated duo-triode
- Noval socket, 6.3V, 365 mA heater
- Maximum anode voltage $130V_{DC}$
- $V_{HKMAX}=150V_{DC}$
- $P_{AMAX}=1.8$ W, $I_{AMAX}=25$ mA
- TYPICAL OPERATION:
- $V_A=90V$, $V_G=-1.3V$, $I_0=15$ mA
- $gm=12.5$ mA/V, $\mu=33$, $r_I=2.6$ kΩ

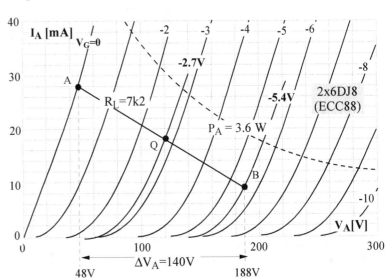

RIGHT: The quiescent point Q, load line, voltage and current swings for two ECC88 in parallel

The bias of the output tubes is only 19 Volts, meaning the RMS grid voltage cannot exceed 19/1.41 or 13.5V, so even a preamp gain of 15 is enough (assuming 1 V_{RMS} at the input).

Due to the relatively low-value grid leak resistor (110k), the coupling capacitance must be increased to at least 360nF (for -3dB frequency of 4 Hz) from the usual 100nF or 220nF value; otherwise, the bass frequencies will be attenuated by such high pass CR filter.

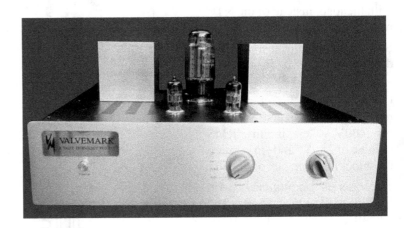

Optimizing the performance of the suppressor grid

While the original Telefunken EL156 had the suppressor and cathode tied together internally, the Chinese octal version has the suppressor grid brought out to its separate pin. Connecting the suppressor to a source of negative voltage instead of the usual strapping to the cathode improves the efficiency and linearity of the output stage and the reliability of the tubes. The most convenient negative voltage source is the bias circuit, typically with around -30V available for suppressor connection.

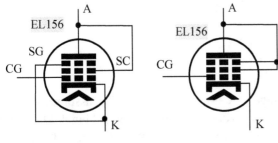

ABOVE: The usual suppressor grid (SG) connection to the cathode and the connection to the anode

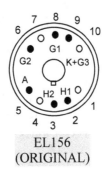

EL156 (ORIGINAL)

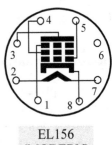

EL156 (MODERN)

Another benefit of a negatively-biased suppressor is the inherent protection it provides in case of a bias failure. -30V on the suppressor will repel many electrons from the anode and limit the zero-bias anode current to approximately half of its value in the case of the suppressor at zero volts, for instance, 200mA instead of 400mA.

This may not prevent overheating, leading to damage or destruction of the power tube(s), should the loss of bias not be noticed and the amp not shut down quickly, but it may protect the output transformer from burning out. The suppressor grid can be strapped to the anode as well. While this doesn't markedly improve the performance, it does result in the lower internal resistance of such pseudo-triode, which is always welcome.

MEASURED RESULTS (triode mode):
- BW: 15Hz - 42 kHz (-3dB, at 1W into 8Ω)
- V_{MAX} = 6.7V_{RMS}, P_{MAX} = 5.6W
- DF_{1kHz} = 3.7 DF_{100Hz} = 3.6

MEASURED RESULTS (ultralinear mode):
- BW: 19Hz - 40 kHz (-3dB, at 1W into 8Ω)
- V_{MAX} = 10V_{RMS}, P_{MAX} = 12.5W
- DF_{1kHz} = 2.6 DF_{100Hz} = 2.4

Design project: PSE ultralinear EL84 amplifier with a separate power supply

The idea and the design approach

These Made-in-China aluminium chassis are sturdy and rigid, with nice contrast, black anodized panels, and natural fascia.

The final impetus to try this configuration came from a box full of EL84 (6BQ5) pentodes, ranging from Australian AWA and Radiotron brands, Mullards, Telefunkens and Valvos, Yugoslav Ei and Dutch Philips.

These lovelies are still plentiful, inexpensive, reliable and long-lasting, and, above all, great sounding. Just as some prefer the sound of 2A3 triodes to their bigger 300B cousins, many audiophiles prefer the sonics of EL84 over larger EL34.

Since there have been enough EL84 push-pull amps, the brief was to design and construct a single-ended one for a change. One thing that EL84 does not have is power. It is a tiny tube by any standard, so we decided to parallel two per channel and operate the output stage in ultra-linear mode to get a reasonable power output.

Plus, since EL84s tend to have a warm and tubey sound (that is why they have been so popular with guitar players), we thought the U/L connection could result in a more neutral tonal balance. The concern was that a triode connection would sound too "colored" and syrupy.

Also, the output power in triode mode would be minuscule even with two tubes in parallel per channel.

The topology

The power supply is on a separate chassis, with the power transformer (1) flanked by two small filtering chokes (2). The rectifier tube (3) is centrally positioned in front of the mains transformer, with a rotary on-off switch on the right (4). The multicore umbilical cord feeds the heater and high voltage supplies to the audio chassis, on which components are also symmetrically arranged, centrally-located volume control knob (5), and the single ECC40 duo-triode (6), with output transformers (7) at the back. The four EL84 tubes are positioned in a V arrangement.

The tiny red neon "power on" indicator (8) is very elegant and not distracting at all during the night listening, in contrast to many China-made amplifiers, which feature very bright (usually blue) LEDs. Since these can be highly annoying, the first modification many audiophiles make is to disconnect them completely.

The output tubes

One EL84 in the SET regime requires an output transformer with high primary impedance, 5-7 kΩ, which is inconvenient. Luckily, two in parallel need only 2.5-3.5 kΩ, which is a standard range for most other triodes such as 2A3 and 300B! With 300V between anodes and cathodes and 76 mA total anode current, at least 4 Watts of output power are obtainable, according to the anode characteristics below, or 5-6 Watts in real life.

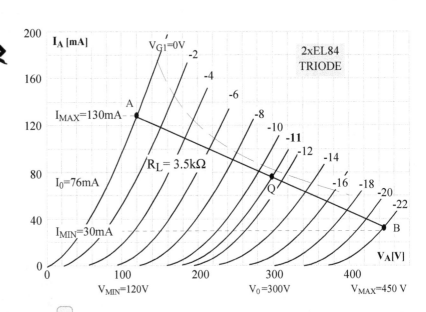

GRAPHIC ESTIMATION

$R_L = 3.5$ kΩ, $V_0 = 300$V, $I_0 = 76$mA
$V_{MIN} = 120$V, $V_{MAX} = 450$V
$\Delta V = 330$V
$\Delta I = 130 - 30 = 100$ mA
$P_{OUT} = \Delta V \Delta I / 8 = 330 * 0.1 / 8 = 4.1$W
$\Delta V_+ = 180$V, $\Delta V_- = 150$V
$D_2 = (\Delta V_+ - \Delta V_-)/2\Delta V * 100\, [\%] =$
$= 30/(2*330)*100 = 4.5\,\%$

TUBE PROFILE: EL84 (6BQ5)

- Indirectly-heated power pentode
- Heater: 6.3V, 0.75 A
- $V_{AMAX} = 300$ V, $P_{AMAX} = 12$ W, $P_{SMAX} = 2$ W
- SE OUTPUT TRIODE OPERATION:
- Bias voltage -11V, load: 3.5 kΩ
- $V_A = V_S = 300$V
- $I_{A0} = 38$ mA, $I_{MAX} = 65$ mA

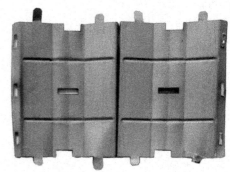

The audio circuit

Simplicity was one of the design's aims, so cathode biasing was chosen for the output stage (next page), with separate cathode resistors and capacitors for each tube, so precisely matched tube pairs are not required since each tube biases itself.

There's only one preamp triode and one capacitor in the signal path, with plenty of headroom in the input stage biased at 3mA of anode current and 3V on the cathode.

The idle current is 38 mA per output tube or 76 mA total. With 300V between the anode and the cathode, the idle power dissipation is $P_0 = 0.076*300 = 23$ Watts, close to the maximum anode dissipation of 24 Watts (for two paralleled tubes).

ECC40 was chosen for the preamp/driver stage, shared between the two channels. One ECC40 triode is more than capable of driving two paralleled triode-strapped EL84.

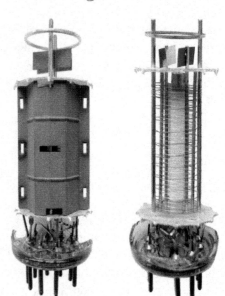

RIGHT: The internal view of the 6BQ5 pentode. Once the anode is removed, three pairs of grid supports are visible, as is the outer, coarsely spaced suppressor grid. The white-coated cathode is in the middle, and the control grid is the dense spiral right in front of it.

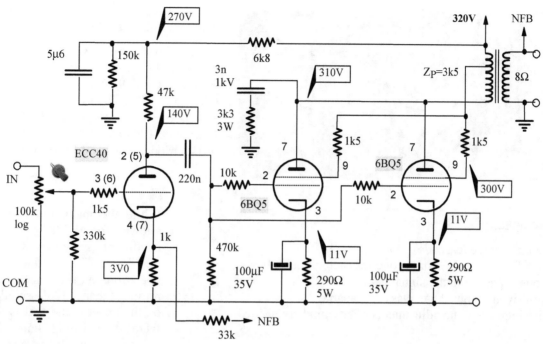

The power supply

The directly-heated 5U4 rectifier supplies high voltage to all (indirectly-heated) audio tubes before they had a chance to heat up (too early), so if you want to prolong their life, use an indirectly heated rectifier such as 5V4G instead. Both can be plugged in interchangeably; the pinout and the heater voltage are the same.

The voltage drop across the 5V4 rectifier is 10-15V lower than across 5U4, so instead of 320V at the output, you will get 330-335V, and the output power will be slightly higher. At relatively low distortion levels, the 10 watts of output power was a pleasant surprise. The maximum power, albeit with high distortion, was 18 Watts per channel!

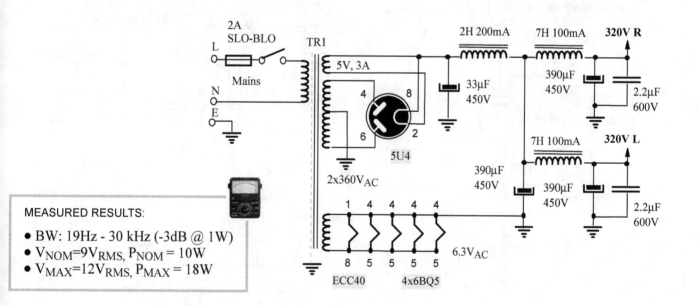

MEASURED RESULTS:
- BW: 19Hz - 30 kHz (-3dB @ 1W)
- $V_{NOM}=9V_{RMS}$, $P_{NOM} = 10W$
- $V_{MAX}=12V_{RMS}$, $P_{MAX} = 18W$

Design project: PL508 PSE "enhanced triode" amplifier

PL508 (17V heater) and EL508 (6.3V heater) are lovely power tubes, robust, cheap, and great-sounding, but two limitations must be overcome in any circuit that uses them. The first is their smallish anode power dissipation, only 12 Watts.

Notice high screen grid dissipation of 4 Watts (33% of the anode dissipation). That gave us an idea that if used in "enhanced triode" mode, where the signal is fed into the screen grid instead of the control grid, the tubes could operate not just in class A1 but also efficiently work in class A2 with a significant screen grid current flowing. That would increase the output power substantially.

AUDIOPHILE VACUUM TUBE AMPLIFIERS

TUBE PROFILE: PL508/EL508 (17KW6/6KW6)

- Indirectly-heated power pentode
- Heater: 17V/300mA (PL508)
- Heater: 6.3V/ 0.825A (EL508)
- P_{AMAX}=12W, P_{SMAX}=4W
- V_{AMAX}= 400V, V_{G2MAX}= 275V
- I_{KMAX}= 100 mA
- As a triode: gm=8 mA/V, µ=9, r_I =1.1kΩ

KEY FEATURES:

- Single input stage (cascode)
- Cathode follower driver stage, one for each output tube
- Only one capacitor in the signal path
- Screen-driven output stage, class A_2 capable
- Self biasingd esign, no adjustments of any kind are needed

While the maximum anode voltage allowed for PL508 is +400V_{DC}, the maximum screen voltage is only +275V_{DC}. That is fine for pentode operation, but if connected as triodes, no more than +300V should be used on the anodes, and that is the second factor that would limit the output power if used as an ordinary pseudo-triode. However, in screen-driven triode mode, that voltage limitation becomes irrelevant; the "enhanced triode" mode solves both problems.

Audio circuit

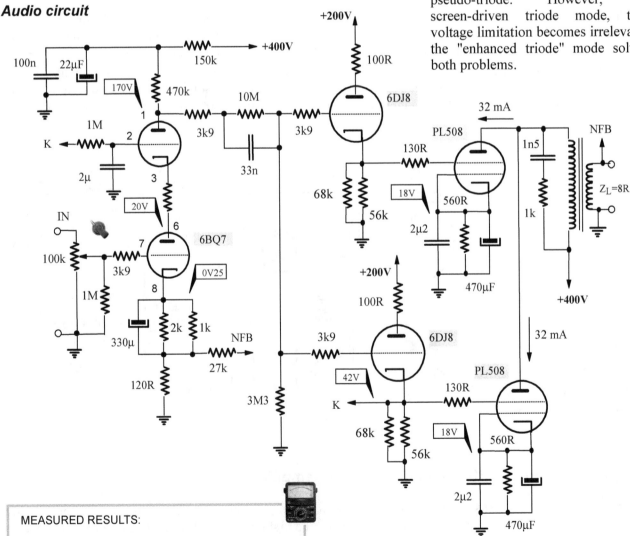

MEASURED RESULTS:

- BW: 17 Hz - 26 kHz (-3dB, at 1W into 8Ω)
- BW: 20 Hz - 24 kHz (-3dB, at 10W into 8Ω)
- V_{MAX}= 14V_{RMS}, P_{MAX} =24W

The cathode follower driver stage needs a tube that sounds good, has a low internal impedance, and can supply more than 10mA of screen grid current that flows in Class A_2. One such beauty is the 6DJ8 (ECC88) duo-triode.

To get even higher output power, we decided to parallel the output tubes. Since the idle DC current is only around 30-35mA per tube, a standard SE output transformer designed for 90-100mA (300B and other power tubes of 40-60W anode dissipation, such as triode-strapped KT88, KT120, etc.) could accommodate even three PL508 in parallel.

A designer has two options, use a more powerful driver tube to drive both output tubes together (their inputs connected), or use two smaller triodes to drive the output tubes separately. We chose the latter option. The cathodes of driver triodes are at +42V, while the self-biased output triodes have a cathode voltage of around +18V, meaning the quiescent bias on the screen grids is around +24V.

Since the voltage amplification factor of the output stage is very low, and the CF drivers provide a mild attenuation of the signal, all the voltage gain must be provided by the first stage. In this instance, we opted for one of our favorite duo-triodes, 6BQ7A, connected here in a cascode configuration to significantly increase the voltage gain.

For those of you who haven't read Vol. 1 of this book, the gain of the cascode is approximately $A \approx gm_1 * R_L$, where gm_1 is the transconductance of the lower triode and R_L is the anode load of the upper one.

In our circuit, due to the low anode current chosen, the transconductance is also relatively low, $gm_1 \approx 1.4mA/V$ and $R_L = 470k\Omega$, so the gain of our 1st stage is $A_1 \approx 1.4*470 \approx 660$ times. The mild negative feedback will reduce that still high gain to the required levels.

Design project: Battery-powered PL508 SET amplifier

In Volume 2 of this book, we discussed the pros and cons of battery power and presented a few designs for low voltage, battery-powered phono stages. Due to their high anode voltages and high power levels, battery-powered power amplifiers are a much more difficult undertaking. Never being the ones who surrender easily, we decided to design and make one. The results were worth the enormous effort.

RIGHT: The rear-view off the battery powered (24V) SET amplifier with PL508 tubes

PL508 - a wonder tube!

PL508 (American equivalent 17KW6) frame pentode was developed to vertically drive the electron beam in cathode-ray tubes (screens) for smaller TVs. A much more powerful tube was needed to drive the CRT horizontally, and its bigger brother, the venerable PL509, was developed for that task.

To power this amp up, we have decided to use two 12V car batteries connected in series, or, for the truckies in our midst, one large 24V truck battery.

To an experienced eye, one look at the PL508 tube profile (previous page) reveals the reasons for our choice. First, look at the heater specs. 17V is under 24V battery voltage, so there are no issues there. 17V+6.3V for preamp tubes' heaters is 23.3V, an ideal match to a 24V battery.

However, preamp tubes usually draw 300mA, and very few power tubes draw such a low heater current. Tubes with standard 6.3V heaters, such as 6L6, EL34, KT88, and others, draw from 1.2 to 1.8A each, so they are out. Luck would have it that our Mighty Midget, PL508, draws the same heater current as preamp tubes of the 12AU7, 12AT7, and 12BH7 variety! In fact, the tube was designed so its heaters can be connected in series with all other tubes in a TV receiver.

Thus, the 300mA was a fixed requirement, and 17V was the consequence, the voltage needed to develop the required anode power! PL509 and PL519 also draw only 300mA, but they need 40V on their heaters due to their higher anode power dissipation (30W and 35W). They could also be used for this project, but we would need four 12V batteries for a 48V supply.

Triode-strapped PL508 has great microdynamics, a detailed, refined treble, an airy and transparent sound, and fast and tight bass.

Has that got anything to do with the fact that this tube worked on a fixed mains frequency (50 or 60 Hz) in TV sets? Nobody knows, but it sounds plausible. Remember, the mains frequencies are audio frequencies!

PL508 are relatively cheap tubes to buy, although they are not plentiful. Connected as triodes, the internal impedance is around 1kΩ, higher than 2A3 or 300B, but the amplification factor µ=7.5 means that PL508 is much easier to drive (more sensitive) than 2A3 or 300B, which is good news.

However, connect two or three of these lovelies in a SET parallel stage, and the internal impedance will drop to 500Ω and 330Ω, respectively, much lower than any "real" triode such as 2A3 or 300B!

Although the low anode dissipation could be a turn-off for many DIY constructors, our chosen circuit is Class A_2 capable (grid current flows in A_2 mode), raising the maximum output power from 6 Watts in Class A_1 to 15 Watts in Class A_2!

KEY FEATURES:

- Capable of low-distortion Class A_2 operation
- Cathode follower driver stage with fixed bias
- Only two coupling capacitors in the signal path
- Battery-powered (fully disconnected from mains during operation)

MEASURED RESULTS:

- BW: 15Hz - 67 kHz (-3dB, 1W into 8Ω)
- BW: 20Hz - 30 kHz (-3dB, 15W into 8Ω)
- Class A_1: V_{OUTMAX}=7V_{RMS}, P_{MAX} = 6W
- Class A_2: V_{OUTMAX}: 11V_{RMS}, P_{MAX} = 15W
- Z_{OUT}= 1.8Ω (DF=4.4)

OUTPUT TRANSFORMERS:

- Turns ratio: TR=1,320/90 = 14.67
- Impedance ratio: IR=215
- Primary impedance: Z_P=215*8Ω = 1,700Ω
- EI114 GOSS laminations
- a=38mm, stack thickness S=45mm
- Center leg cross section A=aS = 17cm^2
- Primary: 1,320 turns, 0.3mm^2 wire
- 6 sections x 220 turns each, series-connected
- Secondary: 1,540 turns, 0.4mm^2 wire
- 7 sections x 90 Turns each, all in parallel
- L_P= 5.6 H

Audio section

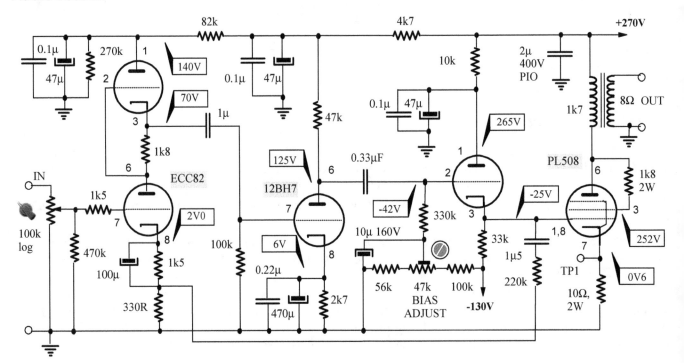

We have already analyzed the cathode-follower driver stage of this circuit in Volume 1 of this book. Since it is directly coupled to the output tube's grid, it can supply significant grid current, making operation in class A2 possible with low distortion.

Driver tubes such as 12BH7, 6CG7, 6SN7, 5687, and many others can be used. A pair of 12BH7 duo-triodes was chosen here since their two heaters can be connected in series across the 24V battery bank.

The 10R resistor is for measuring the cathode current and, indirectly, the bias voltage while adjusting it to the desired value. 0V6 means the idle DC current through PL508 (no signal) is 60mA.

This proven and great-sounding audio circuit (SRPP first stage, common cathode second stage, and DC-coupled cathode follower) was used a few times in Volume 1 for output tubes such as 300B, 6C33C-B, and SV572-10.

As seen from the amplifier's photograph, the connection to and disconnection from the battery is manual, using battery clips, as is the battery charger connection. This can be made switchable, of course, so that as soon as the amplifier is turned off (disconnected from the battery), the battery is switched to the output of the battery charger and vice versa. The charger should never be in the circuit while the amplifier is operating.

We designed and built our own DC-DC converter, operating on 44kHz and providing two voltages, +260V_{DC}, and -130V_{DC}. Since such design and construction is a long and arduous task, it is recommended that you buy a ready-made dual output converter or, if you cannot find a suitable one, a pair of converters. The design can be simplified by omitting the CF driver stage, obviating the need for the -160V converter. However, in that case, the output stage won't be able to cross into Class A_2 operation, and the maximum output power will be reduced.

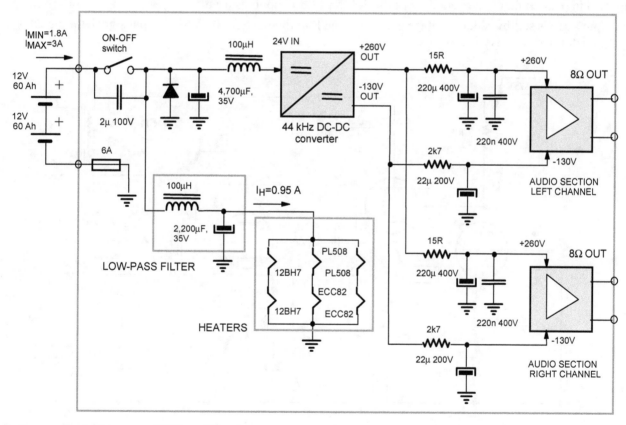

Block diagram of the battery-powered SET amplifier

Yarland FV II and FV35A amplifiers

FV II was one of the first Chinese amplifiers we bought way back in 2005. It looked cute and worked straight out of the box. While not a shocker, it wasn't a delight either. It sounded OK but unremarkable.

Unfortunately, we didn't spend much investigation and modification time to make more meaningful conclusions. The audio PCB was small, messy, and crowded, with components tacked on both sides, making significant modifications very difficult and frustrating. We dumped the PCB in the bin, gutted the whole amp, and converted it to work with PL508 tubes. It sounded much, much better. Its 2A3 tubes were reused in one of our balanced 2A3 push-pull amplifiers.

The output transformers have 2.8 cm center leg and 3.2 cm stack thickness, so their cross-section is A=2.8*3.2= 9cm², which is way undersized.

Single-ended transformers for quality amplifiers in the 7-12 Watt output class should have A=14 cm² or higher, preferably around 18 cm²!

MEASURED RESULTS:
- BW: 10Hz - 33 kHz (-3dB, 1W into 8Ω)
- P_{MAX} = 7.5W (1 kHz)
- DF= 2.75 (Z_{OUT}=2.9 Ω)

The audio section

The first stage has a very high anode resistance (200k), which indicates that the designer wanted to maximize voltage gain. With 6NP's mju of 100, the stage gain of up to 70-75 is possible, albeit with very high distortion. However, notice that one of the two cathode resistors in the 1st stage is not bypassed by a capacitor, meaning there is a mild local negative feedback. The measured 1st stage gain was 50 times (34dB).

Usually, this kind of cathode arrangement is chosen when negative feedback is brought back from the output transformer to the top of the un-bypassed resistor (220R), but in this case, there is no such NFB used. Perhaps the designer changed his mind halfway through or wanted to reduce the gain that way?

6N2 is a Chinese copy of the Russian 6N2P (6H2P). It can be considered an equivalent of 12AX7 (ECC83). The only difference is that 6N2 and 6N2P can only work with one heater voltage (6.3V), while 12AX7 heaters can be connected in parallel for 6.4V operation or in series if a 12.6V supply is used.

The output stage is self-biased. Its voltage gain is around four times, so for 0.1V$_{RMS}$ input signal, the signal on the output transformer's primary is 20V$_{RMS}$.

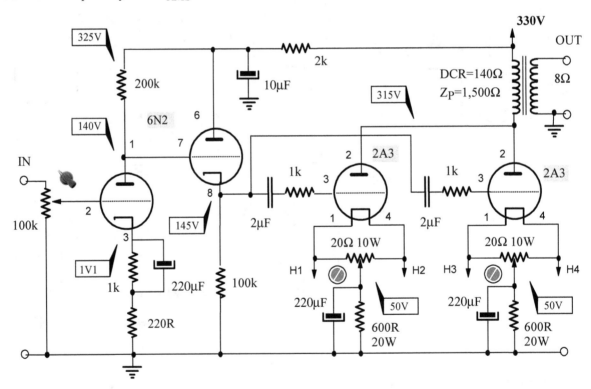

ABOVE: YARLAND FV II amplifier, audio section (one channel only shown), © Yarland

The power supply

A very simple power supply (next page), solid-state rectification with a time delay, LC filters are separate for each channel, the right approach. However, the 2H inductance of filtering chokes is low, as is the 150μF capacitance. The total energy storage in capacitors is 25 Joule.

To increase it and improve the bass, the value of the capacitors should be increased to 390 or 470 μF, and they should be bypassed by quality PIO or film&foil capacitors of 220n or higher capacitance.

On the positive side, the power transformer was a large unit that didn't buzz or overheat like many other Chinese amplifiers.

As a sideline, how did we use PL508 tubes (17V heaters) with this power transformer? Simply, we connected all four 2V5 secondaries in series, together with the 7V secondary, and got precisely the required 17V! Just make sure they are in-phase - connect them in series one-by-one and measure the voltage; the first two should give you 5V; if you get a very low voltage (theoretically zero volts), reverse one connection.

The PCB was removed, Magnoval sockets were installed, and the amplifier was rebuilt point-to-point. The output transformers were reused.

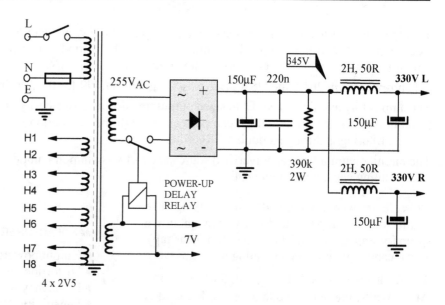

ABOVE: YARLAND FV II amplifier's power supply © Yarland

Audion Sterling EL34 amplifier

Made in France, Audion Sterling is a single-ended amplifier with one EL34 pentode per channel, delivering 12 Watts in ultra-linear mode.

In their promotional blurb, Audion says, "It is auto biased, uses a small amount of feedback in the design, and delivers a conservative 12 CLEAN watts into 8 ohms, which compares to competitors boasting 24-30W." The capital CLEAN is theirs.

However, in the specifications, the THD figure is quoted as "Distortion @ 1Watt: <0.1% No Feedback", which seems to be the identical sentence and figure used to describe most of their other amps and is impossible to achieve with NFB! So the two specification claims contradict each other.

One version uses a printed circuit board; another model is wired in a point-to-point fashion. The basic version has one input, another has five line-level inputs.

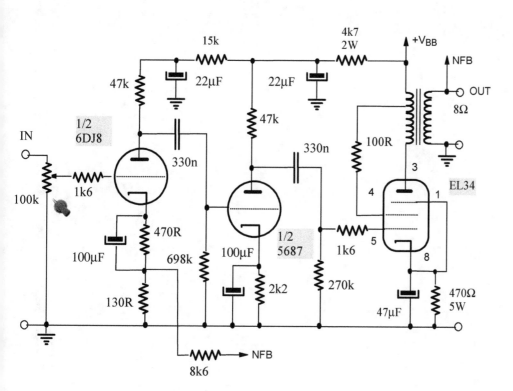

LEFT: Audio section (one channel) of Sterling EL34 amplifier, © Audion

To add to the confusion, some models have the inputs at the very back; other variations have them on the left-hand side, close to the volume control potentiometer. The PCB model uses a Russian 6N1P dual triode as a voltage input/driver stage, while the point-to-point version uses the 6H23N.

When reviewed in 2012, the USA price was $4,098 or $3,499 for the non-integrated version with one input.

The amp is tiny and light. Two EI96 sized output transformers sit side-by-side at the bottom of the chassis, under the toroidal mains transformer, which can be seen through the ventilation slots at the back. No chokes are used; high voltage filtering is through a double CRC filter.

The circuit diagram from an unverified source shows two common-cathode stages capacitively coupled, so there are two coupling caps in the signal path.

DC voltages were not marked on the drawing. The details of the output transformers are unknown (primary impedance and % of the ultralinear tap), so we cannot analyze the operating conditions.

The feedback ratio is $\beta=130/8,730=0.015$ or 1.5%. At 12 Watts, the output voltage on the 8Ω load is $V=\sqrt{(P*R)} = \sqrt{(8*12)} = 9.8V$, so with 0.2V at the input, the voltage gain (with NFB) of the whole amp is $A_F=9.8/0.2=49$ or 33.8 dB, which is quite high.

There is also the local NFB in the output stage which works in "ultralinear" mode.

MANUFACTURER'S SPECIFICATIONS:
- Output power: 2 x 12 Watts (8Ω)
- Distortion @ 1Watt: <0.1% No Feedback
- Frequency Response: 14 to >34KHz ±1 dB
- Sensitivity: Variable >200mV Full output
- SNR: >85 dB (CCIR)
- Tubes: 1 x 6H1N (6DJ8), 1 x 5687, 2 x EL34
- Size - 42x23x19 cm (D/W/H) (excluding tubes)
- Weight: 11 kg

BIG BOTTLES: SET AMPLIFIERS WITH HIGH VOLTAGE TRANSMITTING TUBES

- Audioromy 838 SET amplifier
- CARY CAD805C and 805AE
- VAL MP-211 monoblock amplifier
- BEZ T10B and T10B-5 SET amplifiers
- Mr. Liang LS-845-3 amplifier
- Antique Sound Lab MG-Power 805
- GM70 SET amplifier
- 813 SET amplifier with "enhanced triode" driver stage

"Its poor high-frequency linearity, incipient ultrasonic instability, and the presence of supply hum in its output are all factors that would rule the ATM-211 out for me, ..., the Air Tight amplifier appears to be able to produce a sound much better than its measured performance implies."
John Atkinson, Editor of Stereophile magazine, in his review of Air Tight ATM-211 amplifier
http://www.stereophile.com

Audioromy 838 SET amplifier

The first AudioRomy amplifier we bought was an EL34 push-pull design that looked great but sounded awful. Only after extensive modifications we've managed to get half-a-decent sound out of it. This amp sounded better than the EL34 model, but nothing remarkable.

The design is Class A_2-capable since the driver stage is a cathode follower, directly coupled to the output tube grid. One-half of the 6AS7 duo-triode is used due to its low internal impedance and high driving power, a wise choice. Three preamp duo-triodes are also shared between the two channels; each preamp stage uses a different triode.

One input goes through all three voltage amplification stages, while the second input goes straight to the grid of the 2nd stage. The cathode resistors of the first two stages aren't bypassed, introducing a mild local negative feedback. Global NFB from the speaker output is brought to the cathode of the 2nd stage. There are three capacitors in the signal path.

The power supply is all solid-state; only one little filtering choke is shared between the two channels. The energy storage in the PCB-mounted capacitors is very low for this kind of amplifier.

The output stage operates at the power level of only 51 watts (analysis on the next page). This is way too low for a tube rated at 125 Watts, so the design is highly conservative.

The Audioromy style of construction is instantly recognizable (next page). The whole power supply on a small PCB, cramped and undersized. The output tubes' heaters have their rectifier bridge; both are mounted on the same smallish heatsink, which gets very hot (1).

TUBE PROFILE: 805

- Directly-heated thoriated tungsten triode
- Heater: 10V, 3.25 A
- Socket: 4-pin jumbo
- Anode dissipation: 125W
- Max. anode current: 210 mA

MEASURED RESULTS:

- BW: 4Hz - 35 kHz (-3dB, 1W into 8Ω)
- BW: 12Hz - 24 kHz (-3dB, 10W into 8Ω)
- $V_{OUTMAX} = 17 V_{RMS}$, $P_{MAX} = 36W$

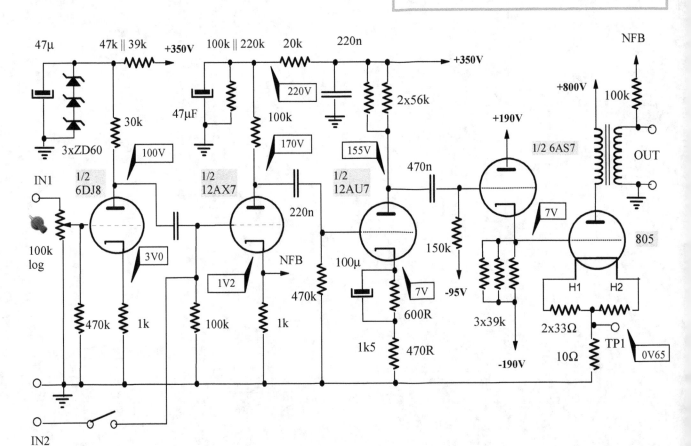

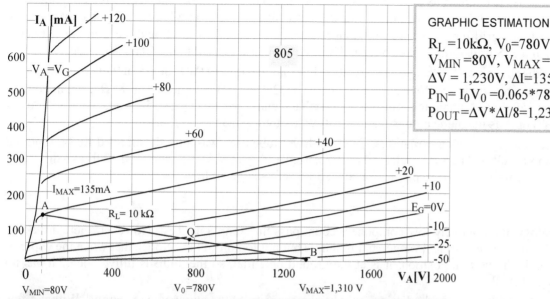

> GRAPHIC ESTIMATION
> $R_L = 10k\Omega$, $V_0=780V$, $I_0=65$ mA
> $V_{MIN}=80V$, $V_{MAX}=1,310V$
> $\Delta V = 1,230V$, $\Delta I = 135-5 = 130$ mA
> $P_{IN} = I_0 V_0 = 0.065*780 = 51W$
> $P_{OUT} = \Delta V * \Delta I / 8 = 1,230*0.13/8 = 20W$

LEFT: The internal view of Audioromy 838 SET amplifier

1) bridge rectifiers for the two output tubes' heater DC supply
2) filtering electrolytic capacitors
3) filtering choke
4) plenty of space for a bigger choke or an additional one
5) bias adjustment trimmer potentiometers
6) 6AS7 tube's octal socket
7) the four preamp tubes and the associated components

The four elcos in the HV filtering circuit (2) are connected two and two in series (to cope with 800V+ anode voltage). A tiny choke is shared between both channels (3), but there is plenty of space to add a bigger one (4). Output transformers are mounted on top of the chassis, above the choke (notice the eight bolts on the photo).

Two trim pots (5) adjust the bias for 805 triodes. The 6AS7 duo-triode is mounted between the two output tubes (6). Finally, the whole audio circuit is cramped around the four preamp tube sockets (7).

CARY 805C and 805AE

The CAD 805 Anniversary Edition (US$15,800 per pair in 2015) are 55 Watts per channel SET mono amplifiers with two 6SN7 duo-triodes in input stages, a 300B driver stage, and a choice of 845 or 211 output triodes, with switch-selectable bias, presumably switching different cathode resistors in and out to suit each tube.

In the 1st stage, two 6SN7 triodes are paralleled, and their output is DC-coupled to the grid of the 2nd stage, a "series anode constant-current Class A amplifier," whatever that means. I would guess that it is an SRPP stage whose output is capacitively-coupled to the grid of the 300B triode.

The 300B driver uses fixed bias (to be set at 60mA of anode current) and is both capacitively and inductively coupled to the output tube's grid (just as in model 805C). This dual-coupling is the main reason for including this expensive piece of hi-end audiophilia as a case study.

The interstage transformer is claimed to have flat response from 30Hz to 23 kHz. The capacitive coupling is in operation for the first 27 watts (Class A_1), while the interstage transformer works as 845 or 211's grid choke. Above that level, the output tube's grid turns positive, its input resistance drops below 2kΩ, and the grid current starts flowing. This is when the IS transformer takes over and provides the grid current in the Class A_2 mode.

The quiescent power dissipation of 211 triodes in Cary 805AE is claimed to be 94W (100mA at 940V_{DC}). While the declared nominal output power is 55 Watts, the maximum is 72 Watts at clipping. The 8kΩ primary output transformer is rated at 150 Watts, although it is not clear what that means, that it can pass 150 Watts of class A power (at what frequencies?) or if its magnetic core is rated at 150 Watts by its size?

Model 805 progressed through a few incarnations. The first driver tube used was a triode-strapped EL34, replaced by a 300B triode in model 805B. The 805C, released in 1998, used a single 6SL7 duo triode and 300B as a driver. Its circuit is available on the web, thus in the public domain, so we will analyze it here.

The audio circuit

A fairly simple topology and relatively low component count pass the Simplicity Test. However, there are a few potential sources of trouble. The biasing arrangement is simple but not foolproof (1). Notice that when the wiper of the bias pot (100k) is in the bottom position, it is at the ground potential, so the 300B is fully "open," it has no negative bias on its grid at all.

The DC anode current would surge and be limited only by the capacity of the anode power supply. Within a few seconds, the anode would start glowing cherry red color, overheat and, unless the mains fuse blows, eventually they would be destroyed. Such a high current through the interstage transformer's primary winding would most likely also overheat its relatively thin wire, which would act as a fuse and fail open circuit at its weakest spot, thus requiring a total transformer rewind or replacement.

But wait, you will say, there is a protective fuse in the 300B's cathode (2). Isn't that going to blow before the tube fails, or the IS primary fails open circuit? Yes, there is a fuse, but it is rated at 250mA. If the user fiddles around and leaves the bias not fully at the ground but at some low level, say -10 or -20V, the current may shoot up to 150-200 mA. After a while, such a high DC current would undoubtedly damage the expensive 300B triode or severely shorten its life.

Can the primary wire of the IS transformer take 200mA of current? So, it is more likely that the primary of the IS transformer would burn out first, thus, paradoxically, protecting the 300B tube before the fuse does.

A fuse in the signal path will raise the eyebrows of puritan audiophiles. Another pair of mechanical contacts, another couple of soldering joints and dissimilar materials in series with the signal. The biasing jack (3) that breaks up the cathode circuit is always a nuisance. In most multimeters, the probe must be unplugged from the V-W terminal and plugged into the mA-A (current) socket.

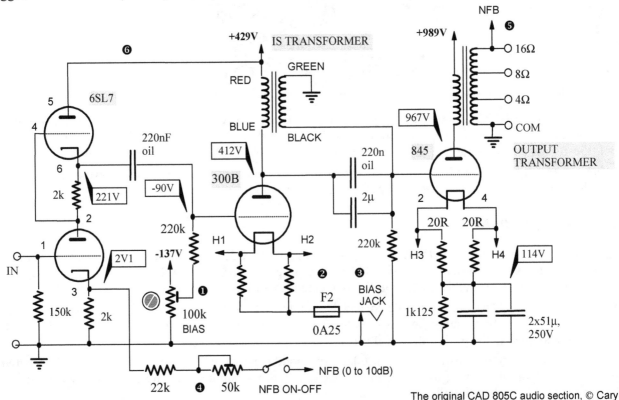

The original CAD 805C audio section, © Cary

BIG BOTTLES: SET AMPLIFIERS WITH HIGH VOLTAGE TRANSMITTING TUBES

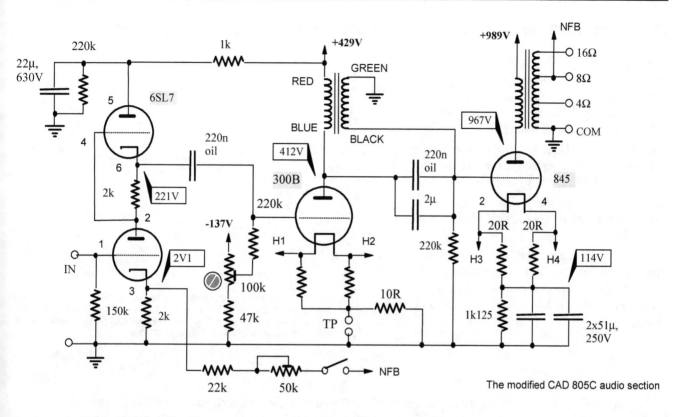

The modified CAD 805C audio section

A much more elegant option is to insert a 1Ω or 10Ω resistor in series with the cathode circuit and measure cathode current indirectly through the DC voltage drop on that resistor.

The manufacturer's specs say that the negative feedback is 0-10dB. With the NFB switch in ON position and the NFB pot at 10dB (slider fully to the right on the diagram), the 50k pot is bypassed, so the 22k resistor determines the feedback strength. The feedback factor is 2k/24k =0.083 or 8.3%. We don't have signal voltages marked on the diagram and don't know transformers' voltage ratios, so we cannot calculate the strength of the negative feedback in dB (decibels), but that seems much more than just 10dB of feedback (4).

With NFB pot at zero (trimmer's slider fully to the left on the diagram), the whole 50k is in series with the fixed 22k resistor, so the feedback factor is 2k/74k = 0.027 or 2.7%! Even that minimum feedback setting is far from zero feedback, so the only way to get zero NFB is to switch it off using the switch, not by turning the pot to "zero."

The next issue (5), still concerning NFB, is that it is taken from the 16Ω terminal, where few modern speakers would be connected. Depending on the speakers used, a more prudent option would be to move it to a 4Ω or 8Ω terminal. How to recalculate the required value of the NFB resistor is covered elsewhere in this Volume. Since NFB is adjustable here, it is not mandatory to replace any resistors; NFB levels can simply be readjusted manually.

There is no power supply decoupling between the first and 2nd stages (6), corrected on the modified diagram (1k resistor and 22μF film capacitor to ground). The chassis is relatively small for an 845 amp. The interior is overcrowded, almost messy, with lots of large-sized film caps, so adding another one is physically difficult.

The most dangerous piece of advice ever!

Cary 805AE user manual says this in its "Troubleshooting Guide"section: "Hum or 'buzzing' in speakers Cause: ground loop Remedy: Install 2-pin adapter in AC cord to float the ground."

You should NEVER "float the ground" by using such "cheater plugs" or by disconnecting the earth (ground) pin from the power cord or the amplifier. This would disconnect the metal chassis from the ground conductor in your electrical installation and make the amplifier illegal and deadly dangerous! If a fault develops and the amplifier starts a fire, a subsequent investigation may result in your insurance claim being rejected.

Should a short circuit develop, the chassis may end up at an extremely high DC voltage of 400-1,200V or an AC mains voltage of 100-240V. Even if a "ground fault interrupter" or "earth leakage circuit breaker" is installed in your switchboard (mandatory in Australia), since there is no connection to the earth conductor, such protection would be ineffective, and touching the chassis would result in an electric shock, probably lethal!

The impact of NFB on the A-f characteristics of an amplifier

The test results by *Stereophile* reveal a few interesting points about this amp and transformer-coupled SET amps in general. You should read the whole review since we will focus only on one issue here.

With the lowest NFB level, the drop in the bass region is the deepest (curve #3), and the f_L (lower -3dB frequency) is the highest.

The overshoot in the treble region (peak around 27 kHz) is very mild, likely to be inaudible.

As the NFB increases, the bass region's dips (the early attenuation of low frequencies) are reduced. However, the peaks in the treble region are increased, a clearly undesirable effect.

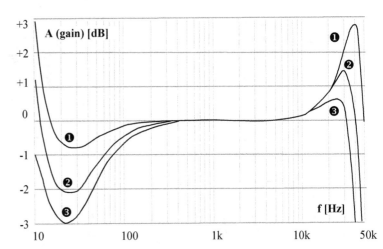

Cary CAD-805, frequency response at 1W into 8Ω with 10dB (1), 5dB (2), and 0dB (3) negative feedback. It is not clear if 0 dB means NFB completely off (true 0dB) or just the NFB pot in the "min" position (not 0dB). Redrawn from Stereophile (www.stereophile.com) with permission.

> "In classical terms, the CAD-805's test bench results cannot be categorized as anything other than mediocre, at best, even for a tube amplifier. ... such a set of measurements raises a question: Does the amplifier sound the way it does, in whole or in part, because or despite of its objective performance?"
>
> Thomas J. Norton (http://www.stereophile.com/content/cary-audio-design-cad-805-monoblock-power-amplifier-measurements)

In his interview with Sasha Matson for "Positive Feedback Online - Issue 19", Dennis Had, the guy behind Cary for many years (now retired), made a few claims that clearly aren't true. For instance, the only way to achieve Class A_2 operation of 211 or 845 tubes is by transformer coupling. We know that the same can be done using a powerful driver tube as a cathode follower, directly coupled to 211's grid. Many hi-end amplifiers use this approach, including the venerable Ongaku (see Vol. 1 of this book).

Regarding direct coupling, Mr. Had also claimed that "The Anniversary 805 is all direct-coupled, from the input tube to the output tube."

The 1st stage of Anniversary 805 uses a paralleled 6SN7 triode, directly-coupled to the next 6SN7 gain stage, which in turn is capacitively- coupled to the 300B driver stage. Since there is a capacitive coupling between the second and the 300B driver stage, and since that 300B stage is both inductively and capacitively coupled to the output stage, only the first two stages are directly coupled, *not the entire amplifier*, as claimed!

VAL MP-211 monoblock amplifier

These monoblock amplifiers were one of the first Chinese "hi-end" amplifiers to hit Australian shops. I remember seeing them a few times in a local dealer's listening room in 1997-1998, selling for $5,000 in their original state and $6,000+ modified. That was very expensive for a Chinese-made product.

Most of the time, different pairs just sat on the floor in the corner or in the small workshop next door. The dealer admitted that their power supply capacitors were blowing like popcorn. One quick look at the power supply circuit diagram (not shown here) reveals the reason.

The HV secondary of the mains transformer provides $740V_{AC}$, which is rectified by a diode bridge and filtered by a single CLC filter. Each leg has two 200µF elcos in series, each cap rated at $500V_{DC}$. The DC voltage before the series choke is $963V_{DC}$, with $950V_{DC}$ after the choke. Although each elco was bypassed by a 220k equalizing resistor, operating a 1,000V rated capacitor bank on 963V is an open invitation to major trouble.

Keep in mind that 963V is a steady-state DC voltage. Before the output tubes heat up and start conducting and thus loading the power supply, the unloaded voltage will be much higher, around 1,100V! As in most situations in electronics and in general, the 80-20 rule of thumb should be adhered to here:

> THE 80-20 RULE FOR POWER SUPPLY CAPACITORS
>
> Ideally, the DC voltage in the amp you are designing or constructing should not exceed 80% of the filtering capacitors' voltage rating. For instance, in a 2A3 SET amplifier with 450 V_{DC} filtering caps, the working voltage should not exceed 0.8*450 or 360 V_{DC}!

Energy storage of the HV power supply

If an amplifier with this problem comes your way, you need to replace the four 200 µF caps with six units, at least 470µF each, rated at 450V_{DC}. Connect three in series on each side of the choke, each bypassed by its own 220k resistor. The voltage ratio will be 950V/1,350V = 70%, meaning your margin will be 30%, and your new elcos will live happily ever after.

The new equivalent capacitance will be 470/3=157µF on each side of the choke. Originally, two 100µF elcos at approx. 950V_{DC} means the maximum stored energy was E=C*V^2/2 = 2*100*10^{-6}*950^2/2 = 90 J (Joule). The new energy storage will be E=C*V^2/2 = 2*157*10^{-6}*950^2/2 = 142 J. This is an increase of 50% and is bound to improve the bass response!

The audio circuit

The audio circuit was redrawn by one industrious owner from the original schematics by the manufacturer and published online. It is a copy of the Audio Note's Ongaku circuit, even down to DC supply voltages. The main difference is the use of different preamp and driver tubes, 6SL7 instead of 6072 and 6SN7 instead of 5687WA.

There is also mild negative feedback from the output tube's cathode to the un-bypassed bottom resistor in the SRPP first stage, while Ongaku uses no NFB. Kondo-San chose 5687 as the driver tube in Ongaku. Each triode inside the 5687 glass envelope is rated at 4.2 Watts, compared to 5.0 Watts for 6SN7, so there are no major issues.

The amplification factor of 6SN7 is around 20, which isn't much higher than µ=16 for 5687.

In this application, working as a cathode follower, µ does not matter much anyway! Both tubes have grid-to-plate (anode) parasitic capacitance of around 4pF, so no difference there either.

In terms of internal resistance, however, when operating around 10mA of anode current the internal resistance of 6SN7 triode is 7-8 kΩ, compared to only 3kΩ for 5687.

The mutual conductance of 6SN7 is around 2.6 mA/V, compared to 5.4mA/V for 5687. So, 5687 is a superior tube on both counts, and we now understand why Mr. Kondo chose it for this important task!

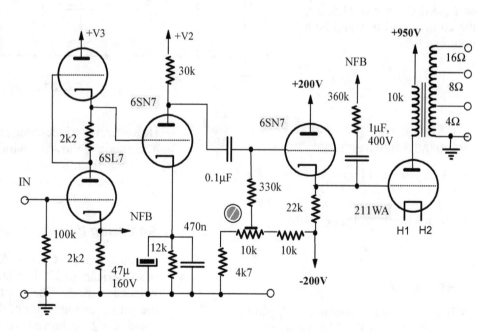

The audio section of VAL MP-211 amplifier

BEZ T10B amplifier

This earlier version of the T10B amplifier could use a variety of pentodes for V1: 6AC7, 6SJ7, 6SK7, 6AG7, 6SH7, 6J4P, and 6J8P. The driver tube is 6L6, but EL34, 6K6, KT66, 6V6, 6P6P, and 6P3P can be used as well. The input pentode is triode strapped, which reduces voltage gain, so the input sensitivity is low, 3.8V for full output power.

The driver power tube is also triode-strapped and works as a cathode follower, providing a low impedance drive signal directly coupled to the output tube's grid. This makes operation in Class A_2 possible.

A grid choke is used instead of a resistor, but since the DC current through it is significant, it had to be wound with a larger diameter wire. Usually, there is no DC current through the grid choke, so a very fine wire is used, and much more turns could be wound. This results in a much higher inductance; 100 or even 200H can be achieved, depending on the permeability of the magnetic material used, of course.

Evaluating a design - a broad view

When analyzing a particular design, the first step is to get an overall feel for what the designer tried to achieve. In this case, all three stages use either real triodes (last stage) or triode-strapped pentodes (the first two stages). Had the designer been a real triode-purist, he would have used a very mild negative feedback or no NFB at all. However, quite a strong global NFB was used, the feedback factor being β=4.12k/(36k+4.12k) = 0.1 or 10%.

In our designs, β is always below 0.02, which usually provides 2-3 dB of NFB, just enough to increase the damping factor and widen the frequency range without affecting the sonics.

The designer of this amp probably had to use a strong NFB to linearize the amplifier. 6L6 beam tetrode is very nonlinear and produces a significant distortion, great for guitar amps, but not in hi-fi service. A similar but more powerful tube, 7027A, makes a much better triode, with only half the distortion of 6L6. So, the choice of 6L6 would be a poor one here.

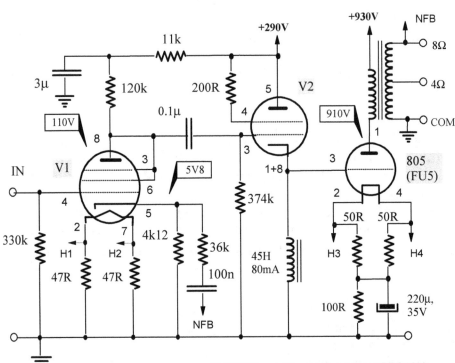

BEZ T10B amplifier, audio section © BEZ 2003

MANUFACTURER'S SPECIFICATIONS:

- BW: 20Hz - 25 kHz (-2dB, 1W into 8Ω)
- P_{OUT} = 25W @ 2% THD
- Input voltage for full power: V_{IN}=3.8V
- Maximum power: P_{MAX} = 40W

TUBE PROFILE: 6SJ7, 6SJ7GT

- Indirectly-heated sharp-cutoff pentode
- Octal socket, 6.3V, 300 mA heater
- V_{AMAX}/V_{AMAX} 300/300 V_{DC}
- P_{AMAX}= 2.5 W, P_{SMAX}= 0.7 W
- TYPICAL OPERATION PENTODE:
- V_A=250V, V_S=100V, V_G=-3V
- I_{A0}=3 mA, I_S=0.8 mA
- gm =1.65 mA/V, μ=1,650, r_I=1MΩ
- TYPICAL OPERATION TRIODE:
- V_A=V_S=180V, V_G=-6V, I_{A0}=6 mA
- gm =2.3 mA/V, μ=19, r_I=8.25 kΩ

However, the triode-connected 6L6, in this case, is used as a cathode follower, which applies a 100% negative feedback from its output to its input and is by definition thus "linearized". So, one could argue that any tube would do here, no matter how nonlinear it is.

Analyzing the operation of the first stage

The choice of the Chinese 6J8P pentode, a close equivalent of American 6SJ7, is also questionable. First of all, it works as a triode. The graphical analysis below shows that the amplification factor of the first stage is only around 15, which can be achieved with many real triodes.

Secondly, although its triode curves look relatively linear, small-signal pentodes make better and more linear triodes, such as EF86, 6AU6, and many others. The anode load of 120kΩ is fine, no issues there, but the anode current is way too low for such a relatively powerful pentode (2.5 Watts anode dissipation).

Notice that the whole voltage swing A-B is in the very nonlinear region, where anode characteristics are very curved and not equidistant (next page) - this means high distortion.

BIG BOTTLES: SET AMPLIFIERS WITH HIGH VOLTAGE TRANSMITTING TUBES

You may say, wait, but the 100% NFB in the cathode follower stage will reduce such distortion. No, the local NFB inherent in cathode followers only reduces the distortion that stage would produce. If the distortion is present at the input, that distortion will be propagated to the output; that is why they are called cathode "followers"!

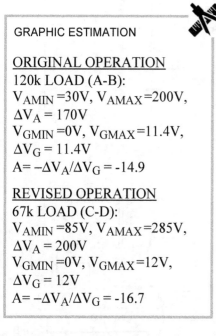

GRAPHIC ESTIMATION

ORIGINAL OPERATION
120k LOAD (A-B):
V_{AMIN} =30V, V_{AMAX}=200V,
ΔV_A = 170V
V_{GMIN} =0V, V_{GMAX}=11.4V,
ΔV_G = 11.4V
$A = -\Delta V_A / \Delta V_G = -14.9$

REVISED OPERATION
67k LOAD (C-D):
V_{AMIN} =85V, V_{AMAX}=285V,
ΔV_A = 200V
V_{GMIN} =0V, V_{GMAX}=12V,
ΔV_G = 12V
$A = -\Delta V_A / \Delta V_G = -16.7$

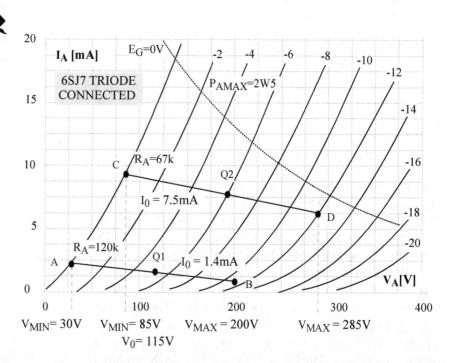

Possible modifications

The experience of many audiophiles is that triode stages with higher anode current levels (5-15mA, depending on the tube used) sound better than the "starved" designs that run on 0.5-1.5 mA. So, why don't amp designers always go for high current stages? The answer will soon become apparent.

Moving the operating point from Q1 to Q2, despite reduced anode resistance (from 120kΩ down to 67kΩ), increases the stage gain from 15 to 16.7. That wasn't our aim, but it is a welcome bonus.

Since the voltage swings around Q1 are equal ($\Delta V- = 115-30=85V$, $\Delta V+ =200-115=85V$), the 2nd harmonic distortion in the original operating regime is zero. This could be a coincidence or the result of the designer's careful planning. The 2nd harmonic distortion in a higher operating point Q2 and with voltage swing C-D would also be zero ($\Delta V- =185-85=100V$, $\Delta V+ =285-185=100V$)!

Sizing the cathode resistor and the required anode supply voltage

Now we need to size the replacement cathode resistor. The bias voltage in Q2 is -6V, so the cathode must be at V_K=+6V, and since $V_K = I_0 * R_K$, we get $R_K = V_K / I_0 = 6/7.5$ mA = 0.8kΩ, so we would use either 790Ω or 820Ω (standard values).

From our graph, we know that the voltage between anode and cathode must be around 190V, plus 6V on the cathode means that anode must be 196V above ground potential, to which we must add the voltage drop the anode current makes on the anode resistor. That will give us the required DC voltage of the power supply. So, $V_B = V_{RA} + V_A = R_A * I_A + V_A = 67k * 7.5$ mA + 196V = 502+196 = 698V!

The voltage drop on the anode resistor must be around 500 Volts, so the power supply must provide around 700V_{DC}, and we only have 290V! This is the problem we only mentioned in Volumes 1 and 2 of this book, and now we must solve it. There are two ways to go about it.

Option 1: 67k anode resistor

We know that the output stage operates at +930V on the output transformer and around +910V on the anode of the 805 tube. So, we could also use that voltage for the first stage after adding an additional filter at that point.

As discussed in Vol. 1, small mains transformers (240V_{AC} to 6.3 or 12.6V_{AC}) make great low current chokes for preamp stages and screen grid circuits. They typically have a DC resistance of around 1 kΩ and 12-18H inductance.

Two-stage filtering would be best. Firstly we will reduce the voltage as much as possible using the CRC filter, followed by the CLC filter.

Three elcos in series are needed on each side of the series resistor R_S, as illustrated. An 8mA of current on 1k choke would create a DC voltage drop of only 8V, so R_S must drop 930-8-700 = 222V. Using Ohm's law, we get R_S= 222V/8mA = 27.8 kΩ, so use the standard value of 27kΩ.

The power dissipated into heat on that resistor is $P=V^2/R_S$= $222^2/27{,}000$ = 49,284/27,000 = 1.83 Watts. The 80-20 rule does not apply to power ratings of resistors; ideally, a resistor should have a power rating of three times the heat dissipated on it, or in this case, 1.83*3 = 5.5 Watts. A standard 5W resistor should be fine here.

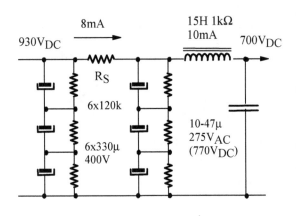

Option 2: Anode choke

Valab 200H/35ma anode/plate chokes are available on ebay and are recommended for tubes with internal resistance under 7kΩ (5842, 5687, 6H30, 6DJ8, 6N1P, 12AU7, 6J5, 6SN7). A mix of two magnetic materials is 50% nickel-iron + 50% Z6 grain-oriented silicon steel. EI66 laminations are used, with a stack thickness of 24 mm.

Choke data: N=14,000 turns, L= 200H@120Hz, L=1,100H@20Hz, 2,330 Ohm DC resistance, Weight: 800g

The anode of V1 must be 196V above ground potential, to which we must add the voltage drop the anode current makes on the anode choke. The DC resistance is 2k3, so the required DC voltage of the power supply is $V_B = V_{CH} + V_A$ = $R_{CH} * I_A + V_A$ = 2.3k*7.5 mA + 196V = 17+196 = 213V.

The DC voltage drop on the anode choke is only 17V. The 7.5mA flowing through the 11k resistor in the decoupling circuit (page 150) drops the supply voltage 82.5V to 290-82.5 = 207.5V. We need 213V, so we could keep that resistor as it is (only 5.5V difference, within a margin of error), or, if you are a pedantic personality, replace it with a slightly lower standard value, say 10k!

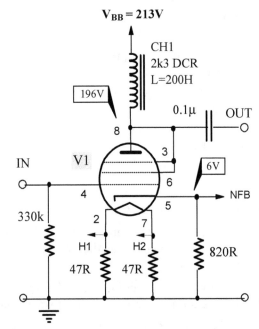

ABOVE: Thanks to the use of anode choke and LC coupling, this version of the modified input stage achieves a high anode current (above 7mA) without using a very high anode supply voltage V_{BB}!

BEZ T10B-5 SET amplifier

In this version of the T10B amplifier (next page), BEZ replaced the triode-strapped 6L6 in the cathode follower with a 300B triode. Cathode biasing is used, and since the idle current through 300B is 67mA, the voltage drop on the 500R cathode resistor is 33.5V.

Another 19.5V_{DC} of cathode bias voltage comes from the DC voltage drop the cathode current makes on the grid choke. Although it's not specified on the diagram, we can easily find out that the DC resistance of that choke is DCR= 19.5V/67mA = 0.295k or 290 Ω.

Since both the anode current of the output tube (120mA) and the voltages on the output transformer's primary are given, we can calculate the DC resistance of the primary winding. The DC voltage drop is 1,015-998V=17V, so R_P=17/0.12 = 142Ω. That is a fairly low resistance for this type of transformer.

To get a high primary inductance at such a high primary impedance (around 10kΩ), a very high number of primary turns is needed, resulting in high DC resistance of 250-350Ω. A low figure of below 150Ω means that a large diameter primary wire was used on a large size transformer core, required to fit so many turns.

This version has a higher input sensitivity (2V compared to 3.8V), due to the higher gain of the first stage, which is now pentode-connected.

MANUFACTURER'S SPECIFICATIONS:

- BW: 20Hz -25kHz (-2dB, 1W into 8Ω)
- P_{OUT} = 25W @ 2% THD
- Input voltage for full power: V_{IN}=2.0V
- Maximum power: P_{MAX} = 40.5W
- V_{HUM}= 1.6mV

BIG BOTTLES: SET AMPLIFIERS WITH HIGH VOLTAGE TRANSMITTING TUBES

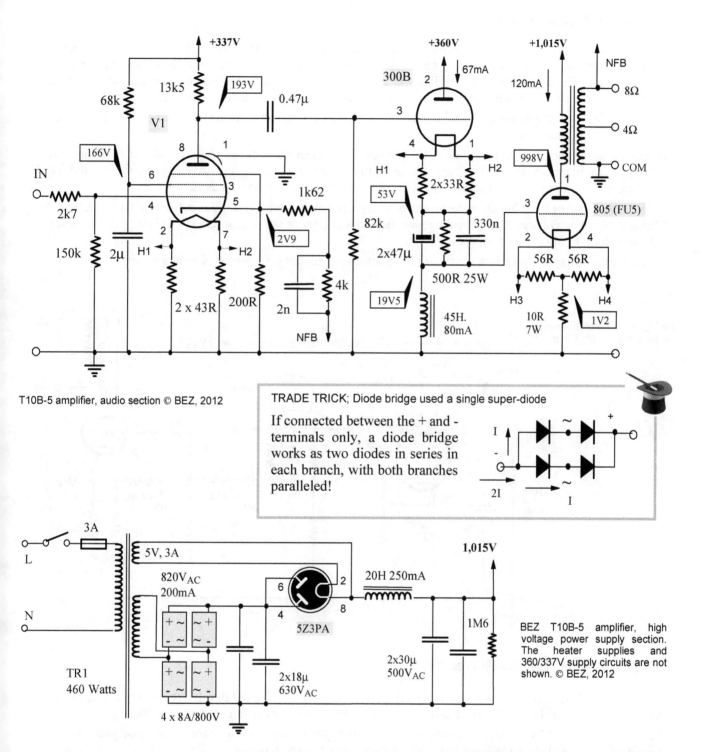

T10B-5 amplifier, audio section © BEZ, 2012

TRADE TRICK; Diode bridge used a single super-diode

If connected between the + and - terminals only, a diode bridge works as two diodes in series in each branch, with both branches paralleled!

BEZ T10B-5 amplifier, high voltage power supply section. The heater supplies and 360/337V supply circuits are not shown. © BEZ, 2012

The power supply illustrates a few interesting points. Notice that instead of one rectifier bridge or four individual solid-state diodes, four bridges are used. Their AC (~ sign) terminals are unused; the only connections are between the + and - terminals! Used this way, each bridge has two branches parallel, with two diodes in series in each branch. That doubles the peak reverse voltage the rectifier circuit can handle.

A dual rectifier tube is used in series with the filtering choke inside the CLC filter. This provides a small "soft-start" delay and also "decouples" the rest of the filter and the load from the mains transformer, preventing back-feeding of the EMF (electromotive force) into the transformer's secondary.

Finally, instead of three $450V_{DC}$-rated elcos in series and the associated voltage equalizing resistors both before and after the choke, motor start film capacitors are used, two in parallel. The ones before the choke are rated at $630V_{AC}$ (meaning they can withstand $2.8*630 = 1,760$ V_{DC}); the ones after the choke have higher capacitance (30μF each) and can withstand $1,400$ V_{DC}.

Notice a large choke; 20h at 250mA is some serious stuff. Overall, this amplifier seems properly designed and well-engineered.

Mr. Liang LS-845-3 amplifier

Another Chinese variation on the same theme, the impressive-looking Mr. Liang LS-845-3 monoblock amplifier, uses a 6J4P pentode input stage with very high gain. The high gain is needed since it is the only voltage gain before the output stage. The directly coupled 300B driver stage is a cathode follower, which reduces the gain. Strangely, its output is capacitively coupled to the grid of 845 tube, so only Class A_1 operation is possible.

At least the first two stages are directly-coupled, eliminating one capacitor from the signal path and the associated time constant and phase shift.

The power supply (not shown) uses four diode bridges for DC heating voltages of 300B and 845 tubes, and another four bridges for the main HV, bias voltage, anode supply voltage for preamp and driver stages, and an auxiliary DC supply for the 555 timer and two relays that delay the application of HV to all stages.

The output stage uses fixed bias from a dedicated solid-state power supply, a fully rectified and CRC filtered source of the negative bias voltage. The AC voltage for the anode supply of the output stage (V1) is supplied by a separate mains transformer, always a good idea.

There were no voltages marked on the circuit diagram, so we cannot do a deeper circuit analysis.

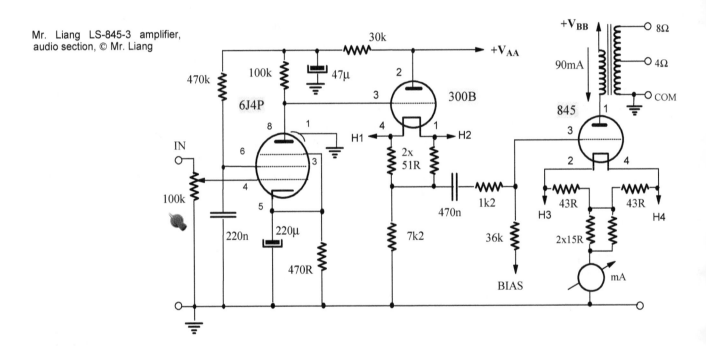

Mr. Liang LS-845-3 amplifier, audio section, © Mr. Liang

Antique Sound Lab MG-Power 805

While Cary's CAD805 operates via capacitive coupling while in Class A_1 and through transformer coupling only in Class A_2, once the grid current starts flowing, this amp works transformer-coupled all the time.

The first two stages (12AX7) are DC-coupled, so there is only one coupling capacitor in the signal path, between the cathode of the second stage (cathode follower) and the grid of the 6L6 driver.

The cathode resistor in the first stage is not bypassed so that the global NFB signal can be brought in from the 16W tap on the output transformer. This feedback is very strong; the feedback factor is $\beta=1k/16k = 0.0625$ or 6.25%.

Most of the voltage gain is provided by the first stage; that is why 12AX7 was used. It is a duo-triode with the highest m. The input sensitivity is 2V for full power.

The use of a cathode follower (CF) between the first and the driver stage is advocated by some designers. Such impedance decoupling (very high input and very low output impedance of the CF) practically eliminates the loading effects between stages, reduces distortion, and widens the frequency range of the amplifier.

As for the sonic consequences, the myth of cathode followers being bad for the sound is just that - a myth! We used the same CF approach in our screen-driven 813 SET amplifier with PL508 driver stage, also featured in this volume.

Notice that the driver stage uses a fixed bias of -30V. That voltage is adjustable with a trim pot but is not regulated.

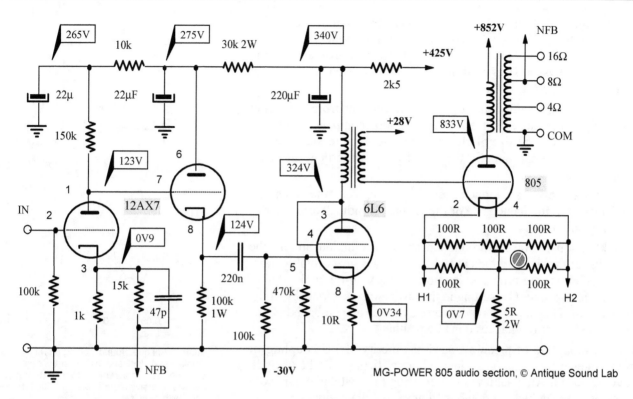

MG-POWER 805 audio section, © Antique Sound Lab

The output tube is biased through the secondary of the interstage transformer, which is connected to +28V power supply, which is both adjustable and somewhat regulated (a simple series bipolar transistor regulator) without any Zener diode to keep the reference voltage stable, just a simple resistor in the transistor's base.

The output stage's idle current is $I_0=0.7V/5\Omega = 140mA$, while the idle current through the 6L6 driver and the primary of the IS transformer is $I_0=0.34V/10\Omega = 34$ mA. The output stage is positively biased, so its grid current flows continuously.

The all-solid-state power supply (not shown) uses a voltage doubler that converts $330V_{AC}$ 500mA secondary voltage to 852 V_{DC} for the output stage. The half-point of the voltage doubler is taken out as $+425V_{DC}$ for the preamp and driver stages. Since that voltage is not well filtered, strong decoupling is required, firstly the 2k5 resistor and 220μ elco, followed by the 30k and 22μ, and finally 10k and 22μ.

The LC-coupling of Explorer 805 DT by The Antique Sound Lab

Explorer 805 DT, a similar model by The Antique Sound Lab, a Hong-Kong company, was reviewed in Stereophile in March 2004. Its circuit is similar to MG-Power 805, but the Explorer's output stage is LC-coupled to the output transformer.

The shunt-fed output stage (or "parallel-fed," thus the name "parafeed arrangement") has the output transformer in parallel with the output tube ("shunting" it to the ground).

The DC anode current I_A does not flow through the transformer's primary winding, so smaller transformer cores can be used. Smaller lamination sizes mean lower parasitic capacitances and higher upper-frequency extension.

There is no DC current flux through the primary winding and the corresponding reduction in the incremental inductance, so much higher primary inductances can be achieved than in series-fed output stages, which translates into better bass.

However, a large-value film coupling capacitor C is now in the signal path to prevent the DC current from flowing through the output transformer's primary.

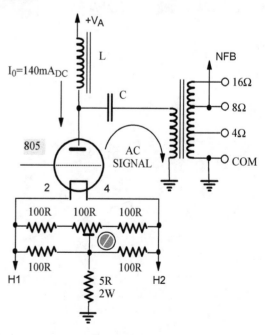

ABOVE: Explorer 805 DT LC-coupled output circuit

Secondly, an additional choke is needed as a series anode load. This choke must be designed to accommodate high anode DC current through its winding (140mA in this case). It must have an air gap and be of a large physical size, so it does not saturate even at the lowest frequencies of interest, 10-20Hz!

So, we have simply shifted the stringent requirements from the output transformer to the choke. Instead of one magnetic component, we now need to wind two! The size of the chassis must be increased. And we have introduced a capacitor in the signal path, a sacrilege in many audiophiles' minds.

Furthermore, now the inductance of the choke, the coupling capacitance, and the inductance of the primary winding form a series resonating circuit.

The 1 kHz square wave reproduction shows oscillation around the rising edge of the signal, which at 10 kHz becomes unacceptably high. These oscillations are primarily due to the less-than-optimal value of the coupling capacitor C resulting in the low resonant frequency of the output circuit. They are then made worse by the strong global NFB; the output transformer resonances don't help either.

While LC-coupling of preamp stages has merits, if LC-coupling of output stages were such a great idea, all amplifiers would be using it, and very few do! All it achieves is increasing the cost and complexity of designs while delivering dubious (if any) sonic benefits.

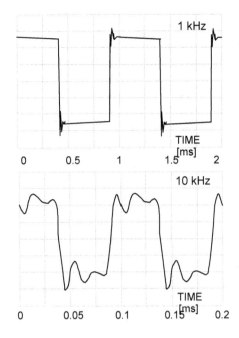

RIGHT: Antique Sound Lab Explorer 805 DT, small-signal 1kHz and 10 kHz square wave into 8 ohms, Redrawn from Stereophile (www.stereophile.com) March 2004 review, used with permission

Design project: GM70 SET amplifier

GM70 is a high voltage transmitting triode of Russian origin. It is slightly shorter than the American 211 but wider and uses a different socket, so the two are not directly interchangeable. The tube was produced in the former Soviet Union in huge quantities, so plenty of NOS tubes are for sale.

There are two versions, with graphite and copper anode. The copper anode version is much more expensive, not because it is rare but because of the claim that it sounds better.

The output power with 5k load and Q-point at a relatively low anode voltage (760V) is only 12.2W, the power that can be achieved much easier using low voltage tubes.

With a 10kΩ load, the operating point must shift towards much higher voltages, 1,000 to 1,100V. The current swing remains roughly the same, around 130-140mA, but the voltage swing is now almost doubled, and so is the output power, 22W instead of 12.2W!

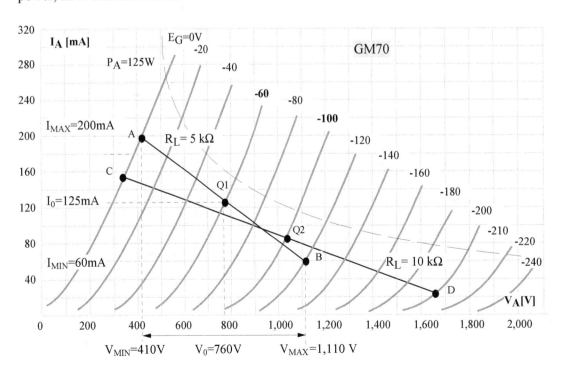

GRAPHIC ESTIMATION

Class A_1 triode with 5k load (A-Q1-B):
V_{MIN} =410V, V_{MAX}=1,110V, ΔV = 700V
ΔI = 200 - 60 = 140 mA
P_{IN} = $I_0 V_0$ = 0.125*760 = 95W
P_{OUT} = $\Delta V \Delta I/8$ = 700*0.14/8 = 12.2 W

Class A_1 triode with 10k load (C-Q2-D):
V_{MIN} =340V, V_{MAX}=1,660V, ΔV = 1,320V
ΔI = 154 - 22 = 132 mA
P_{IN} = $I_0 V_0$ = 0.086*1,040 = 90W
P_{OUT} = $\Delta V \Delta I/8$ = 1,320*0.132/8 = 22 W

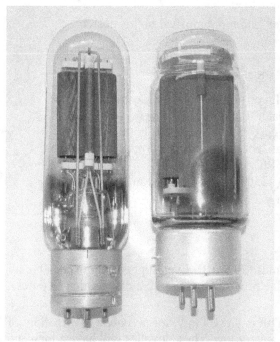
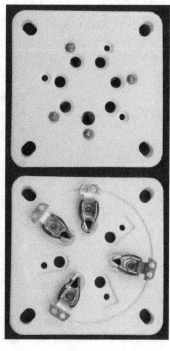

TUBE PROFILE: GM70

- Directly-heated triode
- Heater: 20V, 3 A (60W)
- Plate dissipation: 125 W
- μ= 9.5, r_I= 2,300 Ω

FAR LEFT: Vintage USA-made VT-4C (211) triode with anode dissipation of 100 Watts (left) and GM70 directly-heated triode made in former Soviet Union on the right.

Notice how much bigger (taller and wider) the anode is in GM70 compared to 211, whose anode looks tiny by comparison, which makes us think that this monster of a tube could dissipate much more than its rated 125 Watts!

LEFT: Ceramic tube sockets for GM70 look similar to those that 6C33C-B triode uses but are not identical.

The preamp princess: E86C

The first stage of our design features E86C, an SQ princess with gold pins. Lack of microphony, high reliability, long life, high amplification factor, and high anode current are just some of its technical assets. Its sonic beauty, as always, is in the eyes (or ears?) of the beholder. It sounds transparent, detailed, and uncolored. Nothing is veiled here; there is no artificial sweetness, smudginess, or grainy effect. Yet, its sonic presentation is also dynamic and musical, a very rare combination, indeed.

TUBE PROFILE: E86C

- Indirectly-heated SQ triode
- Noval socket, gold pins
- Heater: 6.3V, 165 mA
- Life: 10,000 hours minimum
- V_{AMAX}=250V, V_{HKMAX}=100V_{DC}
- P_{AMAX}=2.4W, I_{AMAX}=20 mA
- TYPICAL OPERATION:
- V_A=185V, V_G=-1.5V, I_0=12 mA
- gm=14.0 mA/V, μ= 68, r_I = 4.8 kΩ

A GREAT SOUNDING SQ (SPECIAL QUALITY) TRIODE

RIGHT: The quiescent operating point Q on the μ-gm-r_I diagram of E86C.

Notice the very high amplification factor of 67 and low internal resistance (4k8), a rare combination. This means the mutual conductance is very high (14 mA/V), resulting in a dynamic sound.

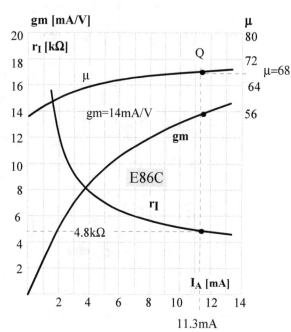

The driver heavyweights: 6CG7 and 6N6P

Russian 6N6P duo-triode is very similar to ECC99 and E182CC, the main difference being that 6N6P operates only on 6.3V heater voltage. Its pin 9 is the shield between the two triode systems, which should be grounded. Pin 9 on ECC99 and E182CC is the heater center tap. Due to these heater wiring differences, the two western tubes are not a drop-in replacement for 6N6P, but there is a tube that is! Our old friend, 6CG7, has the same pinout as 6N6P, and can be substituted in this design, giving tweakers some tube rolling freedom!

The tube profiles reveal that despite their almost identical amplification factors, the two tubes are anything but identical. 6N6P has a much lower internal resistance and much higher transconductance, plus it can pass a much higher anode current. Thus, technically at least, it is a much superior triode.

However, listening tests revealed that 6CG7 sounds better, smoother, and more musical in this design.

TUBE PROFILE: 6N6P (6Н6П)

- Indirectly-heated Noval duo-triode
- Heater: 6.3V, 0.75A, $V_{HKMAX}=100V_{DC}$
- P_{AMAX}: 4.8W (8W both)
- Max. anode voltage & current: 300V/60 mA
- TYPICAL OPERATION:
- $V_A=120V$, $V_G=-2.0V$, $I_0=28$ mA
- gm=11.2 mA/V, µ=22, $r_I = 1.8$ kΩ

TUBE PROFILE: 6CG7

- Indirectly-heated Noval duo-triode
- Heater: 6.3V/600 mA, $V_{HKMAX}=100V_{DC}$
- P_{AMAX}: 3.5 W each, 5W both
- Max. anode voltage & current: 300V/20 mA
- TYPICAL OPERATION:
- $V_A=250V$, $V_G=-8V$, $I_0=9$ mA
- gm = 2.6 mA/V, µ= 20 $r_I=7.7$kΩ

The audio circuit

The high current first stage amplifies the input signal around 30 times (29.5 dB). Even higher gain is possible by bypassing the cathode resistor, but it is not needed since the 2nd stage provides another 22.2 dB of gain (13 times). The total gain up to the GM70's grid is 390 times or 51.7 dB.

Fixed biasing is used; the quiescent DC current through GM70 is $I_0=100$mA for a power dissipation of $P_0=I_0*V_0=0.1*770 = 77$Watts, which for a tube rated at 125 Watts is very low. These lovelies will have a long and stress-free life in this conservative design.

The power transformer gets very hot, it would be better to leave it uncovered, but since we enclosed it in a steel enclosure for aesthetic reasons, ventilation holes under the chassis and in the steel case were mandatory. The steel chassis is powder coated in a black textured finish, acting as a massive heat sink.

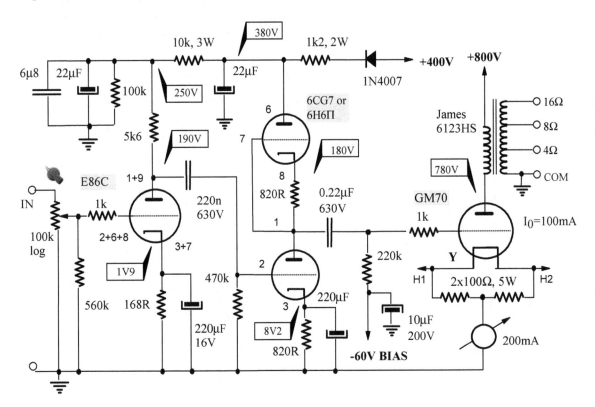

BIG BOTTLES: SET AMPLIFIERS WITH HIGH VOLTAGE TRANSMITTING TUBES

GM70 dissipates 60 Watts of heater power plus almost 80 Watts of audio power (without signal). Adding to that heat from transformer losses and preamp tubes, the heat radiated into your listening room is more than 300 Watts. Make sure this kind of amplifier has plenty of air by placing it on top of a hi-fi rack. Don't place anything in its vicinity. Simply treat it as you would a room heater (away from curtains, carpets, and alike), and you'll be safe!

James 6123H output transformers have 5k primary impedance, but three secondary taps afford us some freedom to experiment and see which connection would sound the best with particular speakers. In our case, that was the 8Ω tap, but our speakers are of relatively high impedance, which would be reflected as 7-12 kΩ to the primary side, depending on the signal's frequency.

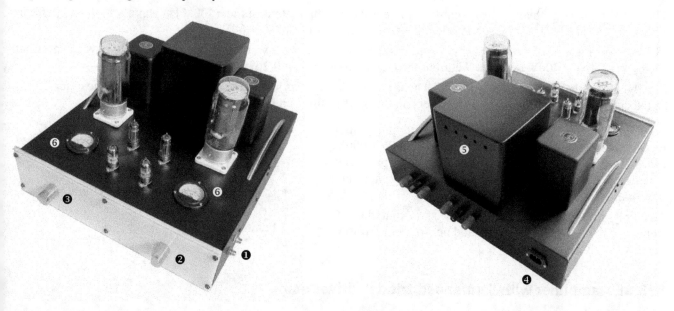

The chassis was the largest size cash drawer we could find. RCA inputs (1) right next to the volume control pot (2), with an on-off switch (3) right opposite the mains inlet at the back (4). Relatively large James output transformers are dwarfed by GM70 tubes and the jumbo-sized power transformer, which could only fit under the largest available Chinese transformer cover (5). It gets very hot despite the ventilation holes at the back and under the chassis for vertical air circulation. Analog meters monitor the DC current through the output tubes (6).

The power supply

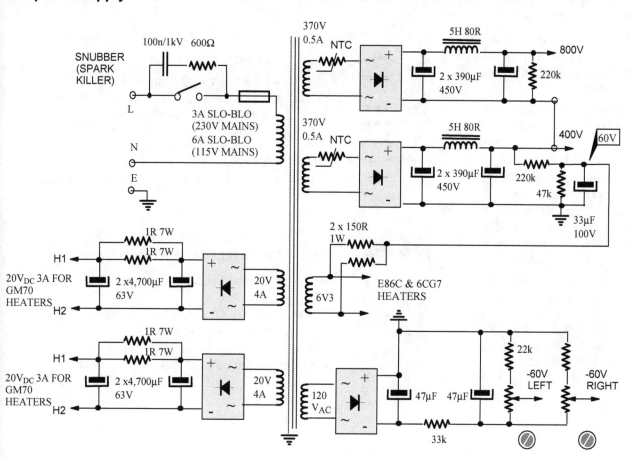

Even without any tube rectifiers, the power supply for this beast is still quite complex. Each output tube requires its own 20V, 4A secondary winding, rectifier, and filtering circuit. Luckily, since the preamp tubes are all indirectly heated with standard 6.3V heaters, only one heater supply is sufficient. However, there is a snag.

Since the 2nd stage is an SRPP amplifier, the cathode of the upper triode is on a DC potential of 180V, so its heaters must not be simply grounded. Here they are elevated to around +60V, meaning there is still 120V between the heater and cathode, which is above the maximum allowed of 100V (for 6CG7 and 6H6П).

After hundreds of operational hours, these robust tubes seem to tolerate this situation well. The E86C input tubes' heaters are also elevated to +60V, but that is within their +100V maximum.

The +60V voltage is obtained by a simple voltage divider at the output of the +400V HV supply, 47k and 220k in series, so 47/(47+220) = 0.15, so the reference voltage is 15% of 400V or 60V.

The resistors must be of a relatively high value so they don't load the HV supply unnecessarily. Here we have a total of 267 kΩ across +400V, so the DC current through them is only 400/267 = 1.5mA.

Finally, notice that the HV supply of 800VDC is two 400V supplies stacked in series, one on top of the other. It can be executed with solid-state or tube rectifiers and simplifies things a lot.

Lower voltage SS diodes can be used, and many tube rectifiers cannot handle such high voltages of 800 - 1,000V. Elcos don't have to be stacked in series with paralleled equalizing resistors. This reduces the parts count and improves reliability.

The only disadvantage of this approach is that the AC ripples on the two power supplies add up, so good filtration in each supply is mandatory.

MEASURED RESULTS:
- BW: 12 Hz-26 kHz (-3dB, 10W into 8Ω)
- V_{OUTMAX}=16V_{RMS}
- P_{MAX} = 32W

813 SET amplifier with "enhanced triode" driver stage

Having already discussed the design that used a screen-driven EL508 or PL508 in a single-ended output stage, let's apply that circuit to drive a large output tube such as 813. Of course, you can use any transmitting tube of your liking, such as 211, 845, or GM70.

The Mighty Mite, otherwise known as PL508 (or 6.3V heater version, EL508), is an incredible tube, both in output service and as a driver. Here it is used in an "enhanced triode mode," which is Tim de Paravicini's marketing speak for a screen-driven triode stage.

The driver stage is capacitively coupled to the output stage that is cathode biased, so only class A1 is possible.

The cathode bias and capacitive coupling obviate the need for a negative bias supply, simplifying the power supply.

There is no cathode follower driver stage that would be necessary for Class A_2 operation, but there is a 6DJ8 cathode follower stage driving the screen grid of EL508.

The first time we built this circuit on a test bench, we used output transformers with a 400:1 impedance ratio (3k2 primary and 8W secondary), meant for 300B triodes.

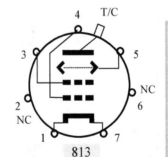

TUBE PROFILE: 813
- Directly-heated beam power tube
- Thoriated tungsten filament, 10V, 5A
- Socket: giant 7-pin bayonet base
- Anode power rating P_{AMAX}: 125W
- Max. anode voltage: 2,250V
- Max. screen voltage: 1,100V
- gm=3.75 mA/V, µ=8.5 (triode connection)

TUBE PROFILE: EL508 (6KW6) & PL508 (17KW6)
- Indirectly-heated beam power tube
- Magnoval socket
- Heater: 6.3V, 1.2A (EL508), 17V, 0.3A (PL508)
- V_{AMAX}=400V, V_{G2MAX}=275V
- I_{K2MAX}=90mA, P_{AMAX}=12W, P_{G2MAX}=4W
- As a triode: gm=8 mA/V, µ=9, r_I = 1.1kΩ

BIG BOTTLES: SET AMPLIFIERS WITH HIGH VOLTAGE TRANSMITTING TUBES

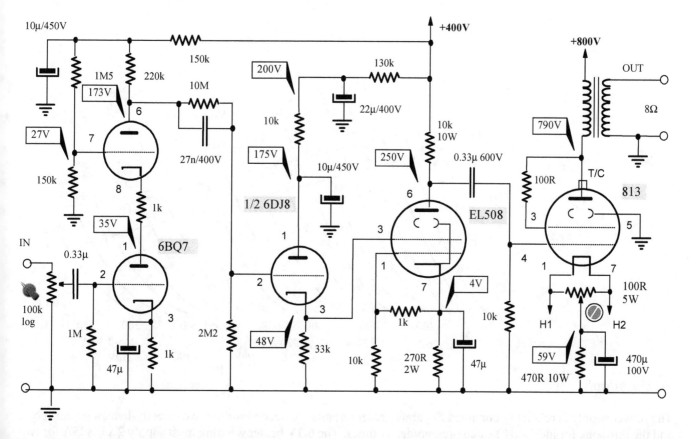

The output power was very high, above 30 Watts, but the lower -3dB corner frequency was also too high, 35 Hz, manifesting itself in a weak bass.

Then we wound a transformer with 4,000 primary and 90 secondary turns. The primary impedance Z_P was 15.8 kΩ, but the primary DC resistance R_P was also high (260Ω) due to many primary turns. Although f_L dropped to 27Hz (a minor improvement), the maximum output power dropped from 32 Watts to 21 Watts!

The conclusion? Primary turns should be reduced from 4,000 to 2,800. Z_P will then be halved to 7k7. Use 0.35mm^2 wire for primary winding instead of 0.25mm. These two measures will lower R_P to below 100Ω and reduce signal and power loss on the high primary DC resistance.

However, due to the larger diameter primary wire, it is impossible to fit 2,800 turns into the window of a 19cm^2, so larger size laminations must now be used, with a center-leg cross-sectional area of around 28cm^2! This will have another undesired consequence; the leakage inductance will increase due to the much larger laminations used, which will lower the upper -3dB frequency f_U and narrow the frequency range of the transformer. If you have ever wondered how come even the venerable Ongaku amplifier only goes up to 22 kHz, now you know.

A third option that would alleviate these problems somewhat (but not significantly!) is to use silver wire for the primary winding. Due to silver's lower specific resistance than copper, smaller diameter wire can be used for the same number of turns, meaning smaller size laminations would be needed and f_U would be raised!

Triode operation of 813

The operating points and load lines for the three experimental cases just discussed are illustrated on the I_A-V_A characteristics of triode-connected 813 (next page). Estimates for cases #1 and #3 are below:

GRAPHIC ESTIMATION

CASE 1 (Q1=Q2, R_L=3k2, cathode bias):
V_{G0CB}=-60V, V_{0CB}=730V, I_{0CB}=125mA
V_A=380V, V_B=1,100V, ΔV=V_B-V_A= 720V
ΔI = 220-40 = 180 mA
P_{IN}= $I_0 V_0$ = 0.125*730 = 91W
P_{OUT}=$\Delta V \Delta I/8$ = 720*0.180/8 = 130/8 = 16.25 W

CASE 3 (Q3, R_L=15k, fixed bias):
V_{G0FB}=-70V, V_{0CB}=790V, I_{0CB}=100mA
V_E=260V, V_F=1,300V, ΔV=V_B-V_A= 1,040V
ΔI = 135-65 = 180 mA
P_{IN}= $I_0 V_0$ = 0.1*790 = 79W
P_{OUT}=$\Delta V \Delta I/8$ = 1,040*0.180/8 = 73/8 = 9.5 W

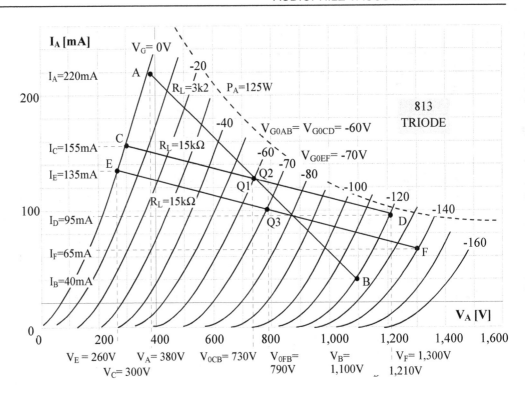

Power supply

The power supply is relatively complex. Separate heater circuits are needed for the two directly-heated power tubes, and the third one for indirectly heated preamp/driver tubes. The 6.3V heater winding must supply 2 x 0.825A for two EL508 power tubes, 2x0.4A for 6BQ6, and 0.37A for one 6DJ8 tube, a total of 2.82A, so as a minimum, the secondary must be able to supply 3A continuous. 4A would be better.

Luckily, since a voltage doubler is used for the anode supply of output tubes, its midpoint can be used to provide half of that voltage for the preamp and driver stages. This voltage is regulated here down to around 400V_{DC}.

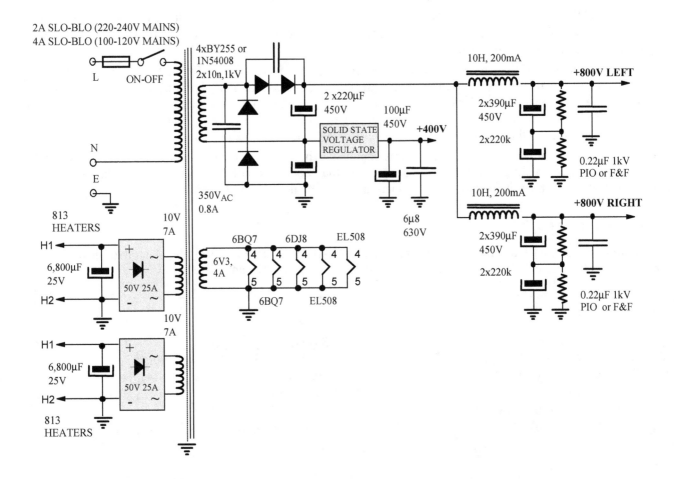

BIG BOTTLES: SET AMPLIFIERS WITH HIGH VOLTAGE TRANSMITTING TUBES

MEASURED RESULTS (15k8 output transformers, 47 H primary inductance, 260Ω primary DCR):

- BW (-3dB): 27 Hz - 23 kHz at 1W, 27 Hz - 23 kHz at 12W
- $V_{OUTMAX} = 13 V_{RMS}$, $P_{MAX} = 21 W$

MEASURED RESULTS (3k2 output transformers, 9H primary inductance, 60Ω primary DCR):

- BW (-3dB): 35 Hz - 39 kHz at 1W, 35 Hz - 28 kHz at 12W
- $V_{OUTMAX} = 16 V_{RMS}$, $P_{MAX} = 32 W$

Fixed biasing trials

In the next experiment, cathode biasing was abandoned and fixed biasing was introduced. The 15k8 output transformers were retained. The maximum output power increased from 21 Watts to 32 watts per channel!

MEASURED RESULTS (fixed bias triode, OPT with 15k8 impedance):

- BW: 17 Hz - 22 kHz (-3dB, 1W into 8Ω)
- BW: 17 Hz - 20 kHz (-3dB, 12W into 8Ω)
- $V_{OUTMAX} = 16 V_{RMS}$
- $P_{MAX} = 32 W$

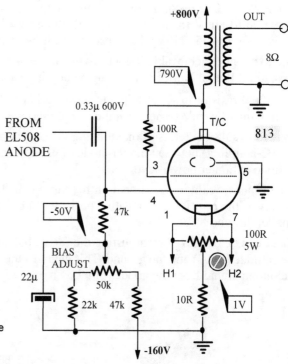

RIGHT: The fixed bias version of the output stage

813 SET amplifier with screen-driven 6EM5 driver stage

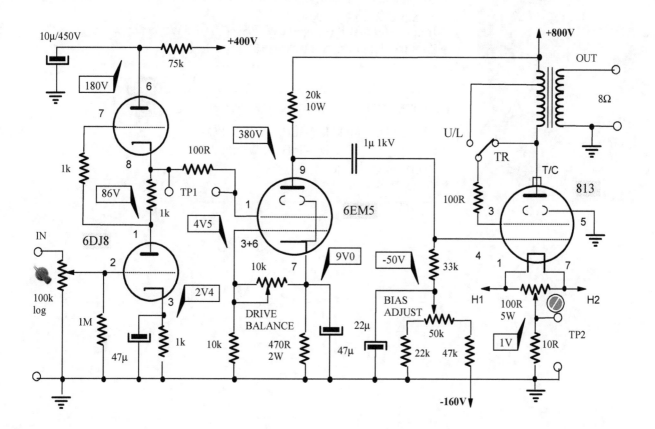

In this version, the second stage (cathode follower) was eliminated, thus simplifying the circuit and lowering the parts count. Secondly, the input stage was changed from a cascode with 6BQ7 to SRPP with 6DJ8 duo-triodes. Finally, 6EM5 was used as the driver tube instead of EL508.

6EM5 was also designed for vertical deflection service in TV sets. The electrical characteristics are similar to those of the European EL508, although EL508 has higher anode and screen dissipation and is generally a superior tube. However, 6EM5 uses a standard Noval socket, and NOS are widely available and thus cheaper.

The first stage has a voltage gain of around 25; the 2nd stage amplifies 2.2 times, and the output stage with triode connected 813 has a gain of 5, for a total overall voltage gain of A=25*2.2*4 = 275!

When the anode supply voltage of the second stage was +400V, with a 10k anode load, the DC voltage on 6EM5's anode was 240V, resulting in maximum output power of around 21 Watts. When the $+V_{BB}$ of the 2nd stage was increased to 800V (as per the diagram), the anode DC voltage increased to 380V. This was above the maximum allowed for 6EM5 (315V), but these robust tubes seemed fine and worked well for extended periods.

Due to a higher achievable voltage swing at the anode of the 2nd stage, the maximum output power increased from 21W to 28 Watts! Notice only a single coupling capacitor in the signal path; the first two stages are DC-coupled.

The 100Ω resistor in series with the screen grid of 6EM5 enables the measurement of screen current, which ranged from 0.3mA (30 mV$_{AC}$ between the test points) with no signal, and 2 mA (0.2 V$_{AC}$) at full output power.

For best results (in terms of maximizing the output power and minimizing distortion), connect an oscilloscope (set on AC-coupling) between pin 9 (anode of 6EM5) and ground, and observe the waveform of the signal. This should be done at higher power levels.

For instance, with 50V$_{AC}$ signal on the input (pin 1, the screen grid), you will get around 110V$_{AC}$ on the anode. However, the positive and negative peaks will not be equal - the positive may be 110V while the negative is much smaller, around 80V.

These can be equalized by adjusting the "drive balance" potentiometer, part of the voltage divider network between the cathode (pin 7) and the ground. This changes the DC voltage (bias) on the control grid (pins 3 and 6) and centers the output signal from the anode (pin 9).

TUBE PROFILE: 6EM5

- Indirectly-heated miniature beam power tube
- Noval socket, heater: 6.3V, 0.8A
- V_{AMAX}=315V, V_{G2MAX}=285V
- I_{KMAX}=60mA, P_{AMAX}=10W, P_{G2MAX}=1.5W
- As a triode: gm=5.8 mA/V, µ=8.7, r_I = 1.5kΩ

MEASURED RESULTS (fixed bias, 40% U/L tap):

- BW: 30 Hz - 30 kHz (-3dB, 1W into 8Ω)
- BW: 25 Hz - 19 kHz (-3dB, 12W into 8Ω)
- V_{OUTMAX}= 18V$_{RMS}$
- P_{MAX} = 40W

THE WAY IT USED TO BE: VINTAGE PUSH-PULL AMPLIFIERS

- Vintage EL84 (6BQ5) push-pull amplifiers
- Heathkit UA-1, UA-2 and AA-61
- PACO SA40
- Stromberg Carlson ASR-444
- Quicksilver 8417
- Dynaco (Dynakit) Mark III
- Fisher SA-1000
- 50C-A10 PP amplifier by YS-Audio
- Bell 2145-A amplifier with combined capacitive and transformer coupling
- New Golden Ear monoblock amplifier
- The Williamson amplifier
- QUAD II
- Eico ST40 and ST70

"The greatest challenge to any thinker is stating the problem in a way that will allow a solution."
Bertrand Russell, philosopher, mathematician, Nobel prize winner

Vintage EL84 (6BQ5) push-pull amplifiers

If you like the sound of EL84 but don't want to fork out thousands of bucks for Manley Stingray, Leben 300C, and similar contemporary amplifiers, there are dozens of vintage models (mostly of mono persuasion) using these tubes in push-pull arrangements. Together with amps using 6V6 lovelies, these were the cheapest amps one could buy at the time, but that does not mean that with a bit of improvement, they cannot sound as good or even better than the modern ones.

Most used tube rectifiers, usually EZ81 or similar, since they are mono units, and all have a phono stage, something many modern units do not incorporate, so there is a significant saving for you, right there. Most of these phono stages use feedback-type RIAA filtering, but some, like Heathkit EA-2 and EA-3, feature passive filtering, which is generally considered better performing and superior sounding.

As with any candidate for improvement, look for amps with a solid, sturdy, non-resonant chassis, as large as possible. There should be plenty of room in the chassis for additions such as chokes, larger filtering caps, and if you want to push the envelope, interstage transformers. Decent (large) mains and output transformers are a must since you are retaining those.

Bogen DB-115 is one such EL84 option, pictured below after our modernization. A fuse holder and a 3-core mains cable were added, new feet, new speaker binding posts, and a few improvements under the chassis. It is a good-looking mono amp with variable damping adjustment. Still, despite having the largest chassis of all amps featured here, due to its topology and the spread-out array of connections and controls, there is no room for a filtering choke or additional capacitors to be (easily) added! Despite their diminutive size, there is enough room inside the chassis of Eico HF-12 to add filtering chokes, something none of these budget units has.

Heathkit EA-2 and EA-3 mono amplifiers offer large transformers and plenty of room on the chassis for additions. The preamp tubes differ between the two models, but the topology is identical.

Dynaco STA-35, a more modern stereo amp, has a phono stage on a separate PCB. If you remove two large can-style capacitors and replace them with smaller modern elcos under the chassis, there is room for a filtering choke and larger filtering capacitors.

ABOVE: Bogen DB-115 with its top cover removed

ABOVE: Eico HF-12 sounds much better than its looks would suggest!

ABOVE: Heathkit EA-2 and EA-3 look very similar but some preamp tubes are different
RIGHT: The inside view of Dynaco STA-35

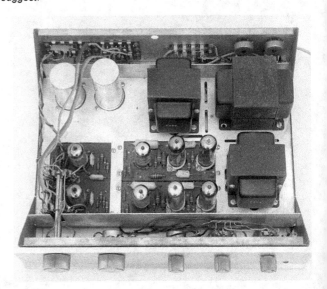

Heathkit UA-1, UA-2 and AA-61

While EA-2 and EA-3 are Heathkit's mono integrated amplifiers (with phono stage and tone controls), UA-1 and UA-2 are their power amp equivalents. The difference between the UA-1 and UA-2 are minor.

UA-1 had a 130W common resistor in the cathode, way too low. With such high anode currents, the output tubes were running at their maximum dissipation levels, making their lives short and stressful, so Heathkit increased the value of R_K to 210Ω. With 13 Volts on the cathodes, the quiescent cathode current was now 13/210 = 62 mA or 31 mA through each tube.

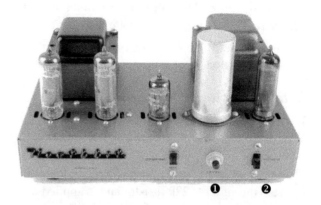

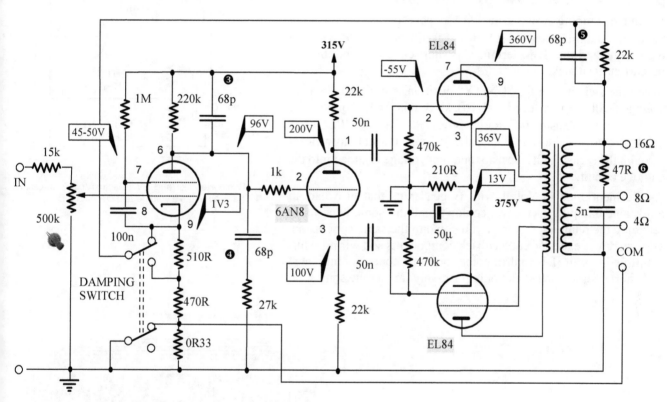

ABOVE: The audio section of Heathkit UA-2

AA-61 was the amp's last reincarnation, with a more modern-looking black & white chassis, instead of a gold one for UA-1 and UA-2.

The chassis is small, but there is plenty of room both on top (between the power and output transformers) and underneath for a filtering choke or anything else you want to add. The -3dB power bandwidth at the rated 12Watt output level is very wide, from 8 Hz to 65 kHz, indicating quality output transformers.

The amps are well designed and built, and apart from adding a choke and larger filtering capacitors in the power supply, the audio section doesn't require any significant modifications.

Compared to poorly designed (from the ergonomic point-of-view) Dynaco amplifiers, all the controls are at the front. The level control wasn't meant to be used as volume control, but it can be; just add a control knob (1). Replace the shonky slider-type power switch (2) and install a rotary one so that you will have two nice knobs at the front.

Three high-frequency compensation and filtering measures were used. The first is the 68 pF capacitor in parallel with the anode resistor of the first tube (3). Then there is the 68pF + 27k RC network between the anode of the input pentode and ground (4), and finally, the 68pF HF compensation capacitor bypassing the 22k NFB resistor (5).

Notice also a Zobel network at amplifier's output, the 5nF + 47Ω RC network between the 16Ω and COM/GND terminals (6).

The time constant of the RC biasing network (210Ω + 50μF) is a bit low, resulting in a relatively high -3dB frequency of around 15Hz. The 50μF elco should be replaced by a 100μ film capacitor for better and cleaner bass. Alternatively, use separate cathode resistors and capacitors, 390Ω and 47μF in each cathode.

You may remove the 15k resistor in series with the signal input and add 100-330Ω resistors in series with screen grids of the output pentodes.

Damping factor control

We included these amps here as a recommendation for a promising modification platform and educational reasons due to their dual NFB scheme.

The standard global negative voltage feedback is taken from the 16Ω tap through the 22k resistor and is applied across the series of the two cathode resistors, 510R + 470R. The feedback ratio is $\beta = R_2/(R_1+R_2) = 0.98/22.98 = 0.0426$ or 4.26%.

When the damping switch is the "MAX" position (as illustrated on the circuit diagram), the 0.33Ω current feedback resistor (between the ground and the COM terminal) is short-circuited, and there is no current feedback.

When the switch is in the "UNITY" position, the global negative voltage feedback is reduced (halved), since it is applied only to the lower cathode resistor (470R), so $\beta = R_2/(R_1+R_2) = 0.47/22.47 = 2.1\%$. However, the output (speaker) signal current now flows through the series 0.33Ω resistor and that voltage is added into the cathode circuit.

As we mentioned in Vol. 1 with Bogen amps that also had an adjustable damping factor, we couldn't find any sonic benefits of reducing the damping factor. If you come to the same conclusion, you can remove the current feedback components and wiring. Thus simplified circuit (DF control potentiometer's contacts eliminated and shorter signal paths) can only positively impact the sonics.

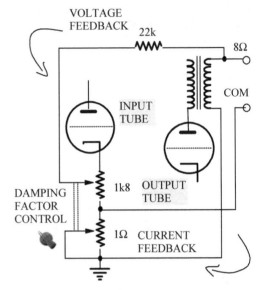

Some Heathkit and EV (Electro Voice) amplifiers have adjustable voltage and current NFB.

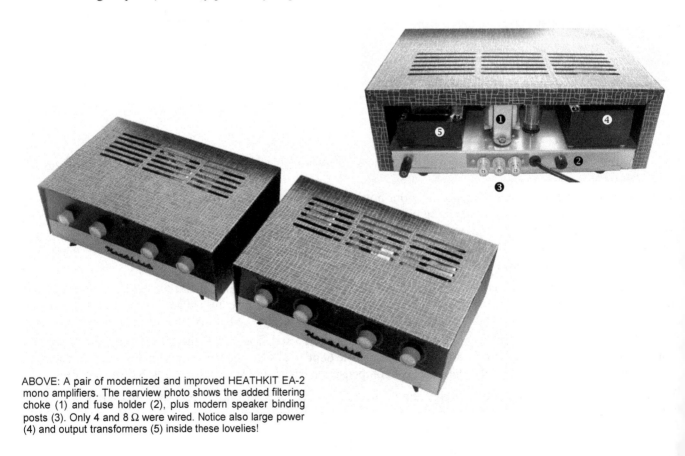

ABOVE: A pair of modernized and improved HEATHKIT EA-2 mono amplifiers. The rearview photo shows the added filtering choke (1) and fuse holder (2), plus modern speaker binding posts (3). Only 4 and 8 Ω were wired. Notice also large power (4) and output transformers (5) inside these lovelies!

PACO SA40

SA40 was PACO's (Precision Apparatus COmpany) entry into hi-fi under its own name in the 1960s. Sometime earlier, the company had bought Grommes, a relatively obscure 1950's amplifier maker. Since SA40 was also its last hi-fi model, it seems PACO had not benefited from Grommes brand name or goodwill. The instrumentation firm (also known for its well-regarded tube testers) quickly abandoned the idea, the field was too crowded already, and transistor amps were taking over. Pity, since SA40 was well made, and it sounds good.

The audio circuit is almost identical to Dynaco MkIII and many other push-pull amps of the era. 7199 pentode section provides all the voltage gain, DC coupled to its triode section as a split-load inverter, with a fixed-biased output stage.

As in Heathkit UA-2, two high-frequency compensation and filtering measures were used. The first is the 68pF capacitor between the 2k2 feedback resistor and the lower output tube's anode; the second is the 47pF + 15k RC network between the anode of the input pentode and ground.

7189 output tubes work as pentodes, with 300V on their screen grids and 395V on the anodes. At the output, bias balance around the fixed -14V grid bias is provided by the 100Ω trimmer potentiometer between two cathodes.

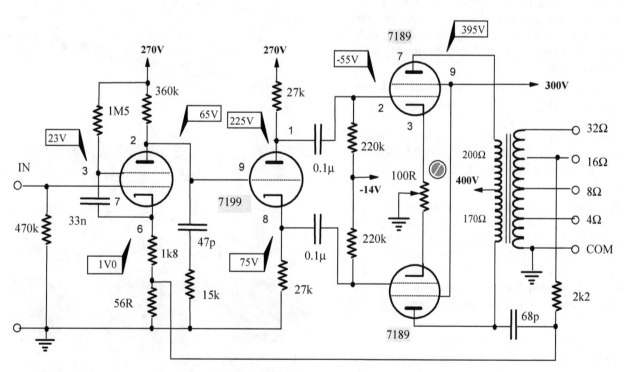

ABOVE: The power amplifier section of PACO SA40 (one channel only)

Notice unequal DC resistances of the two halves of the output transformer's primary (anode) winding. This means it was wound using fast, el-cheapo methods without much sectionalizing. In Vol. 2 of this book, we explained a superior way of sectionalizing and connecting the primary sections, resulting in equal DC resistances without the need for bifilar winding.

Compared to Heathkit EL84 amplifiers, PACO's output transformers are tiny toys, and for that reason alone, Heathkit amps are a much better candidate for modernization and improvement.

Stromberg Carlson ASR-444

ASR444 is the big brother of the similar-looking but lower powered ASR433, which used two 6BQ5 pentodes in each channel for a 12 Watt output. Here, 7027A tubes in pentode push-pull configuration produced 30 Watts per channel, so ASR44 is a much better vintage buy. Of course, 30 Watts from such power tubes working as pentodes isn't much because the output stage is biased to work mainly in Class A operation.

The phono stage was included, of course, since it was a fully integrated amplifier. One 12AX7 was employed in each channel's phono stage, followed by 6AV6 in the preamplifier, 6U8 triode/pentode as the voltage amplifier and phase inverter, and finally a pair of 7027 beam power tubes.

The variety of controls is mind-boggling. Apart from the usual volume, balance, and tone controls (bass and treble), there is also a "master gain" potentiometer, normal-reverse switch, and the output control selector, with five positions: A, B, stereo, mono, and crossover. Why somebody would need anything apart from "stereo" and perhaps "mono" for bridging the amp into a monoblock is beyond comprehension.

There are also switches for turning the loudness control on and off and for rumble filters. Since these aren't and are very unlikely to ever become highly desirable collector items, there is no need to keep them in their original condition.

You have two options: keep all the controls operational or strip them away and keep only the basic modules, the phono stage, and the power amplifier, with only the input selector switch and volume control operational.

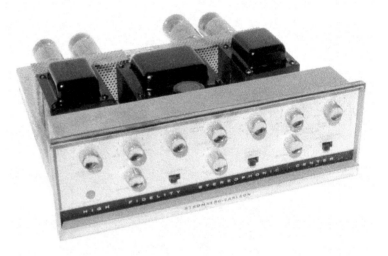

There are too many switches, controls, and contacts in the signal path, which is incredibly long. The shielded cables for all inputs are strapped together into a thick spaghetti bundle and routed across the back, along one side, and then across the amp's front, the worst topology I have seen.

It seems SC engineers had never heard of or seen extension shafts used at the time in quality gear. Using one on the amp's selector switch would have eliminated the whole bundle of thick shielded cables!

The power and output transformers are of decent size and quality, so a more drastic option is to salvage the transformers and tubes and throw everything else in the bin. Then build a better amp on a modern chassis. The golden knobs and trim look incredibly dated, a polite word for tacky or kitschy.

The driver circuit uses a very common topology of the time, popularized in Dynaco and many other mid-fi amplifiers. The pentode provides all the voltage gain, directly coupled to the split load triode inverter (no voltage gain), both in one physical tube (6U8 triode-pentode in this case). There is only one pair of coupling capacitors per channel.

The output cathode currents are $35V/247\Omega = 142$ mA or 71mA per 7027 tube, which is quite high, meaning the output stage is biased to operate mostly in class A, crossing into class B only at the highest output power levels.

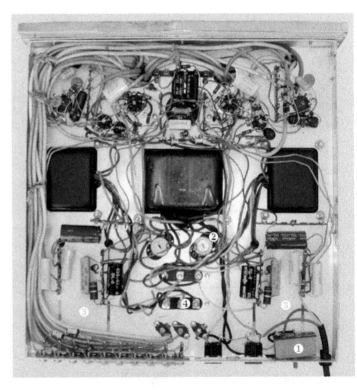

RIGHT: The under-the-chassis view of the restored ASR-444 amplifier:
1. Resettable circuit breaker
2. Hum balance rheostats
3. Output stage cathode biasing components
4. Speaker connections

Indeed, the listening tests confirmed the charming sound of this amplifier (after power supply improvements, of course, outlined on the next page).

Interestingly, the global NFB is taken from its own, dedicated tertiary winding.

Notice how close the screen voltage (+420V_{DC}) is to the anode supply voltage (+450V_{DC}) and the voltage on the anodes themselves (+430V_{DC}). This means it would be very easy to convert the output stage to triode operation. Simply remove the +420V connection from pins 4 and add a 100-220R resistor between pin 4 (screen grid) and pin 3 (anode) of each output tube.

The amplifier's input sensitivity will be reduced by this conversion to the triode operation, and the output power will be lowered as well.

The output transformer's primary impedance will be a bit high for triode mode, but that will reduce distortion.

ABOVE: To keep the height of this sleek amplifier to a minimum, the power tubes were mounted horizontally. This created a severe shortage of real estate on top of the amp, despite its large footprint. Notice how close the output binding posts are to the 7027 hotties! Connecting and disconnecting speaker cables to this amplifier is a hazardous undertaking.

BELOW: Stromberg Carlson ASR-444, the original audio section

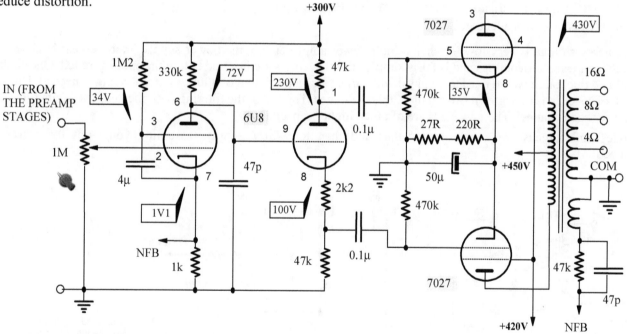

Alternatively, experiment by moving your speaker connection (customarily connected to the 8Ω output) to the 16Ω output instead. That would reduce the reflected load impedance to the primary side.

Notice a very low bypass capacitance of 50μF in the cathode circuit. The time constant it determines together with the cathode resistance of 247Ω is equivalent to the -3dB frequency of the output stage of 12.9 Hz. This is too high for good bass reproduction, so the cathode capacitor should be increased to 100 μF (for -3dB of f_L=6.5Hz).

Improving the power supply

On the positive side, while most amps of the time (late 1950's, the early 60s) did not even have a mains fuse, ASR444 had a resettable circuit breaker. The rest of the power supply is pretty standard or, more accurately, *substandard*.

The output of the voltage doubler, containing a very high AC ripple, is fed straight to the output stages (1). The conventional "wisdom" is that such ripple is present on both output tubes' anodes. Since the ripple propagates through the output transformer's primary in opposite directions, it ends being of opposite phase on the anodes, so it should cancel.

However, such cancellation is not complete even at quiescent conditions, without a signal, due to the always-present slight imbalance in tube parameters and between the two halves of the primary winding.

In dynamic conditions, with the presence of a music signal, the hum (ripple) isn't canceled at all; in fact, it interacts with and modulates the audio signal and creates all kinds of unpleasant sounding harmonics and intermodulation artifacts, making the amplifier sound harsh and irritating.

This sonic degradation is almost always erroneously ascribed to the pentode operation of the output stage. In fact, if properly designed and implemented, a pentode stage can sound as good as triode stages. Pentodes' screen supplies need to be rock stable.

A simple RC filter, 3k9 resistor, and 20 µF elco (2) is grossly inadequate.

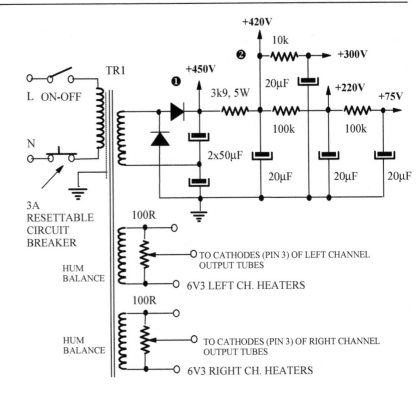

The original power supply.

The energy storage is very low for such a high-power amplifier. So, the power supply improvement is aimed at correcting these shortcomings and is fairly simple. Two LC filters are inserted before the anodes are fed. One choke would be fine, but it would need to be physically large since a minimum of 400mA current rating is needed. Due to a lack of space, it is easier to fit two smaller 200mA chokes. The rest of the filtering is duplicated, a separate filtering chain for each channel. The new elcos are of much higher capacitance.

Anodes and screens are connected together and then to +430V supply (assuming you'd opt for a triode connection).

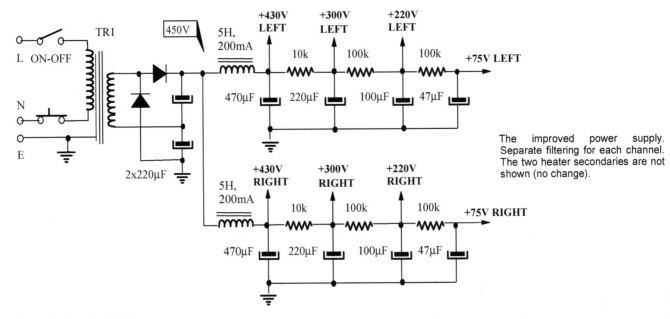

The improved power supply. Separate filtering for each channel. The two heater secondaries are not shown (no change).

Quicksilver 8417

Quicksilver manufactured the "Philips 8417" monoblock amplifiers from 1984 to 1988 when Philips stopped producing the 8417 tubes. Its successor, the "GE 8417" monoblocks, were made from 1990 to 1992. Many owners reported reliability issues, amps blowing fuses, and being unstable. Still, these, for some reason, gained a good reputation and circuit diagrams are online, so, assuming the circuits are correct, let's look at their design.

Two different preamp tubes are used, but only a half of each; the other half is just sitting there, idling. The explanation may lie in the vintage Bogen M120 PA (public address) amplifier, on which this design seems to be based

The 150 Watt Bogen beast had a quad of 8417 driven by a single 7247 duo-triode. This tube has two dissimilar triodes in one glass envelope, one equivalent to 12AX7, the other to 12AU7. Probably thinking that such a rarely used tube is soon to become obsolete (and they were right), the boys from QS simply took a 12AX7 and a 12AU7 and left half of each unused!

8417 beam power tubes

One of the last vacuum tubes developed, the Sylvania datasheet says that 8417 is "a beam-power pentode featuring a T-12 bulb and an octal 6-pin base."

Such lax use of language will frustrate audio puritans. There is no such thing as a beam-power pentode. A tube is either a "beam power tube" or a pentode, but it cannot be both! A pentode has a third (suppressor) grid; a beam power tube has beam-forming plates.

Secondly, an octal socket means that it has eight pins (octo=8), not six! However, looking at the bottom of an 8417 tube, Sylvania's meaning is apparent - two pins are entirely missing, so this is a 6-pin tube on an 8-pin base.

8417s have a very high mutual conductance (around 23 mA/V) and a high amplification factor.

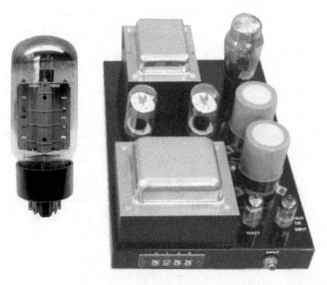

ABOVE LEFT: The 8417 is an elegant-looking but troublesome and unreliable tube!
ABOVE RIGHT: The rear view of Quicksilver 8417 monoblock. Proper topological positioning, the RCA input is right next to preamp tubes, and the output transformer is right in front of the speaker terminals.

TUBE PROFILE: 8417

- Indirectly-heated beam power tube
- Octal socket, 6.3V, 1.6A heater
- V_{AMAX}=660V, V_{G2MAX}=500V
- I_{K2MAX}=200mA, P_{AMAX}: 35W, P_{G2MAX}=5W
- gm=23mA/V, r_I = 16kΩ, bias -12V
- As a triode: μ=16.5, r_I = 660Ω

This means that a very low bias is needed, around -12V, compared to -20 to -30 V for 7027A, 6L6, EL34, and especially the hard-to-drive KT88. This means that 8417 tube is very easy to drive, so with a 2V signal from a typical CD player or DAC, any triode can drive it, even 12AU7 whose μ=17!

8417 beam tetrodes were never popular with designers, primarily because of their poor reliability. To achieve high gm, the screen grids and anodes had to be placed very close together.

Looking at the comparison table on the next page, you'll notice a very high idle cathode current of 146mA, which at maximum power output swings up to 350mA! Since the screen current increases with rising power levels, such high screen currents heat the thin screen wire, which expands and causes internal tube shorts (screen-anode).

The audio section

FIRST STAGE: The voltage drop on the anode resistor is 440-160=280V, so the anode current is 280V/300kΩ = 0.93mA, and the DC voltage at the cathode is 0.93mA*(820+220) = 1V.

The anode resistor is of an extremely large value, meaning the stage's voltage gain is high. On the other hand, its un-bypassed cathode resistor introduces local negative feedback and lowers the gain. Also, the global NFB from the output transformer's secondary to the lower cathode resistor is very strong, further reducing voltage gain.

The gain before the global NFB is $A_1=-\mu R_L/[r_I+R_L+(1+\mu)R_K]$ = -100*300/(63+300+101*1) = -100*300/464 = -65

The DC-coupled 2nd stage is a cathodyne or split-load phase inverter. The anode current is (440V-300V)/23,700 = 5.9mA and the abode and cathode voltage gains are $A_A=A_K=-\mu R_L/[r_I+R_L+(1+\mu)R_K]$ = -16.5*23.7/(5.3+23.7+17.5*23.7) = 0.88

With the output power of 60 Watts, the maximum voltage on the 16Ω tap is $V_{MAX}=\sqrt{(PR)} = \sqrt{(60*16)} = 31V$

NEGATIVE FEEDBACK: β= 220/(220+5,000) = 0.04215 so at full power the AC feedback voltage on the 220Ω cathode resistor is $V_F=\beta*V_{MAX}$ = 0.04215*31= 1.31V

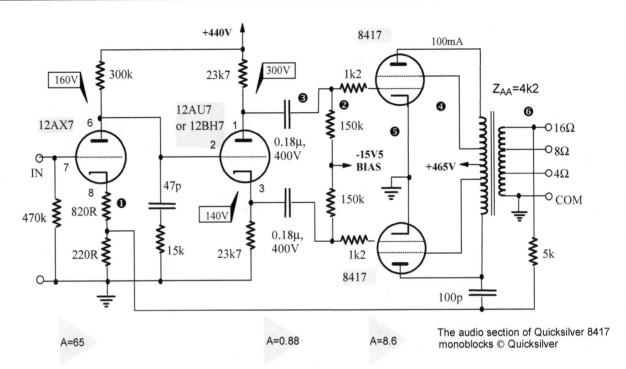

A=65   A=0.88   A=8.6

The audio section of Quicksilver 8417 monoblocks © Quicksilver

TUBE PROFILE: ECC83 (12AX7)

- High μ duo-triode, Noval socket
- Heater: 6.3V/300mA or 12.6V/150mA
- V_A= 300V_{DC}, V_{HKMAX}=180V_{DC}
- P_{AMAX}=1W, I_{KMAX}=8 mA
- TYPICAL OPERATION:
- V_A=250V, V_G=-2.0V, I_0=1.2 mA
- gm = 1.6 mA/V, μ=100, r_I = 62.5 kΩ

TUBE PROFILE: 12AU7 (ECC82)

- Medium μ dual triode, Noval socket
- Heater: 6.3V/300mA or 12.6V/150mA
- V_{AMAX}=300V, V_{HKMAX}= 100V
- P_{AMAX}=2.75 W, I_{AMAX}=20 mA
- TYPICAL OPERATION:
- V_A=250V, V_G=-8.5V, I_0=10.5 mA
- gm=2.2 mA/V, μ=17, r_I = 7.7 kΩ

Modification options

Since half of each input tubes is unused, the modification possibilities are almost endless. You could convert the input stage into SRPP or a common cathode stage with an active load or a constant current sink. A balanced operation throughout is also possible, the 12AX7 stage would become a differential amplifier and phase splitter, and the 12AU7 would work as a balanced driver stage.

We will show the most straightforward option here, paralleling both input triodes (1), which simply means installing wire jumpers between pins 1 & 6, 2 & 7, and 3 & 8. This will increase the gain of both stages slightly.

The maximum grid resistance of 8417 with fixed bias is 100kΩ, so grid resistors of the output stage should be reduced (2). To maintain the lower -3dB f frequency of around 6 Hz, the coupling capacitors must be increased by the same factor (3).

PARALLELED TUBES

REDUCED INTERNAL IMPEDANCE: r_I/N
INCREASED TRANSCONDUCTANCE: gm*N
AMPLIFICATION FACTOR DOES NOT CHANGE
r_I = internal resistance of one tube
gm = transconductance of one tube
μ = amplification factor
N= number of paralleled identical tubes

Since 8417 tubes are rare and matched NOS pairs are next to impossible to find, a separate bias adjustment for each output tube should be incorporated to allow less-than-perfectly-matched tubes to be used. Even if you manage to find a matched pair, tubes age differently, and what is a matched pair now will not be matched after one or two years of operation!

Screen resistors in 220R - 2k range should be inserted in the screen circuits of 8417 tubes. To determine the lowest value, some experimentation and listening tests are needed. This should stabilize the output stage (4).

There is no provision for checking the cathode current through the output stage, so 10R resistors should be installed between the two cathodes and ground (5).

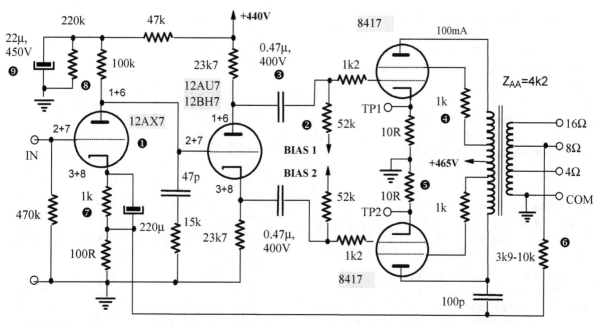

The modified audio stage of Quicksilver 8417

Since the 8Ω tap is more likely to be used than the 16Ω one, the NFB was repositioned. This required the NFB resistor to be reduced to 3k9. Should you wish to reduce the strength of NFB, use the original 5k resistor on the 8Ω tap or even increase it to 10k (6).

The value of the upper resistor in the first stage's cathode was increased to 1k. It was also bypassed for AC signals by a 220μ elco. This will increase the bias voltage, allowing larger signals at the input and increasing the gain of the 1st stage (7).

The anode resistor was reduced to 100k, which will reduce the gain, effectively countering the increase achieved by the previous modification. However, it will improve the sonics (8).

There was no power supply decoupling between the first and second stages. Thus, a series voltage dropper resistor of 10k was installed, together with the associated 22μF capacitor to ground and the 220k bleeder resistor which helps regulate the anode voltage (9). This will reduce the anode supply voltage of the 1st stage a bit, but since it is very high (440V), even a reduction to 400V will not change things significantly.

MOVING THE GLOBAL FEEDBACK RESISTOR FROM ONE IMPEDANCE TAP TO ANOTHER

Most vintage amps have global negative feedback taken from the 16Ω tap of the output transformer, yet most modern speakers are of nominal 4Ω or 8Ω impedance. Ideally, the NFB should be taken from the tap to which the speaker is connected. Here are two formulas to help you calculate the value of the NFB resistor and its bypass capacitor when you reposition the NFB point: TRADE TRICKS

QUICKSILVER 8417 MONOBLOCK EXAMPLE: A 5 kΩ NFB resistor is taken from the 16Ω tap. To change to the 8 ohm tap, its value should be $R_8 = R_{16}/\sqrt{(16/8)} = R_{16}/\sqrt{(2)} = R_{16}/1.41 = R_{16} \times 0.707 = 3.55$ kΩ, so we would use a standard E12 value of 3k9 or E24 value of 3k6.

To move NFB to the 4Ω tap, R_{NEW} needs to be exactly half of the 16Ω tap resistor or 2k5.

If a frequency compensation capacitor is connected across the NFB resistor, its new value should be $C_8 = C_{16}*\sqrt{(16/8)} = C_{16}*1.41$ or 41% larger capacitance!

Operating points of the first two stages and H-K issue of the 2nd stage

The DC voltages aren't marked on the drawing, but we can figure them out since we know the anode supply voltage. To get the DC operating points, we have to draw the load line and construct the bias lines for each stage. To construct the bias line for the 1st stage with ECC83, choose two current levels, say 0.5mA and 2mA, and calculate the bias voltage in each point.

With 0.5mA through the cathode resistors (total of 1,040Ω or 1.04k) the cathode voltage will be $V_K=0.5*1.04 = 0.52$V, which is -0.52V bias, and for 2mA $V_K=2*1.04 = 2.1$V, so $V_G=-2.1$V!

Draw these two points on the graph X and Y), connect them with a straight line, find the intersection of that bias line with the load line, and voila, you have your Q-point for the 1st stage, exactly at $V_G=-1V$ and $I_A= 1mA$!

From the graph, we read that the DC voltage between the anode and cathode is 135V, which means that the anode of the first stage is at +136V above ground.

If we draw the bias line of the 2nd stage (which is almost a horizontal line!), we get Q2 at -7 V bias and around 5.4mA anode current.

-7V bias means pin 3 (cathode) is 7V above the grid, which is at +136V, so the cathode of the second stage (pin 3) is at $+136+7 = +143V_{DC}$ above ground!

The heater-to-cathode limit for ECC82 is only +100V, so the designer has significantly exceeded that limit! This will put the heater-cathode insulation of these tubes at considerable stress and shorten their life.

Thus, we have to elevate the referent potential of the 6.3V heater winding to at least +43V.

Improving the power supply

Issue #1: The first capacitor (1) is too big; 330µF exceeds the limit, around 60µF per rectifier, so we replaced it with a 100 µF one. This decreased the energy storage significantly, so the 330µF elco (2) should be replaced by a 470µF unit to compensate. The two original elcos are mounted vertically outside the amp, retain them to preserve the looks, and install new, smaller elcos inside the chassis.

Energy before: $E=C*V^2/2 = 660*10^{-6}*439^2/2= 64$ Joule

Energy after: $E=C*V^2/2 = 570*10^{-6}*439^2/2= 55$ Joule

Issue #2: The unfiltered output of $465V_{DC}$ for the output tubes is taken *before* the CLC filter. The AC ripple at that point is too high even for a guitar amp, let alone a hi-fi one.

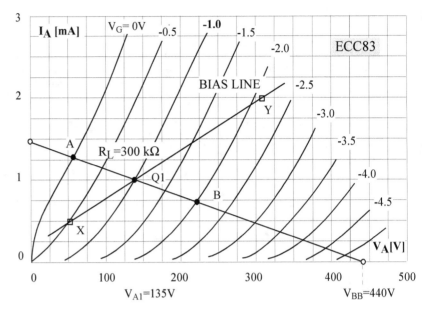

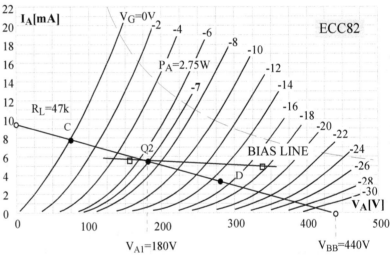

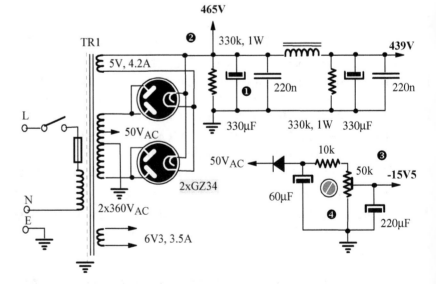

ABOVE: The original power supply

A high inductance choke of 5-10H, good for 250 or 300mA costs less than a pair of matched output tubes or one "hi-end" coupling capacitor, so skimping on such an essential makes no technical, financial or sonic sense! The existing tiny choke becomes the second choke in the CL_1CL_2C filter, and a new choke is inserted here, resulting in much better filtration for all stages, a quieter amp, and improved sonics.

The common bias circuit should be separated for the two power tubes. In the upper position of the bias trimpot the top resistance of the voltage divider is 10k, the bottom 50k, so the output voltage (wiper) is at 50/(10+50) = 0.83 or 83% of the -50V supply or V_{BIAS}= -50*0.83 = -41.5V. That is fine, the output tubes would be fully cut off.

However, at the other extreme, with the pot's wiper or slider at the ground potential, the grids of 8417 tubes would have no negative bias voltage, the DC current through them would surge and be limited only by the capacity of the anode power supply! Within a few seconds, anodes would start glowing cherry red color, tubes would overheat and, unless the mains fuse blows, be destroyed.

Such a high current through the output transformer's primary will overheat its relatively thin wire, which would act as a fuse and fail open circuit at its weakest spot, thus requiring a total transformer rewind or replacement!

Many mains transformers burn out before the output tubes because their primary fuses are of too high a rating (say 3A when 1A fuse should have been used).

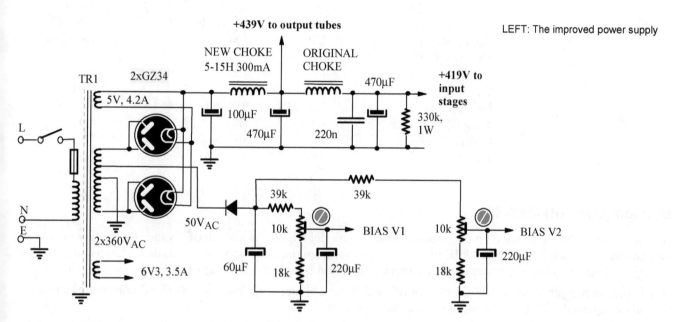

LEFT: The improved power supply

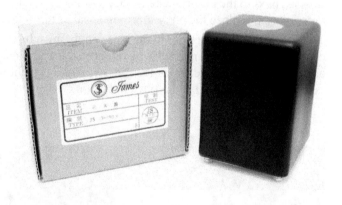

These great-looking James 10H 250 mA chokes (right), made in Taiwan, sell on eBay for US$95 + US$33 airmail (Sept. 2015). If you have an amplifier that is basically sound (good output transformers, decent design, etc.) but has a wimpy power supply with poor filtering and regulation, replace its series resistor or the tiny, grossly inadequate choke with these babies.

Just make sure you have the space on top of the chassis since their footprint is 80x90mm, and the height is 110mm. Each weighs a hefty 3.35 kg!

The DC resistance is only 86 Ω, so their insertion won't reduce the voltage too much (with 200mA, the DC voltage drop will only be $\Delta V=0.2*86$ = 17 Volts.

Replacing 8417 tubes

8417 power tubes are rare, unreliable, and generally undesirable in audiophile amplifiers. Luckily, there are quite a few candidates for their replacement, although all require some changes to operating conditions, either socket rewiring, bias changes, or both.

If 8417 are replaced by KT88, KT90, KT100, or even KT120, no socket rewiring is needed, but higher bias voltage is needed, and a higher gain is required from preamp stages. With 12AX7 at the front, this should not be a problem. Reducing NFB will also increase gain.

7027A are the best replacement, albeit at a reduced output power level. The heater load on the power transformer will reduce by 1.4A, and the anode load on the high voltage winding will also reduce somewhat so that the mains transformer will run cooler and quieter. The distortion will go down; the amp will sound cleaner and more transparent.

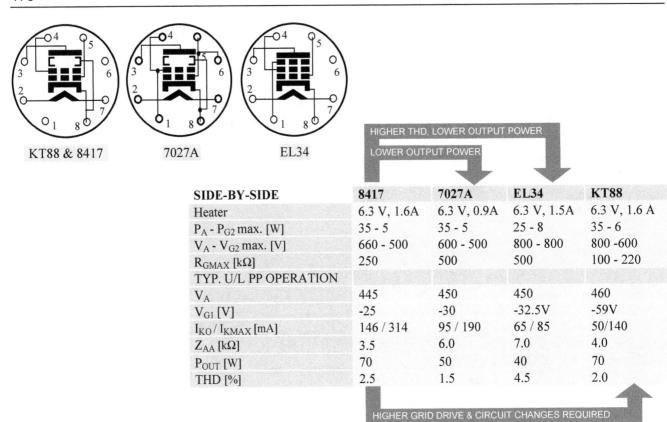

Dynaco (Dynakit) Mark III

In Vol. 2, we analyzed Dynaco's most popular amplifier, model ST-70. As a result of savage cost-cutting measures, Dynaco monoblocks, be they Mark I, II, III, or IV, all suffer from the same deficiencies as their stereo cousin: a tiny chassis, a joke of a choke, inadequate power supply, common biasing for both power tubes and many others.

Not in any particular order, a few simple upgrades that are obvious from the photos. The 2-core mains cable (1) should be replaced with a 3-core one and the chassis grounded.

The internal fuse should be removed and replaced with an external fuse holder in location (2) at the back of the chassis, next to the on-off switch. That will free up some valuable space under the chassis. While you are still on the outer works, replace the poxy screw-in speaker terminals with quality binding posts (3).

The aging selenium rectifier in the bias supply must be replaced by a modern silicon diode (4). The outside-mounted can type multi elco (5) must also be replaced due to its age or retained just for cosmetic purposes and a bank of higher capacity elcos mounted inside the chassis. There is plenty of room (6), but an additional choke must also go there, so you will very quickly fill the tiny chassis up.

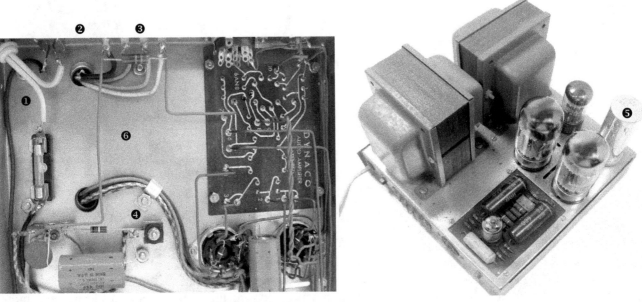

ABOVE: Dynaco MkII, under-the-chassis view of the original wiring

Power supply improvements

Move the fuse to live feed after the power switch, install a 3-core mains-approved power cable or IEC mains socket and ground the chassis (1).

The filtering elcos are rated at $500V_{DC}$, but the voltage is 490V. Before output tubes warm up and start drawing anode current, the voltage goes up to 550V! Any replacement elco will not last very long under those conditions.

Replace the first two filtering elcos with two capacitors in series (2), bypassed with 100-220k resistors. If possible (space permitting, since these are physically large), use a 47µF high voltage film capacitor ($630V_{DC}$ or higher) or a motor start capacitor instead.

Replace the 6k8 resistor with a filtering choke. If possible, replace the existing 1H5 choke with a larger one; 7-10H would be best, or add a smaller choke (2-4H) in series with the existing one (3)!

Always replace aging selenium rectifiers with silicon diodes (4).

The bias voltage is not regulated at all. Install a negative voltage regulator of your choice, a Zener diode as a minimum (5)! Separate the shared bias adjustment into two independent adjustments (6).

Point A supplies the input and phase splitter stages. Instead of an electrolytic capacitor in its filtering circuit, use a high voltage film capacitor such as Solen 47µF $630V_{DC}$ or a cheaper motor start polypropylene capacitor (7). That is the simplest and best value-for-money sonic improvement you could ever make.

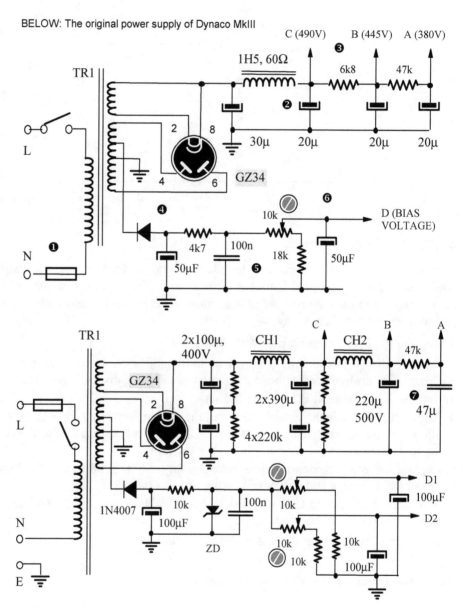

BELOW: The original power supply of Dynaco MkIII

Dynaco MkIII: The modified power supply

The original audio section

The input stage uses a pentode section of the 6AN8 tube. Its anode resistance is very high, 270k, and so is its gain. However, both cathode resistors are un-bypassed, so local NFB is introduced, reducing the gain. Into the 680R and 47R junction enters the global NFB signal from the 16W output tap and the high-frequency compensation feedback from the ultra-linear tap via the 390pF capacitor. There is also an additional local HF NFB to the grid of the 2nd stage via the 12pF capacitor.

The coupling between the 1st and 2nd stages is direct, so there is only one pair of coupling capacitors in the signal path, 0.25mF. There is no provision for balancing the concertina or split-load phase splitter or individual biasing of the output tubes.

As was almost always the case in the 1950s and 60s, the global NFB is taken from the 16Ω output tap, while today, most speakers are of the 4, 6, and 8Ω persuasion.

Dynaco MkIII: The original audio section

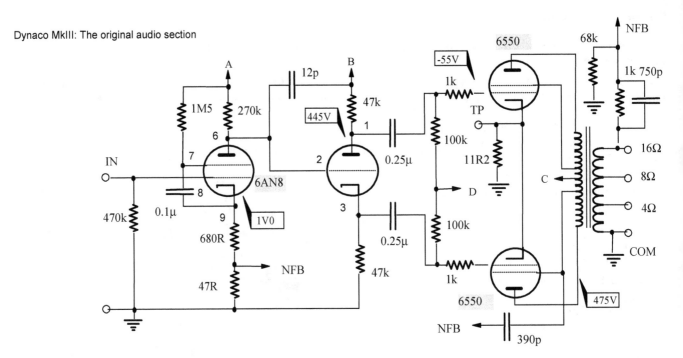

Suggested improvements

The rationale behind this approach is to reduce the gain of the first stage and implement a different kind of negative feedback, a cathode coupled output stage instead of the overall or global NFB. If you don't want to implement cross-coupling in the output stage, at least move NFB to the 8Ω tap (without changing the value of NFB resistors) and reduce its strength that way.

1. Replace the 270k anode resistor with 100k.
2. Remove the NFB from the junction of 680R and 47R. Bypass both with a good quality 220-330µF elco. Completely un-bypassed cathode capacitors usually negatively impact the sonics.
3. Remove 12pF local NFB from the 2nd stage.
4. Add a 10k trimpot in the cathode circuit of the phase inverter and change the 47k cathode resistor to 39k. This will allow for AC (signal) balance adjustment. Observe 1 kHz signals after the coupling caps on an oscilloscope and adjust 10k trimpot until their amplitude are the same!
5. Replace coupling capacitors with better quality 330n film&foil units. Paper-in-oil caps will most likely be physically too large and will not fit on the tiny & cramped PCB.
6. Separate the two joined 100k grid leak resistors and take each to its own bias supply, as indicated in power supply improvements on the previous page.
7. Add 100Ω 3W screen grid resistors to the output tubes.

Dynaco MkIII: The audio section after a few easy modifications.

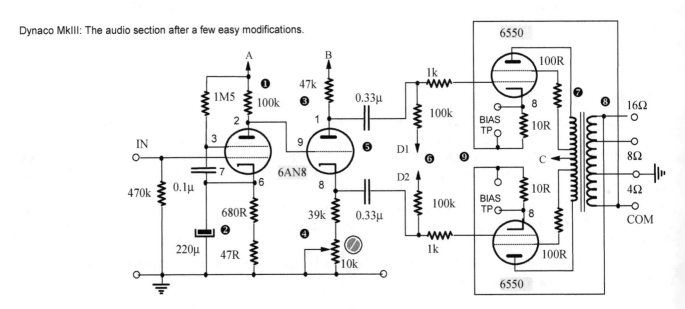

8. To implement partial cathode coupling, remove the 11.2Ω cathode resistor from the output stage and add two 10Ω resistors in series with each cathode (to measure cathode currents). Disconnect the speaker COM terminal from the ground and ground the 4Ω terminal, and cross-couple the tubes as per the diagram.
9. Install two pairs of test points, one pair per output tube, as illustrated. Each 10mV of the DC voltage on these TPs corresponds to 1mA of cathode current. Notice that none of the test points are grounded! They are at a very low voltage with reference to ground and thus safe to touch; there is no danger of high voltage shock.

Fisher SA-1000

Released in late 1963, Fisher K-1000, the kit version, and SA-1000, the factory-built version, sold for US$279.50 and $329.50, respectively. The power output was specified at 75 watts per channel into 8W (music power) and 65 Watts of RMS power, with 0.11% THD. The power bandwidth was 10Hz to 30kHz, and the damping factor was above 30. The input sensitivity was variable by adjusting the position of the input attenuator (separate for each channel); it ranged from 0.7V to 2.75V. The physical dimensions were on a smallish side, 385 x 300 x 200 mm.

You can tell a lot from a vintage amplifier's weight. These beasts weigh almost 32kg, so there is plenty of transformer iron in there. Unfortunately, these days, this criterion does not work, especially with Chinese-made amplifiers. Most of the weight comes from their thick and heavy aluminium and steel chassis, not the usually tiny transformers.

Behind the hinged bottom section of the front panel (1) are the less-often used controls such as the subsonic filter switch and screw adjustments for biasing the output tubes and for AC balance of the phase splitter.

The actual selector of the output tube to be biased is in the top section (2), together with the small bias meter (3) and two input attenuators (4).

Inside, most resistors and capacitors were mounted in-line on one long terminal board and a shorter vertically-mounted board, the construction style more commonly used in guitar amps of the era. McIntosh is another notable hi-fi manufacturer that preferred that method.

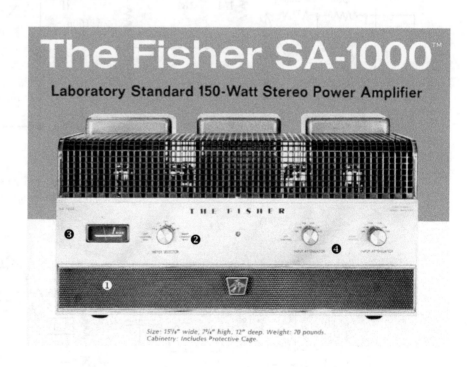

Five elcos are all mounted on a vertical divide, so access is easy. Solid-state rectification was used for high voltage (voltage doubler), bias, and screen voltage supply; see the four discrete diodes on the small terminal strip.

The amplifier was released too late to become a commercial success. The solid-state craze was spreading through the audiophile community, and tubes were becoming passé. Neither the Fisher name nor the use of marketing ploys such as "laboratory standard" and specifying the amp as a 150 watter could change perceptions. Nevertheless, we are not interested in history but in some educational aspects of this design, and there are quite a few.

Circuit diagram

The input attenuator (next page) uses a 6-position single-pole switch (separate for each channel), with positions marked "0", "-3dB", "-6dB", "-9dB", "-12dB", and "OFF". The signal goes through two cascaded RC (high-pass) filters, 2.2nF & 1MΩ, and 5nF & 1MΩ. Normally, there is no need for two such filters, but the circuit forms part of the rumble filter, which is the reason behind this cascade.

The first stage is a cathode follower using ECC808 duo-triode, a low-noise relative of the ECC83. It has the same amplification factor as ECC83 (μ=100), but there is a slight variation in the pinout. Pin 9 is not the center of the heater but a screen between the two triodes.

Thus, ECC808 heaters can only operate on 6.3Volts (pins 4 and 5). ECC808 has a lower anode dissipation than ECC83 (0.5 Watt compared with 1 Watt of ECC83).

The CF output feeds a common cathode stage, which is DC-coupled to the grid of the third stage, the other half of the same ECC83 tube, strapped as a split-load (concertina) phase splitter. The AC balance adjustment is provided in the high DC voltage anode circuit; a safer option would be to move it to the cathode side of the splitter.

The 4th stage is a differential amplifier using ELL80 (6UH8), twin audio power pentode strapped as a triode. The first and the third stage provide no voltage amplification; most of it comes from the second stage, to which the global NFB is applied, and from the third stage. The low bias of 8417 output tubes (-24V) means that very modest voltage amplification is required of each of those two amplifying stages.

Due to the rarity of these amplifiers, we've had no pleasure of subjecting one to the measurement and listening tests and cannot comment on their sonic strengths and weaknesses. From a pure engineering viewpoint, the design fails the Simplicity Test. There are too many capacitors, resistors, filters, switches, and gain stages in the signal path.

Four types of negative feedback are used: the cathode follower at the input, the un-bypassed cathode resistors in the second stage, the local cathode feedback of the output stage, and the global feedback from the speaker output back to the 2nd stage.

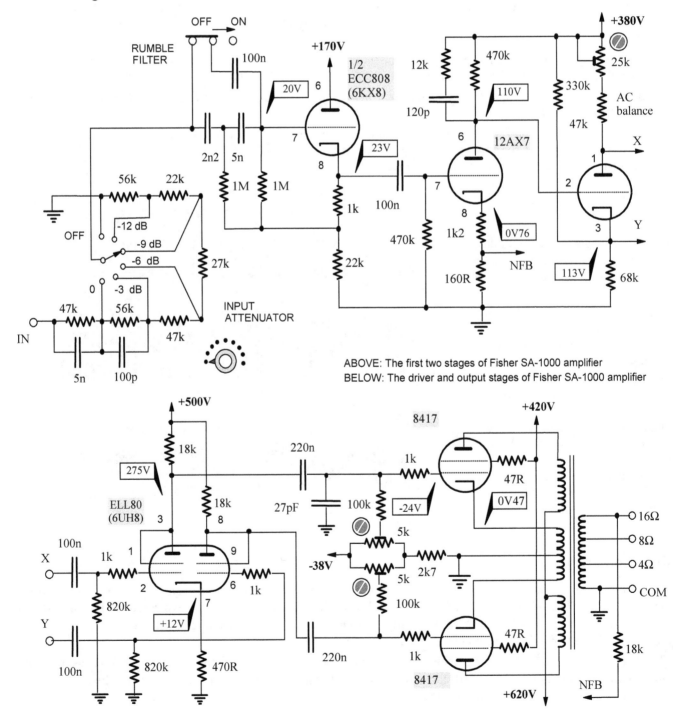

ABOVE: The first two stages of Fisher SA-1000 amplifier
BELOW: The driver and output stages of Fisher SA-1000 amplifier

The output stage uses fixed bias (-24V) and operates in the pentode mode with +620V on the anodes and +420V on the screens.

There is also a dedicated primary winding that provides cathode feedback. It is unknown how strong that negative feedback is since no winding data was provided on the original circuit diagram.

The power supply is unremarkable. Voltage doubling using two solid-state diodes provides the high voltage, no choke in the CRC filtering.

Half-wave rectification for the negative bias supply (-38V) is provided by the third diode, while the fourth diode is a positive half-wave rectifier, this time for the screens of output tubes. This seems the weakest point of the whole amplifier. Only full-wave rectified and regulated DC voltage can keep the sensitive scre

> **KEY FEATURES:**
> - Frequency-compensated input attenuator (five steps)
> - Unusual choice of input (ECC808), driver (ELL80) and output tubes (8417)
> - Cathode follower input stage
> - DC-coupling between the 2nd and 3rd stage
> - Cathode feedback via tertiary winding on the output transformer

> **TUBE PROFILE: ELL80 (6HU8)**
> - Twin audio frequency power pentode
> - 9-pin Noval socket, heater: 6.3V, 550 mA
> - Maximum anode/screen voltage 300/300 V_{DC}
> - P_{AMAX} = 6 + 6 W, V_{HKMAX}=100V, I_{KMAX}=40+40 mA
> - As a pentode: V_A=250V, V_S=250V, V_G= -4.6V
> - I_A=24mA, I_S=4.5mA, gm =6 mA/V, r_I=80kΩ
> - As a triode: gm =6.5 mA/V, μ=17, r_I=2.6kΩ

BELOW LEFT: Some vintage tube data books, such as "Tubes and Transistor Handbook" presented basic tube parameters in a typical circuit each tube would be commonly used in. This short-form summary for ELL80 is from the 10th edition, published in 1963.
BELOW RIGHT: The internal construction of Telefunken ELL80 from two angles.

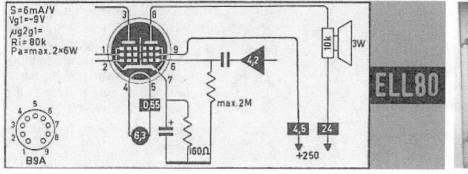

50C-A10 PP amplifier by YS-Audio

Chui Yat Sang (Nelson Chui) is the designer behind YS-Audio in Hong Kong. His amps and preamps are sold under the "Audio Experience" brand name. We had one of his line stages in for analysis; it was well designed and professionally constructed. This early 1990s design by Mr. Chui was chosen not because of the obsolete and hard-to-get 50C-A10 tubes but for its elegant simplicity. First, let's make a historical detour and look at another commercial amplifier that used 50C-A10.

Vintage benchmark: Luxman MQ60 and KMQ60

KMQ60 was a kit version of Luxman's MQ60, a 60 Watt (30+30W) push-pull amplifier. The circuit (quite ordinary, so it will not be reproduced here) reminds one of the British school of valve design from the same era - pentode preamp stage directly-coupled to common cathode phase splitter, driving the fixed-bias output stage. NFB from the 16W tap back to the pentode's cathode resistors. The output triodes are biased at -55V, so a high gain is needed from the first two stages, for which 6267 (an equivalent of the ubiquitous EF86 pentode) and 6AQ8 were used.

The 50C-A10 output triodes (there was also the 6C-A10 variant) were produced only by NEC in Japan, specially for use in Luxman amplifiers. The static parameters are very promising, high amplification factor (double that of 300B), meaning high sensitivity and easier drive, and low internal impedance of only 620Ω!

MANUFACTURER'S SPECIFICATIONS:

- BW: 15Hz - 60 kHz (-1dB)
- THD: less than 0.5% @(1kHz @ 30W) or less than 1% (55Hz-10kHz @ 30W)
- Damping factor: 15, Output hum: 0.7 mV
- Input voltage for rated power: 0.85V
- Rated power: P_{OUT} = 30W+30W
- Weight: 13.8 kg

ABOVE RIGHT: The top view of MQ60. The chassis is so small, and the major components (elcos and transformers) are so close together that very few terminal strips were needed. Diodes, resistors, and capacitors are simply connected in a true point-to-point fashion between sockets, transformer- and capacitor-terminals.

TUBE PROFILE: 6C-A10, 50C-A10

- Indirectly-heated power triode
- Compactron socket (12-pin)
- Heater: 6.3V/1.5mA (6C-A10)
- Heater: 50V, 0.175A (50C-A10)
- V_{HKMAX}: 200V_{DC}, P_{AMAX}=30 W
- V_{AMAX}: 450V, I_{AMAX}:200 mA
- Static data:
- V_A=250V, V_G=-22V, I_0=80mA
- gm=14 mA/V, r_I=620Ω, μ = 8.7

GRAPHIC ESTIMATION

Q-POINT: R_L=2k6, V_{bias}=-25V, V_0=278V, I_0 =90mA
VOLTAGE SWING: $\Delta V=V_B-V_A$ = 433-88 = 345V
CURRENT SWING: ΔI = 160-30 = 130 mA
INPUT POWER: $P_{IN}= I_0 V_0$ = 0.09*278 = 25W
$P_{OUT}=\Delta V\Delta I/8$ = 345*0.13/8 = 45/8 = 5.6 W
DISTORTION (2nd harmonic): ΔV_+=190V, ΔV_-=155V
$D_2 =(\Delta V_+ - \Delta V_-)/2\Delta V*100$ [%] = 35/(2*345)*100 = 5.1 %

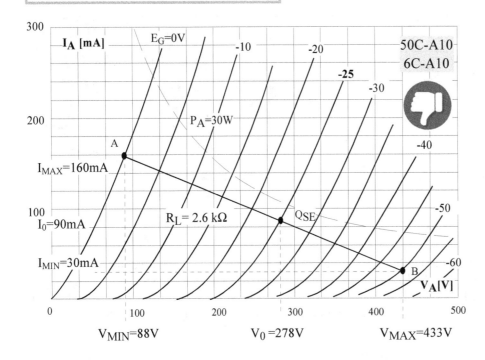

The 50V heater voltage seems a weird choice, until one realizes that the mains voltage in Japan is 100V. So, Luxman simply connected two power tubes' heaters in series across the mains voltage of 100V_{AC}!

Nowadays, such practice is illegal.

Shuguang in China copied the tube and started production, but the tubes are not widely available, are expensive and users have reported a high failure rate.

As a SET, the distortion is above 5% (Q_{SE} on the left), so not a linear tube anyway!

Triode connected F2a pentode has a similar internal resistance of around 800Ω and could replace 6C-A10, or 50C-A10 if heater voltage is changed. Of course, tube sockets and wiring would need to be changed in any case. F2a is more sensitive as a triode, with a μ=17 (double that of 50C-A10), so lower drive voltages would be needed. However, F2a is grossly overpriced due to its use in Shindo amplifiers, so we are back to square one.

Ideally, the Compactron sockets should be retained, so a candidate comes to mind. 6LB6 is a 30W beam tube for horizontal deflection service. It has a top cap, but the main limitation is its low screen rating of only 200V. 6JB5 beam tube has no top cap and a screen rating of 300V, but its plate dissipation is only 15Watts.

Sadly, that means there is no direct replacement for 50C-A10 triodes.

The YS-Audio design

The vintage UTC A-20 transformer acts as a phase splitter at the amplifier's input, and from there onward, the whole circuit is fully balanced. Both stages use identical topology, a differential amplifier and are directly coupled. Using an input transformer and DC coupling between the two stages eliminates two pairs of coupling capacitors from the signal path and their associated time constants and phase shifts, a very clever approach.

Another benefit of this topology is that both balanced and unbalanced inputs can be used. For a balanced input, the + and - connections are simply brought to the ends of the primary winding, neither of which is grounded. The bottom transformer terminal is grounded for unbalanced (RCA) input, and the top one is "hot."

Two different preamp duo-triodes were used, 396A for the first and ECC82 for the driver stage. 396A is a Western Electric name for 5670, which in turn was copied by the Chinese under the 6N3P name, although audiophiles claim that 396A has a smoother sound. We have discussed this duo-triode in detail elsewhere in this book.

The output stage uses cathode bias with a clever way of balancing (compensating) for any mismatch between the two power tubes. As an exercise, assuming perfectly balanced tubes (trimmer potentiometer in the mid-position), calculate the voltage in point X (across the 250R cathode resistor).

Since the two currents in the first stage are identical (differential amplifier), they both flow through the common cathode resistor. We can find the DC current from the cathode voltage through the 1st stage: $1.5V = 2I_1 * 1,500$, so $I_1 = 0.5$ mA. That is a very low current for 5670, a design mistake in my view. At that point, the tube's internal resistance is very high (above 30kΩ), and the amplification factor is very low, around 24 (instead of 36 or more). See the μ-r_I-gm curves for 5670 elsewhere in this book. Thus, the amplification of the stage will be low, only $A_1 = -24*27/(30+27) = -10.5$!

We don't know the cathode voltage of the second stage, but we know that the voltage drop the anode current makes on the 27k anode resistor is $285-198 = 87V$, so $I_2 = 87/27 = 3.2$ mA. The cathode voltage of the 2nd stage is thus $V_K = 2I_2*15 = 96V$. Since grids are at 85V DC potential, the grid-to-cathode (bias) voltage of the 2nd stage triodes is $85-96 = -11V$!

Without consulting the μ-r_I-gm curves for ECC82 for precise values of parameters, the voltage amplification of the 2nd stage is roughly $A_2 = v_{OUT}/v_{IN} = -\mu*R_L/(r_I + R_L) = -17*27/(7.7+27) = -13.2$!

The primary impedance of the output transformer wasn't specified, so we cannot estimate the voltage gain of the power stage, but we aren't interested in 50C-A10 tubes anyway, so our analysis will stop at this point. You can use the same topology (the first two stages) with many power tubes of similar sensitivity (bias -25 to -40V).

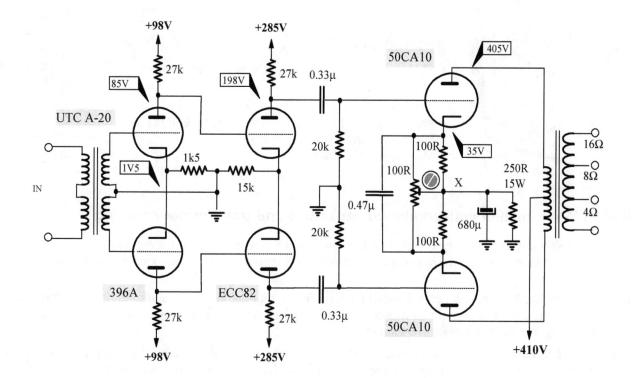

50CA10 P-P Power Amplifier, audio section, © YS-Audio

CALCULATING THE IDLE CATHODE CURRENT OF THE OUTPUT STAGE

There are two ways to solve this problem, the exact way and the approximate fudging. First, the exact way. We draw the circuit assuming the wiper is in the middle position, and the tubes are perfectly balanced, so $I_{K1} = I_{K2} = I_K$. The equivalent resistance of 50 and 100Ω in parallel is 33.3Ω.

Now we write the voltage loop equation as $35V = I_K*33.3 + 2I_K*250$, or $35 = I_K*533.3$, and from here we calculate $I_K = 65.6$ mA.

The voltage in point X is $2I_K*250 = 32.8V$.

The quick way is to disregard the 50Ω and 100Ω parallel combination (33.3Ω equivalent), since it is much smaller than the other 250Ω cathode resistance. This means we assume that the voltage in point X is actually 35V (the voltage on the cathodes). In that case $2I_K \approx 35V/250Ω = 140mA$, so $I_K \approx 70$ mA. The relative error is only $\varepsilon = (70-65.6)/65.6 = 6.7$ %!

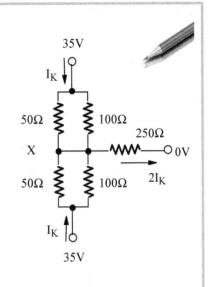

The way it used to be: UTC A-20 transformer

A20 is a universal multi-impedance input transformer. This vintage model hasn't been in production for decades, but salvaged units are still available on eBay. There are also similar vintage models by Sescom, Triad, and other transformer makers of the past.

In audio dBm (also called dBmW) is referenced to the 1 mW of power that a sine source of 0.775 V_{RMS} dissipates on a 600Ω resistive load. Don't confuse it with dBV (or simply "dB"), the unit we have been using for voltage ratios such as amplification and attenuation factors.

The reference power level for dBm is $P_0=0.001$ W, so P [dBm] = 10log(P/0.001). Since UTC A-20 maximum power level is +15dBm, this means 15 = 10log(P/0.001). We can calculate P in watts from here, log (P/0.001) = 15/10 = 1.5, or $P/0.001 = 10^{1.5} = 31.62$, so finally P = 31.6 mW. This is truly a flea-sized transformer!

SPECIFICATIONS:

- Made in USA by UTC (United Transformer Corporation)
- Primary: 50, 125/150, 200/250, 333, 500/600 Ω
- Secondary: 50, 125/150, 200/250, 333, 500/600 Ω
- Frequency range: 10 Hz - 50 kHz
- Max. level: +15 dBm

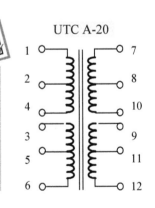

Bell 2145-A amplifier with combined capacitive and transformer coupling

Bell Sound Systems, Inc. was one of the most prominent USA tube gear manufacturers of the early 1950s. Unlike Brooks, they had survived into the next decade, but they had to compete with even more budget amplifier makers in the sixties.

In the early 1950s, 2145-A was Bell's flagship model. The circuit diagram of the audio section is redrawn here the same way it was drawn by the manufacturer. While the driver and output stages are drawn in the conventional (mirrored) way, the input stages seem a total mess.

After passing through the decoupling capacitor (0.25μF) and the 100k volume control pot, the signal goes through the 6k8 grid stopper resistor to the grid of the bottom triode (1/2 of 12AX7 tube). Its cathode is at +1.4V, and its anode is at +135V. The anode resistor has a high value (220k). So, a typical high gain common cathode stage with an un-bypassed cathode resistor introducing a mild local NFB.

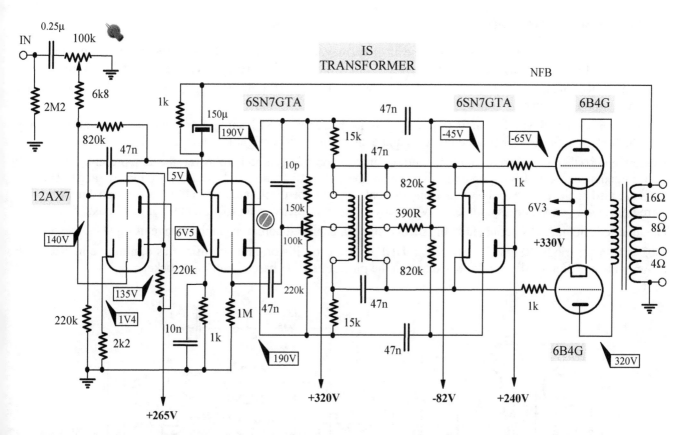

ABOVE: Bell 2145-A push-pull triode amplifier with combined capacitive and transformer coupling

From the anode of the bottom triode, the signal goes straight to the grid of the upper half of 12AX7, whose anode is at the anode supply voltage (+265V), no anode resistor, and whose cathode is at +140V. So, it's a cathode follower whose output is capacitively coupled to the grid of the upper half of a 6SN7 tube. Likewise, its anode is capacitively coupled (47n) to the grid of the cathode follower driver stage (upper half of another 6SN7). Its output also feeds the primary of the IS transformer via a 15k resistor.

A resistive divider network spans the two 6SN7 anodes, from where the signal goes via another 47n capacitor to the grid of the bottom 6SN7 triode, which means the two triodes form a floating paraphase inverter.

From the anode of the bottom triode, the signal then feeds both the other end of the IS transformer's primary and the grid of the bottom 6SN7 in the balanced cathode follower stage.

Notice bypass capacitors between the IS transformer's primary and secondary. The capacitive coupling is operational in Class A₁. Once larger signals cross into class A₂, the transformer coupling takes over.

BELL Custom HIGH FIDELITY AMPLIFIER
with complete Remote Control
Model 2145-A

This unique, new Bell High Fidelity Amplifier adds the master touch to any type built-in or custom audio system. Its remote controls, in an attractive unit for table or console mounting, are amazingly selective. Include single selector for phono—AM, FM radio (or TV), all domestic and foreign records, 78, 45, and 33-1/3 rpm; also compensated volume control; plus smooth, continuously variable bass and treble with boost and cut. All-triode amplifier, 30 watts maximum, reproduces sound at all levels, naturally clear and life-like.

Frequency response plus or minus .25 db to 20 to 30,000 cycles. Distortion at normal listening levels is less than .2 of 1%. Elements can be separated 5 to 25 feet with only .1 db loss at 20 kc. All inputs (3 phono, 1 mike, 2 radio) connect directly to the main amplifier. For more details, write.

Also Ask About Bell's lower cost Hi-Fi Amplifier Model 2122-A.

 BELL SOUND SYSTEMS, INC.
599 Marion Road, Columbus 7, Ohio
Export Office: 401 Broadway, New York 13, N. Y.

Bell advertisement from March 1952, © Audio magazine

New Golden Ear amplifier

This design was published in the Jan 1954 issue of *Audio* magazine. Apart from its excellent measurements and sonic performance (I have never heard a 6AR6 amplifier I didn't like!), it also makes a very educational case study.

A pair of 6AR6 lovelies will deliver 20 Watts in Class A triode push-pull operation, while 6L6, 5881, and KT66 tubes will produce around 15 watts.

The amp was designed with two inputs, so it is ready for balanced or XLR input. For unbalanced operation, simply omit the bottom 470k resistor (marked with *) and connect that grid straight to GND.

The first audio stage is the cross-coupled inverter. Since its input is a cathode follower, the amp presents a very high input impedance to any driving preamplifier, which means low distortion and extended frequency range. The two goals are also achieved due to the voltage and current negative feedback inherent in the cross-coupled inverter. For more information on the cross-coupled or Van Scoyoc inverter, please refer to Volume 2 of this book.

The inverter's output is DC-coupled to the second stage, a balanced common cathode stage using 9002 triodes.

9002 is a single triode designed to work as a UHF detector, oscillator, and amplifier. It's diminutive in size, with 17mm wide and 35mm tall glass bulb, excluding the pins of its B7G base (7-pin mini socket).

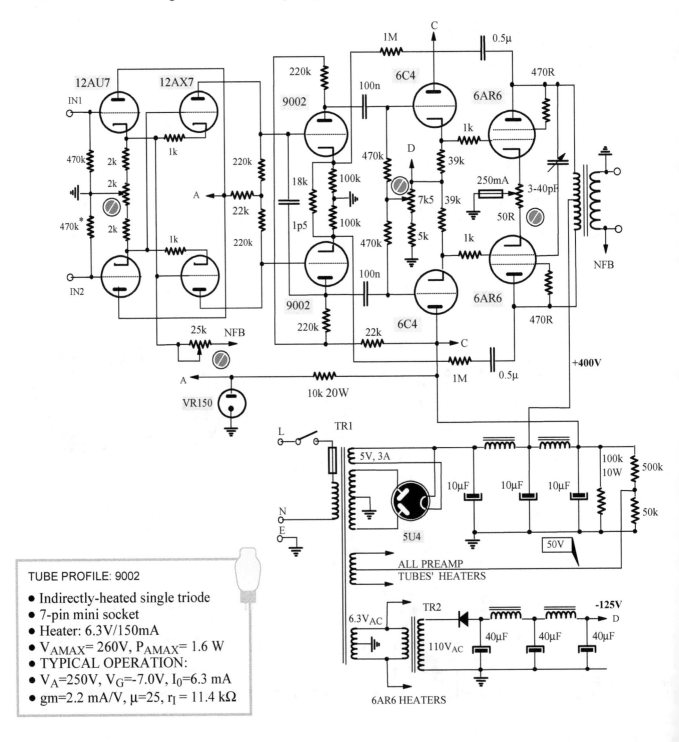

TUBE PROFILE: 9002

- Indirectly-heated single triode
- 7-pin mini socket
- Heater: 6.3V/150mA
- V_{AMAX}= 260V, P_{AMAX}= 1.6 W
- TYPICAL OPERATION:
- V_A=250V, V_G=-7.0V, I_0=6.3 mA
- gm=2.2 mA/V, μ=25, r_I = 11.4 kΩ

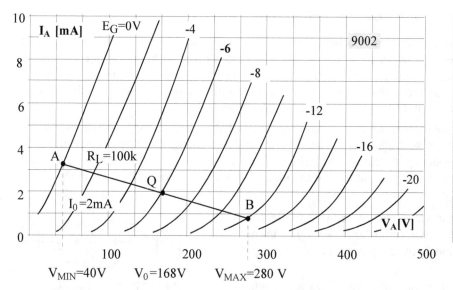

12AU7 can be used instead of 9002 triodes, but with a slightly reduced stage gain (12x versus 17x with 9002). 6SN7 can also be substituted, but the upper-frequency extension may be reduced due to its higher parasitic capacitances.

The operating point in the graphic estimation above is not that of The Golden Ear amplifier; it is simply a general investigation of the 9002 triode's capabilities. The internal resistance is relatively high, and its transconductance is low. There are no substitutes for tube rolling, and since the distortion with a modest anode voltage swing is relatively high (above 3%), we would not recommend this tube; there are much better choices around, such as ECC40, E86C, 6SN7, and 6DN7.

The output of the second stage is capacitively coupled to a balanced cathode follower using a pair of 6C4 single triodes (half of 12AU7) per channel. You should use one 12AU7 nowadays instead. Why did Joseph Marshall use so many single triodes is mind-boggling when duo-triodes were widely available.

The power supply

Due to predominately DC coupling between the stages, the cathodes of tubes are at high potentials, so the heater secondary is not grounded but connected to a 50k/500k voltage divider.

The anode supply voltage to the cross-coupled inverter is stabilized with a $150V_{DC}$ voltage regulator tube. This tube has a very low impedance for AC hum and, as such, is a very effective decoupling device.

The designer did not mark any DC voltages on the diagram or the values of chokes, for that matter, a major faux-pass.

Notice an example of our favorite trick, back-to-back transformers. The designer did not have a 100-120V secondary winding for the bias voltage, so he piggybacked a small filament transformer (110V to 6.3V) onto the 6.3V winding supply heater power to the two output tubes (6AR6) and obtained the 110V that way. After rectification and double CLC filtering, that circuit supplied $-125V_{DC}$ to the cathodes of cathode followers.

The values of the three elcos in the HV section are too small, only 10μF each, which means the energy stored in this power supply is very low. The first elco should be 47μF and the next two 220-330μF! How the designer achieved what he claimed as a decent bass with such a weakling of a power supply is questionable.

The fuse "protection" in the cathode circuit

In fixed (external) biased output stages, if the bias supply fails, output tubes would conduct fully and quickly overheat and be destroyed. The designer thus added a 250mA fuse in the common cathode circuit between the slider of the DC balancing potentiometer (50Ω) and ground. Soon after the sum of the two cathode currents exceeds 250mA, the fuse would blow, and the current would be interrupted. So far, so good.

However, the anode supply voltage would still be present. The electrostatic field established between the anode and cathode would start pulling electrons out of the cathode, and the process known as cathode stripping would start. If discovered early, for instance, during a listening session, where a blown cathode fuse would result in no sound at all, and if the amp is quickly turned off, all would be well. Otherwise, if such a regime goes on undetected, the output tubes would be permanently damaged.

The use of negative feedback and neutralization techniques

The local NFB in this design is inherent in the cross-coupled inverter, as mentioned, but also notice un-bypassed cathode resistors of 9002 triodes. There are two global NFB loops. The first is taken from the output transformer's secondary back to the grid of the bottom triode in the inverter stage (and the cathode of its upper triode pair). The second type of feedback is two balanced loops that capacitively couple the output stage's anodes with the second stage's cathodes brought back to the aforementioned un-bypassed cathode resistors.

One of the design goals was to "eliminate, neutralize, or minimize all positive feedback loops which might cause oscillation, regeneration, hangover or other transients." The inverter and the cathode follower stage needed no intervention to improve the high-frequency response and achieve that design goal; both had an excellent HF response. The neutralization measures to compensate for the dreaded Miller effect were introduced in the 2nd and the output stage.

A fixed 1.5 pF capacitor couples the anode of the lower 9002 triode with the grid of the upper one and a variable (adjustable 3-40pF trimmer capacitor operates in the same fashion between the anode of the upper output tube and the grid of the lower one.

One consequence of the Miller effect is the reduction of gain at higher frequencies, limiting the upper -3dB frequency of triode amplifiers. Since that capacitive coupling from the anode back to the grid due to the Miller effect is negative feedback, it can be neutralized by using positive feedback. We need to provide feedback current, equal in magnitude but opposite in phase, to cancel the feedback current through the Miller capacitance. In push-pull amplifiers, that is easily done by capacitively cross-coupling the grid of the upper tube to the anode of the lower tube and vice versa since their signals are always of the opposite phase. This positive feedback increases gain at higher frequencies.

As the frequency increases, the reactances of C_{F1} and C_{F2} decrease, causing more signal voltage to be fed back, so the shunting action of parasitic capacitances C_{P1} and C_{P2} is compensated for.

If there are phase-shift problems and the NFB becomes positive, the opposite will happen to the neutralizing circuit, so the two capacitive currents between anodes and grids will always be of opposite phase!

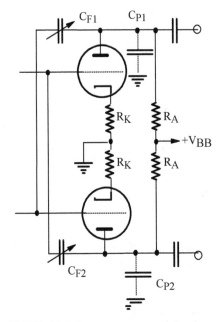

ABOVE: High frequency compensation by capacitive cross-coupling

The Williamson amplifier

This amplifier design by D.T.N. Williamson appeared in articles in April and May 1947 editions of the Wireless World magazine, followed by an updated version with minor changes in August 1949 and additions in October and November 1949. The first version used four L63 singe triodes, while the second version specified preamp tubes as 4xL63 (or 6J5) or two 6SN7 (or B65) duo-triodes.

Williamson's design was not novel even in its day. There is nothing in it that hadn't been seen before, except, perhaps, the triode connection of the output tubes. However, it combined a few clever design choices, resulting in a relatively simple yet (for the time) well-performing package.

It eliminated multiple interstage transformers so widely used in the 1930s and 40s and the associated HF resonant peaks and other irregularities. It also DC-coupled the first two stages and thus eliminated the low-frequency phase shift that an additional RC-coupling would have caused. This was important because Williamson also included the output transformer into the global NFB loop, and every phase shift inside such loop increases the likelihood of oscillation at frequency extremes.

The phase shift caused by the output transformers was especially critical, demanding transformer specifications far exceeding anything seen before this design. The primary inductance was specified as greater than 100H, while the leakage inductance had to be under 33mH, meaning the quality ratio of the transformers was above 100/0.033= 3,030!

The output transformers, made by the British Partridge Company, had two identical coils, each with five primary and four secondary interleaved windings. They were hermetically sealed and potted, as was the practice at the time. The plate-to-plate (anode-to-anode) impedance was 10kΩ (a load of 2.5kΩ on each output tube).

A transitional phase-shift network consisting of R26 and C10, which was previously recommended as a temporary measure, has been added as a permanent feature to increase the margin of stability at high frequencies.

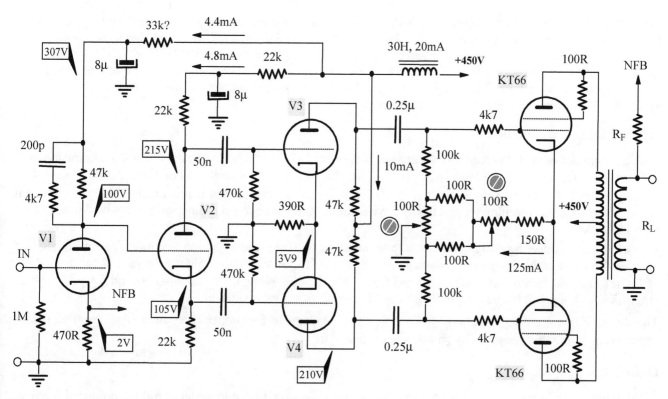

The audio section of the Williamson amplifier, the new version (1949)

The first stage is a common cathode amplifier directly coupled to a split-load inverter. Since that type of inverter provides no voltage amplification, there was not enough gain to drive the grids of output tubes directly, so the designer added another balanced common triode stage. Although the voltage gains of the first and third stages were relatively modest (low-value anode resistors and low-to-medium mju triodes used), there was more than enough gain, so cathode bypass capacitors were omitted. This introduced a mild local NFB in both stages, improving linearity and reducing distortion.

Discrepancies and inaccuracies on the original circuit diagram

The original circuit diagram specifies the DC current of the 1st stage as 4.4mA and the anode supply voltage (top of its 47k anode resistor) as 320V. A simple calculation, multiplying the current by that resistance, gives us the DC voltage drop across that resistor as 47*4.4=207 Volts. When the specified anode voltage of 100V means the anode supply voltage is 307V. Close enough, you might argue, but still a bit sloppy.

The decoupling resistor was specified as 33k, but with 4.4mA flowing through it, its other end should be at 33*4.4= 145V above 307V or at 452V. The HV supply is 450V, and we still have a voltage drop across the choke, at least 5-10V (depending on its DC resistance).

The DC current through the 2nd stage was specified as 5.25mA, but notice the cathode at 105V, meaning the DC current is 105/22 = 4.8mA, not 5.25mA!

Since 215V on the anode was specified, the voltage drop across its anode resistor must be 4.8*22=105V, meaning its anode supply voltage must be 215+105= 320V. When a voltage drop on 22k decoupling resistor is added (4.8*22 = 106V), we get 426V, for the same point for which the 1st stage calculation yielded 452V, a discrepancy of 26V.

Perhaps the anode characteristics of L63 or 6SN7 tubes (both yield the same results) will clarify the discrepancies. Assuming 4.4 mA of anode current is correct for the 1st stage, for -2V bias, we get V_{AK}= 90V. The diagram shows 100-2 = 98V, so close enough.

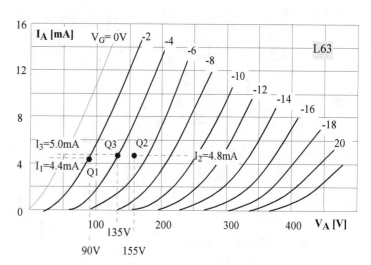

With 10 mA through the 390Ω common cathode resistor of the 3rd stage, the cathodes are at 3.9V to ground, so the bias is -3.9V and 5mA flows through each tube (Q3). The voltage drop across the tube should be around 135V. However, if we calculate the voltage drop across its 47k anode resistor, we get 47*5 = 235V, which, when subtracted from 445V after the power supply choke, gives us the anode voltage of 210V, as marked on our diagram (no stage 3 DC voltages were specified in the original article!) Again, a very significant and frustrating contradiction, 210V versus 135V, so something is definitely amiss here.

AC conditions and issues

The time constant of the first coupling RC network is $\tau_1 = 50nF * 470k\Omega = 23.5$ ms, so the lower -3dB frequency is $f_{L1} = 1/2\pi\tau_1 = 6.8$Hz, which is too high. Also, notice that the time constant of the second coupling RC network is $\tau_2 = 250nF * 100k\Omega = 25$ ms, so $f_{L2} = 1/2\pi\tau_2 = 6.4$Hz, which is also too high, but is practically identical to the -3dB cutoff of the 1st stage, meaning we practically have a double pole in the transfer function and doubled phase shift at such frequency.

A much better engineering design practice is to separate the two frequencies, so the first modification should be to increase the value of the first pair of coupling caps from 50nF to 330nF and the value of the second pair from 250nF to 330nF. The grid resistors should be changed from 100k to 180k! Now, $f_{L1} = 1/2\pi\tau_1 = 1$ Hz and $f_{L2} = 1/2\pi\tau_2 = 2.7$ Hz, which will improve the bass reproduction and the stability at low frequencies.

The decoupling capacitors were only 8μF, which was inadequate for proper reproduction of low frequencies; anyone replicating this design today should use at least 47μF.

QUAD II

Judging purely by its design, it is hard to see why QUAD II became such a legendary product, produced for almost two decades (from 1953 to 1970). A wimpish power supply, pentode operation of all tubes, and double negative feedback are three strikes in the books of triode purists.

The global NFB is taken from a dedicated part of the secondary winding (section Q-P) back to the cathodes of the two input pentodes, configured as a floating paraphase inverter. A tertiary transformer winding is between the cathodes of KT66 output beam power tubes, providing cathode feedback. The output stage is not unity coupled, only partially cathode coupled, and works only in class A, so the maximum output power is relatively (for a PP amp with KT66) low, only 15 Watts.

Ultimately, it is hard to think of even one good reason to choose to go to all this trouble - imperfect phase splitter, push-pull output stage, and high parts count - when one relatively simple single-ended triode amplifier would easily achieve this output power and significantly better sonics! Alas, this was a brainchild of the early 1950s, when SET amps weren't in fashion anymore (or not yet!)

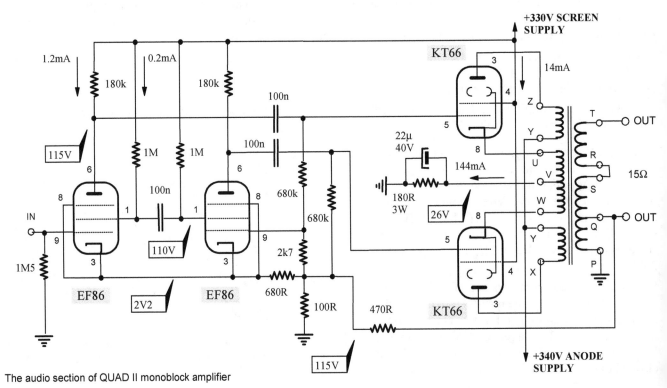

The audio section of QUAD II monoblock amplifier

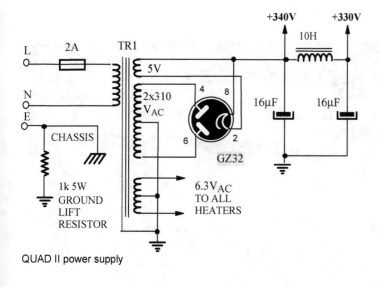

QUAD II power supply

The power supply is QUAD II's weakest point, with very low capacitance and stored energy. Furthermore, the anodes of the output stage are supplied by a high-ripple DC voltage taken before the choke. This results in diabolical levels of intermodulation distortion in dynamic conditions (with the audio signal present).

The first elco should be increased to 47μF, the maximum allowed for GZ32 rectifier, the second elco to 100-220μF.

The self-biased output tubes share the 180Ω cathode resistor and its bypass capacitor, so closely matched tubes are mandatory!

Notice the 3 Watt rating of the cathode resistor, which dissipates $P = V^2/R = 26^2/180 = 3.8$ Watts! No wonder this resistor is usually the first to burn out in these vintage amps! Even a 5 watter would be highly stressed out here, a 7W rated resistor should be substituted!

Negative feedback

The output transformer winding and voltage ratios give us a clue about the strength of the negative feedback. Since $N_{ZX}/N_{PQ}=28$ (the whole primary versus the feedback tertiary winding) and $N_{UW}/N_{PQ}=3.1$ (the cathode winding versus the feedback tertiary winding), the cathode feedback is $N_{UW}/N_{ZX}=3.1/28 = 0.11$ or 11%.

The tertiary voltage as a percentage of the anode-to-anode voltage is 1/28 = 3.57%.

A few bloopers

The grid resistor's value (680 kΩ) is way too high; the datasheet warns that it should not be larger than 500 kΩ! Secondly, notice the absence of screen resistors (between pins 4 of the KT66 tubes and the 330V screen supply). Again, the datasheet warns that "for the prevention of parasitic oscillation, a series resistor of 100-300 Ω should be connected close to the screen grid terminal of the valve socket."

And finally, as for the common cathode resistor of KT66 tubes, the datasheet also says, "The use of common auto-bias resistor is not recommended." Perhaps QUAD's highly esteemed designer, Sir Peter Walker, wasn't such a genius after all?

The output tube options

While the original GEC KT66 tubes are rare as hen's teeth these days, there are currently produced copies, brands such as JJ, Gold Aero, Genalex, Tungsol, and Shuguang. Alternatively, you can modify the output stage to work with a variety of other octal tubes. The use of 6L6 will result in the output power reduction of 25%, plus these aren't very linear tubes anyway, so that option does not seem rational. The cleaner and more detailed sounding 7027A would work well, as would EL34 pentodes. Both require only minor rewiring.

The best option is to rewire the output sockets for 6AR6 lovelies. They are robust, reliable, well-matched, and much better-sounding tubes.

Since vintage QUAD II specimens are pretty expensive to purchase, they are not recommended as an improvement platform, so we will not go deeper into various modification details. After studying this book, you should be able to assess various improvement paths should you find yourself in possession of a pair of QUAD IIs.

The modern descendants

Such is the allure of QUAD II amplifier that various versions are being manufactured again, in China, where else. The "Classic" is a modern version of the original amp, with only superficial changes, while QUAD II-eighty is a higher power descendant, using a quad of KT88 beam tetrodes for a whopping 80-watt output. Its retail price in Australia was AU$14,999.- when reviewed in the March/April 2009 issue of Australian Hi-Fi magazine! There is also QUAD II-forty, using a pair of KT88 in the output stage.

However, the KT88 models have nothing in common with the original QUADs; they use different topologies and power and preamp and driver tubes (6SN7 and 6SL7).

Eico ST40 and ST70

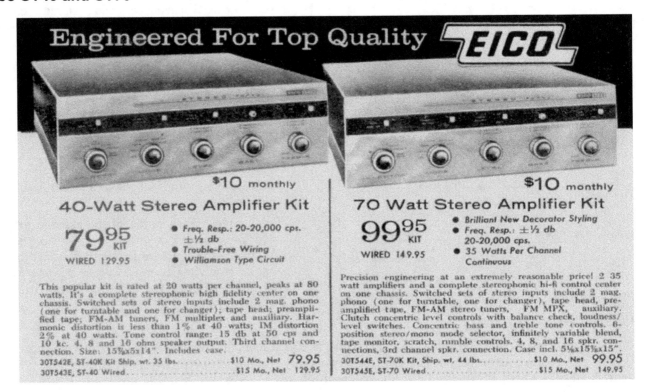

These two were among the last and thus most modern of Eico's integrated amplifiers. Most were sold as kits to DIY constructors (notice a significant difference in price between the kit and factory wired versions). However, their complexity far exceeded relatively simple Dynakit designs, their primary competitor.

You either love or hate the looks of these two amps. Personally, I find them ugly. It seems that most audiophiles prefer the "open" looks of the older amplifiers (with black transformer covers and exposed tubes) to these "boxed-in" attempts to streamline and modernize the looks in the late 1950s and early 1960s. Paradoxically, these seem more visually dated from today's perspective, and since tubes aren't visible at all, they could easily be mistaken for the abhorrent solid-state amplifiers of the era.

Both basic designs are sound, but what lets these amps down is their complexity due to the array of unnecessary controls (the fashion of the day). There are too many switches, contacts, and controls in the signal path. Once those are bypassed or removed completely, the sound improves significantly. Once the power supply is improved, the sound improves further. As in most vintage PP amps, the anodes of the output tubes are supplied with a high ripple DC voltage at the output of the GZ34 rectifier tube (pin 8) before the smoothing filters!

ST40, the smaller brother of ST70, uses the same output tubes, yet its power per channel is only 20 watts, compared to ST70's 35 watts.

How did Eico's designers manage to get almost double the power in the ST70 model? Hint: study the comparison table for clues!

ST70 uses fixed bias, while ST40 uses self-bias (also called cathode bias).

SIDE-BY-SIDE	ST40	ST70
Output stage	push-pull pentode	push-pull pentode
Output stage bias	self- or cathode bias	fixed (external) bias
Anode & screen voltage	395 / 310 V	435 / 390 V
I_0 per tube	56mA	38mA
Output power per channel	20 W	35 W
Load impedance (Z_{AA})	8.2 kΩ	6.5 kΩ
Input/driver tubes	12DW7	1/2 12AU7 + 6SN7
Phase inverter	Cathodyne (split load)	long-tailed pair

All other things being equal (same tube, same circuit, same plate voltage), fixed bias results in increased power since there is no power loss on the cathode resistors. The voltage on the cathodes of the 7591 power tubes is 15 Volts on ST40, while on ST70, the cathodes are at approx. zero potential (0.38V).

In ST70, the plate voltage on 7591 output tubes is higher, 440V compared to 395V in ST40, ditto with the screen voltage (390V versus 310V). The previously mentioned difference of 15 volts on the cathodes should be added to these voltage differences, for a total of about 60 V_{DC} (45 + 15) more on the plates in ST70!

The idle current of ST40 is much higher, 56mA versus only 38mA in ST70. ST40 will operate primarily in class A, crossing into class B only at the highest power levels. In contrast, ST70 will cross much earlier into class B, increasing its power output further.

ST70's output transformer has a much lower primary impedance than the one used in ST40, 6.5kΩ versus 8.2 kΩ. Lower impedance means higher output power but also a higher distortion.

The first stage is a common cathode amplifier with a 12AX7 triode, DC-coupled to the phase inverter in both models. However, ST40 uses the 12AU7 half of 12DW7 as a split-load inverter, while ST70 uses a 6SN7 duo-triode in a cathode-coupled inverter, pejoratively called "long-tailed pair."

And finally, the output transformers on ST70 are much larger than the ST40 units. A larger core means the bass response of ST70 will be better. It will saturate at a lower frequency and higher power levels.

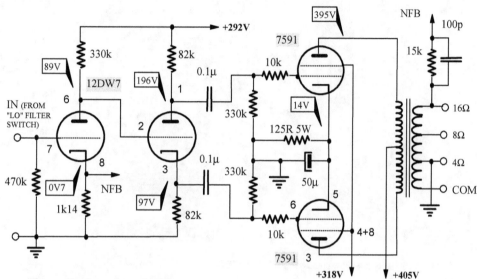

ABOVE: Eico ST40 power amplifier section (one channel only)

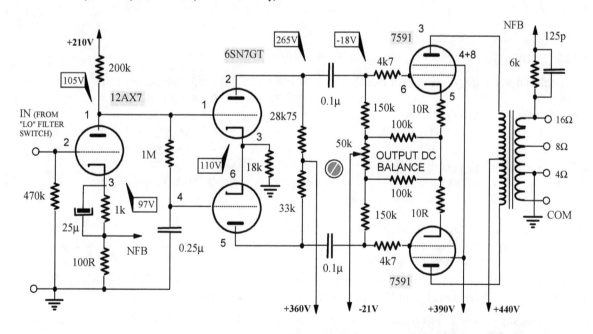

ABOVE: Eico ST70 power amplifier section (one channel only)

RIGHT: Looking deceptively small in photos, almost as it would fit on an ordinary bookshelf, ST70 is a huge and heavy beast with a very large footprint in real life. Many modern hi-fi racks are way too small to accommodate it!

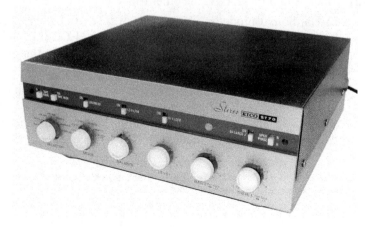

The under-the-chassis (component side) view of Eico ST40 in its original condition:
1. Power transformer
2. 1k8 5W series resistor in the CRC power supply filter
3. Rectifier socket
4. Cathode capacitors (output tubes)
5. Cathode resistors (output tubes)
6. Coupling capacitors to the grids of output tubes
7. 100R "hum adjust" trimmer potentiometers across 6V3 heater secondaries
8. 16Ω 20W resistor used as a load (between the 16Ω terminal and COM) during the "balance check" test.
9. 30nF "death capacitor" between one end of transformer primary and chassis - must be removed and chassis grounded (earthed)!

The tubes used

TUBE PROFILE: 12DW7 (7247)

- Indirectly-heated duo triode (dissimilar triodes)
- Noval socket
- Heater: 6.3V/300mA or 12.6V/150mA
- SECTION 1: V_{AMAX}= 330V, P_{AMAX}= 1.2W
- gm=1.6 mA/V, µ=100, r_I = 62.5 kΩ
- SECTION 2: V_A=330V, P_{AMAX}= 3.3W
- gm=2.2 mA/V, µ=17, r_I = 7.7 kΩ

TUBE PROFILE: 7591

- Indirectly-heated beam power tube
- Octal socket, 6.3V/0.8A heater
- V_{HKMAX}=100V_{DC}
- P_{AMAX}=19 W, P_{SMAX}=3.3 W
- V_{AMAX}: 550V, V_{SMAX}: 440V, I_{KMAX}: 85 mA
- AS A TRIODE:
- µ = 16.8, gm=12 mA/V, r_I=1.4kΩ

7591 was initially designed to operate as an audio tube with high efficiency and high power sensitivity, meaning very low grid signals were needed for maximum power output. While it uses an octal socket, it is not pin-compatible with the 6V6-6L6 family of octal beam power tubes.

Physically, the vintage 7591s are relatively small, so good ventilation is needed for efficient cooling. The modern replacements produced in Slovakia (JJ) and Russia (Electro-Harmonix) are taller.

12DW7 or 7247 contains two different triodes in one glass envelope; the first triode is half of 12AX7, the other triode is half of 12AU7.

Electro-Harmonix produces its replica, 12DW7EH, while JJ makes ECC832, a clever name that combines ECC83 and ECC82! Their internal construction looks vastly different, so one would predict that they would also sound different with a fair degree of certainty.

If you don't have confidence in currently produced tubes, you can rewire the input/driver stage and share one 12AX7 and one 12AU7 between the two channels. Then you can use a range of brands and tube types, flat anode, ribbed anode, boxed anode, black-gray-silver anode, etc.

RIGHT: How is it possible that two tubes with completely different construction and a huge difference in anode size be electrically equivalent?

7591 in triode connection

The G_2G_1 product or triode amplification factor of 7591 beam power tube is quite high, around 17, as is its transconductance as a triode (around 12mA/V). Two tubes connected in parallel would bring the overall power dissipation (anodes+grids) to almost 45 Watts, internal resistance down to 700Ω and transconductance up to 24mA/V, superior to the equivalent figures of the 300B triode!

The internal grid-to-anode capacitance of one tube is 0.25 pF, while the input capacitance (grid to K/H/G2/G3) is 10pF. Assuming an amplification factor of the output stage of around 12, the dynamic or Miller's capacitance of one triode connected 7591 would be $C_{IN} = C_{GK} + C_{AG}(1+A) = 10+0.25(1+12) = 13.25$pF or 27pF for a pair.

As a comparison, the input capacitance of a 300B SE stage (voltage gain of around 3.5) would be $C_{IN} = C_{GK} + C_{AG}(1+A) = 8.5+15(1+3.5) = 76$pF, which is 76/27 = 2.8 times higher!

This means that, compared to 300B, two paralleled 7591 triodes would need 17/3.9 = 4.4 times smaller voltage signal to drive them. Since the input capacitance is 2.8 times lower, the required current that needs to be provided by the driver stage is also 2.8 times lower (to charge the input capacitance). Thus created a "super triode" would be a much easier load to drive. Let's see what we can expect in terms of power output and distortion.

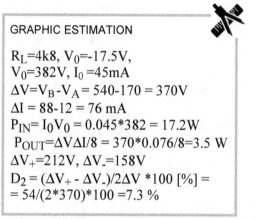

GRAPHIC ESTIMATION

$R_L = 4k8$, $V_0 = -17.5V$,
$V_0 = 382V$, $I_0 = 45$mA
$\Delta V = V_B - V_A = 540 - 170 = 370V$
$\Delta I = 88 - 12 = 76$ mA
$P_{IN} = I_0 V_0 = 0.045 * 382 = 17.2W$
$P_{OUT} = \Delta V \Delta I / 8 = 370 * 0.076 / 8 = 3.5$ W
$\Delta V_+ = 212V$, $\Delta V_- = 158V$
$D_2 = (\Delta V_+ - \Delta V_-)/2\Delta V *100$ [%] =
  = 54/(2*370)*100 = 7.3 %

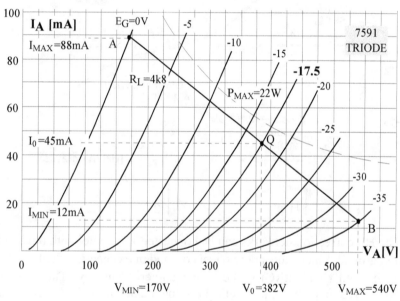

Two tubes in parallel would produce only 7 watts with a load of around 2.5kΩ, hardly worth getting out of the bed for. To make things even worse, the distortion is almost as high as that of the triode connected 6L6, another tube that seems only suitable for push-pull operation.

From our experience, for SE output stages using pseudo-triodes (pentodes and beam power tubes strapped as triodes), the best choices are 6AR6, 7027A, F2a, and EL152/153 lovelies.

HF correction networks

Many of the vintage amplifiers analyzed in this chapter used very strong negative feedback levels. Due to the increasing phase shift at frequency extremes, such strong negative feedback can turn into positive feedback and cause oscillations, turning the amplifier into an oscillator.

The compensation reduces the gain at high frequencies so that at critical frequencies, the product Aβ becomes smaller than 1, in which case oscillations are not possible, according to Nyquist's Criterion, which we studied in Volume 1 of this book. As always, "A" is the gain of the amplifier without feedback, and "β" is the feedback ratio.

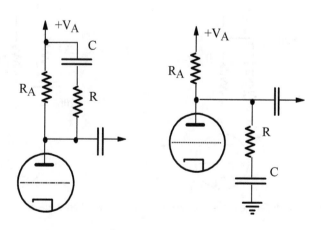

ABOVE: Two ways of implementing HF amplitude correction networks

The Nyquist diagram on the right is for the original, uncompensated amplifier. This amplifier has a high gain in the frequency range between points f_1 and f_2, its Nyquist loop encloses the critical point (1,0) or 1+0j, and oscillations will result at those high frequencies.

The Nyquist diagram below right is for the same amplifier, but this time the amplifier has been compensated. Its gain between high frequencies f_1 and f_2 is reduced, so although its negative feedback still becomes positive at those high frequencies, the product $A\beta$ is now smaller than 1 (the length of the $A\beta$ vector), and there are no oscillations since point (1,0) is not encircled. The amplifier remains stable.

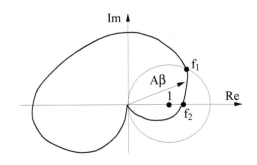

ABOVE: Nyquist diagram of the original, uncompensated amplifier, unstable at high frequencies

BELOW: Nyquist diagram of the same amplifier, this time compensated and stable

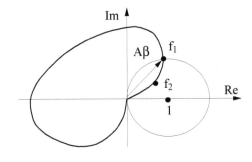

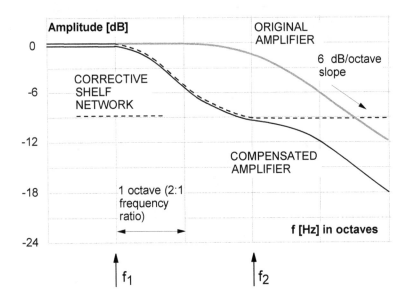

The HF correction networks can be connected in parallel with the anode load (anode resistor) or between the anode and ground. Since there are two resistors and one capacitor, the filter has two frequencies, $f_1=1/[2\pi(R+R_A)C]$ is the lower, and $f_2=1/(2\pi RC)$ is the higher corner frequency. Thus, the amplitude vs. frequency characteristic has a distinctive shelf shape, just as the building blocks of the RIAA filter, as discussed in Volume 1!

To preserve the clarity of the drawing, in the amplitude-frequency diagram above, frequency f_2 is only two octaves higher than f_1 (4x higher frequency); normally, it is much higher. See the two examples below, Williamson's amplifier and Marantz 1000.

$f_1=1/[2\pi(R+R_A)C] = 15.4$ kHz $f_1=1/[2\pi(R+R_A)C] = 2.75$ kHz
$f_2=1/(2\pi RC)= 169$ kHz $f_2=1/(2\pi RC)= 110.5$ kHz

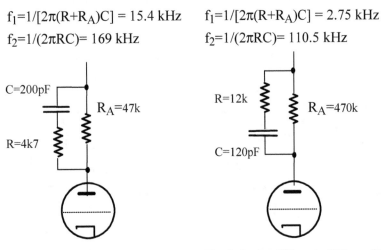

The 1st stage in Williamson's amplifier   The 2nd stage of Marantz 1000 amplifier

NEW? IMPROVED? MODERN PUSH-PULL AMPLIFIER DESIGNS

- Classic 16.0
- Leben CS300
- Design project: EL84 amplifier
- Design project: EL12N amplifier
- Shindo Sinhonia
- Twenty preamplifier tubes in a push-pull output stage?

"If the midrange isn't right, nothing else matters."
J. Gordon Holt, founder of *Stereophile* magazine

Classic 16.0

The Classic No.16.0 is made by Beijing Yushang Electronic Audio Technology Co., Ltd. A very elegant-looking and extremely heavy amplifier.

Behind the power transformer centrally-mounted at the front is a single 1.5H choke (for both channels), flanked by two pairs of EL34 tubes.

At the back, between the two output transformers are two 12AT7 and two 6SN7 duo-triodes.

Each of the four output tubes is biased separately; a selector switch connects a moving coil meter to the circuit so the bias reading can be taken. A very user-friendly solution, no need to turn the amp upside down, remove the bottom cover and search for the bias test points.

There is also a digital hours-run meter, a rather lovely marketing drawcard. However, it can easily be reset, so the display should not be relied upon if buying this amp from online sellers as a secondhand unit.

The primary impedance of its output transformers is specified as 5kΩ anode-to-anode; the claimed primary inductance at 50Hz is incredibly high, 234H, while the specified leakage inductance is very low, 2.27mH.

The frequency bandwidth of the output transformers is claimed to be 5Hz-342 kHz (-3db points at 5V output level), while the frequency range of the whole amplifier is specified as 8Hz -82 kHz (at 10W out). The rest of the specs are in the table.

The friend who owned this amp could not bias two of the output tubes. However, when we powered the amp up, it kept blowing the mains fuse.

Cold checks with an ohmmeter revealed that the high voltage secondary winding of the 400VA toroidal mains transformer was shorted.

We tried to reveal the HV winding and see if we could pinpoint the location of the short and repair it, but had no success.

A few resistors on the PCB were overheating, and one was completely burned out. Did that happen first, which is why its owner could not bias it, or was he overzealous with the biasing screwdriver and damaged the amp as a result? It's hard to tell.

In any case, the power transformer had burned out and needed replacing.

SPECIFICATIONS:

- Output power: 40W (UL mode), 20W (triode mode)
- BW: 15 Hz - 42 kHz (-3dB, at 1W into 8Ω)
- Open-loop gain 45db, gain with NFB 38 dB
- THD: below 1.2% (at 10W out, 1kHz signal, triode mode)
- SNR: 90db (unweighted)
- Output noise: 1.2mV
- Negative feedback: 5.72 dB
- Input sensitivity: 0.22V for full power

Checking discrepancies in specifications

The manufacturer's specifications claim that the open loop gain is 45 and that the NFB is 5.72db. The overall gain should then be 45-5.72 = 39.28 dB, but 38 dB is claimed. The input sensitivity of 0.22V is specified, so let's see what that tells us. Since $P_{OUT}=V_{OUT}^2/Z$, assuming $Z=8\Omega$ and $P_{OUT}=40W$, we get $V_{OUT}=17.9V$. Now, A [dB] = A [dB] = $20*\log A = 20*\log(17.9/0.22) = 38.2$ dB.

So it seems that the 38dB gain is correct after all, meaning that one of the other two figures is not right, either the 5.72dB of negative feedback or open loop gain of 45.

The internal wiring

One huge T-shaped PCB accommodates the power supply and audio circuitry. Smaller PCBs are dedicated to specific functions. The on-off switch is brought to the front (1), the round biasing meter (2) and the PCB with four biasing selector switches (3) are on one side, while the volume control pot (4) and the input selector switch (5) are on the other. The selector switch activates relays mounted on the back PCB (6).

A dual timer on a separate PCB uses a NE556 integrated circuit and controls the power-on delay (7), flanked by the hours-run meter digital circuitry on another small board. Overall, a well-conceptualized and neat layout, but repairs and mods are challenging.

ABOVE: The internal view of Classic 16.0
BELOW LEFT: The burned-out, overheating resistors and charred PCB on both channels. The leg of one resistor somehow fell out or unsoldered itself, most likely due to overheating and vibration.
BELOW RIGHT: Despite the front-mounted power transformer, the amp looks sleek. Moving the transformer to the back and bringing the four preamp tubes to the fore would make it even more appealing, although more conventional.

The audio circuit

The first stage is 12AT7 (ECC81) duo-triode in an SRPP configuration, DC- coupled to the inputs and also DC-coupled to the grid of the phase splitter, the cathode-coupled or "long-tailed pair."

Without negative feedback, the voltage gain of this stage would be roughly half of the μ of the tube, and since μ=70 for ECC81, that would be around 35 times or 3.5V for the 0.1V signal input. However, global negative feedback taken from the 8W tap reduces that gain to around 18.

The phase splitter has a voltage gain of $A_2 = 12V/1.8V = 6.7$ or 16.5 dB.

NOTE: The DC voltages marked are after the replacement of the power transformer. The original DC voltages were higher, as marked on the power supply schematics (next page).

The output stage can operate in UL or triode mode and uses fixed biasing, individually adjustable for each EL34 tube (22k bias trimmer potentiometers).

The metering circuit is not shown n the diagram; biasing switches connect the shared analog meter to one of four test points, the cathodes of output tubes, so the meter displays the voltage drop across 10Ω cathode resistors.

The claimed primary inductance of the output transformers is 234 H (?), but we measured only 10H! Their anode-to-anode DC resistance was 180Ω and the primary impedance was 4k5.

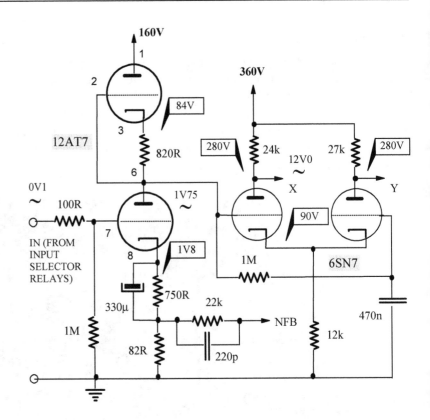

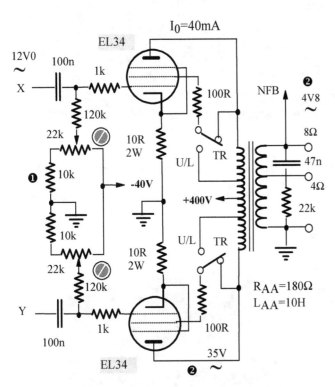

The amplifier is very heavy, not due to substantial output and power transformers or large chokes, but its thick aluminium chassis. Having heard that many prospective buyers judge tube amps' quality by their weight, Chinese manufacturers often use unnecessarily thick & heavy aluminium chassis.

After all, aluminium is much cheaper than grain-oriented silicon steel laminations or copper winding wire, both used in mains and output transformers. Plus, it makes their usually mediocre amplifiers look "hi-end."

Notice fixed 10k resistors (1) in series with each bias trimmer potentiometer. This is a prudent inclusion, meaning the bias (negative DC voltage on the power tube's control grid) cannot be set all the way up to zero volts, which would leave the tube "fully open" and damage or destroy it within minutes.

Notice also (2) that the output stage (in triode mode) amplifies around 3 times (35V_{AC} signal at the anodes, 12V_{AC} at the grids)!

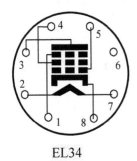

EL34

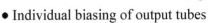

MEASURED RESULTS (U/L MODE):

- BW: 11Hz–41kHz (-3dB @10W)
- V_{OUTMAX} = 19V_{RMS}
- P_{MAX} = 45W
- Gain: 53.6 dB

KEY FEATURES:

- Individual biasing of output tubes
- Bias meter, hours run meter
- Switchable triode - ultralinear mode
- Start-delay timer

Capacitive compensation of inductive loads

Notice the series RC pair (22k resistor and 47n capacitor) across the 8Ω amplifier output. Known as Boucherot cell, this simple filter dampens HF oscillations and improves stability of amplifiers that use strong negative feedback.

It is usually referred to as a "Zobel network", but strictly speaking an RC pair should be called Zobel network only when used in parallel with the woofer (bass driver) of a loudspeaker, in order to correct its characteristics.

The inductive reactance of a dynamic (moving coil) speaker increases in a linear fashion with rising frequency f ($Z_L=2\pi fL$), so this simple type of RC compensation aims to compensate for this rising impedance and make the load independent (or less dependent) on frequency. If L/R_S is made equal to CR ($L/R_S=CR$), then the load as seen by the amplifier's output is purely resistive and thus independent of frequency. Usually R is made equal to R_S, so then $R=R_S=\sqrt{(L/C)}$!

Of course, if the loudspeaker used is of predominantly capacitive nature (such as electrostatic speakers), then this RC network must be removed.

Likewise, since the NFB is below 6dB in this case, this kind of correction should not be needed, so if you own one of these amplifiers you can experiment by removing this RC pair from both outputs and comparing the impact of this modification on its sound.

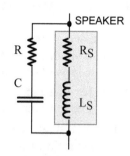

ABOVE: RC compensation of an RL (inductive) load

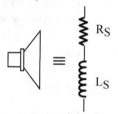

A simple model of a dynamic speaker driver

The power supply

The energy storage is relatively low for an amplifier of such a high output power. Also, a single 1.5H choke is dismally inadequate. The bias voltage is not regulated.

It seems there are or have been two versions of this amplifier; one features better quality elcos, which are then bypassed by film capacitors. The electrolytic caps in our cheaper Australian version were not bypassed at all.

The tube sockets on our unit were extremely loose, all the tubes wobbled in their sockets, resulting in poor contacts and occasional pops and crackles from the speakers. So, from reliability and modification potential, this model fails the grade.

Sound-wise, it sounded decent, much better than the cheaper AudioRomy or Music Angel amplifiers we auditioned, but far from impressive or engaging. Thus, depending on how much you pay for it, the sonics and value-for-money aspects may or may not get a passing grade.

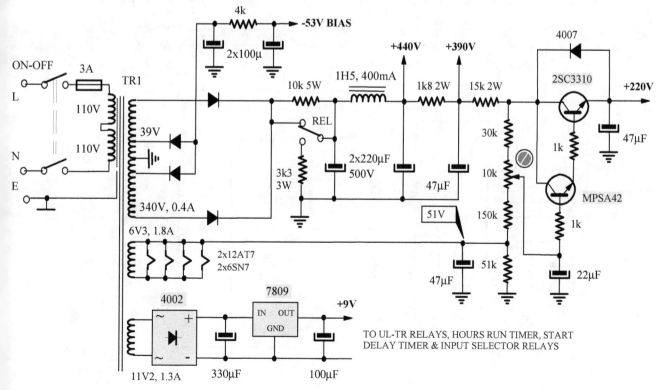

Classic 16.0 power supply

Leben CS300 & CS300X

Now discontinued, Leben CS300 was an old-fashioned-looking push-pull amplifier made in Japan using PTP (point-to-point) wiring method. While CS300 was rated at 2x12 watts, the more expensive variant, CS300SX, was rated at 15 watts per channel.

When reviewed in Stereophile magazine in November 2011, it was selling for US$3,395. For that money, you get a single preamp/phase inverter stage driving a push-pull pair of EL84 in pentode mode.

The output stage is self-biased, and global negative feedback is taken from the 8-ohm tap back to the first stage. The feedback ratio of the global NFB used is $\beta = R_2/(R_1+R_2) \times 100\% = 100/(3{,}000+100) \times 100\% = 0.0323$ or 3.23%, meaning 3.23% of the output voltage is fed back to the cathode of the first triode.

The -2dB frequency range is specified as 15Hz - 100 kHz, the THD distortion at 10 Watt output level is claimed to be 0.7%. According to the results of the *Stereophile* review, most of these claims turned out to be overly optimistic; in other words, the amplifier failed to meet those specs.

Finally, and perhaps most importantly, the weight is only 10.5 kg. That means small output transformers were used, and, as the circuit diagram confirms, there are no filtering chokes of any kind, two major weaknesses of this amp.

Audio circuit

The first stage is a floating paraphase inverter. The input signal from the selector and tape monitor switch (not shown) goes through the volume and balance potentiometers and enters the grid of the upper 5751 triode. Its anode output feeds the grid of the upper EL84 pentode through 470n coupling capacitor.

The grid signal for the lower triode is taken from the joint of two 1M resistors, floating between the outputs of two coupling capacitors. The floating paraphase inverter is a poor choice in hi-fi. It is never perfectly balanced, plus the signal for the lower output tube passes through two triodes, while the driving signal for the upper power tube passes through only one triode.

The bias is $V_G = -1.8V$, which does not agree with the value on the schematics. Either the graphs published by GE are wrong or the tube used in the actual amplifier was very different from the average or bogey value.

Anyway, the Leben designer's choice of the operating point seems wrong. The load line enters the region of low anode currents, where the characteristics are curved and not at all equidistant.

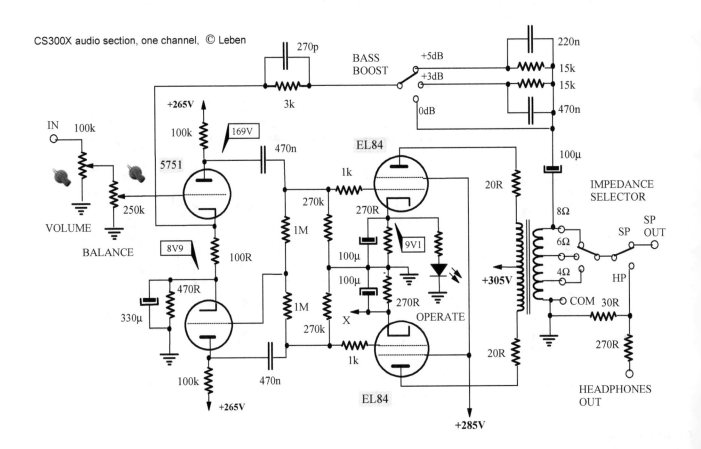

NEW? IMPROVED? MODERN PUSH-PULL AMPLIFIER DESIGNS

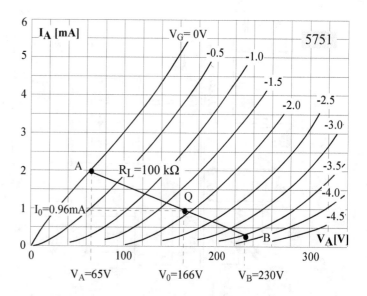

GRAPHIC ESTIMATION

$V_0=166V$, $I_0=0.96mA$
$\Delta V_G = 3.6V$, $\Delta V_A = 230-65 = 165V$
$A_V = -\Delta V_A / \Delta V_G = -165/3.6 = -45.8$
$\Delta V+ = 101V$, $\Delta V- = 64V$
$D_2 = (\Delta V_+ - \Delta V_-)/2\Delta V_A = 37/330 = 11.2\%$

One look at its anode characteristics forecasts high levels of distortion. The Q-B swing is much, much shorter than the Q-A swing. Surely enough, more than 11% 2nd harmonic according to the estimation above!

The voltage on the common 470Ω cathode resistor is 0.9V, so assuming the currents to be identical, each triode pulls $I_A=0.9/2*470 = 0.96mA$. Since the anode supply voltage is 265V and anodes are at around 169V, a voltage drop of 96V, the anode current is $I_A=96/100 = 0.96mA$. The results match, so it seems we can trust whoever measured this amplifier and posted the results online. Thank you.

5751 duo-triode

RCA specifies 0.8W plate dissipation for 5751, GE 1.1 Watt. No matter who is right and who is wrong that is one weak tube, you would think. The mju of 70 reminds us of 12AT7, another tube not on my list of favorites.

Then come the real shockers. A very low transconductance of only 1.2 mA/V and a very, very high internal resistance of 58 kΩ! With so many lovelies around, why would someone use this tube is perplexing.

TUBE PROFILE: 5751

- Indirectly-heated octal duo-triode
- Heater: 6.3V, 350mA or 12.6V, 175mA
- $V_{AMAX}=330V$, P_{AMAX}: 0.8W
- $V_{HKMAX}=100V$
- $\mu=70$, $gm=1.2mA/V$, $r_I = 58k\Omega$

The output stage

Since cathodes of EL84 output tubes are at $9.1V_{DC}$, the cathode currents through 270Ω resistors are $I_K=9.1/270 = 33.7mA$ We don't know (yet!) how much of that is the anode and how much current is drawn by the screen grids. The trick is to look at the power supply and calculate that ratio from there.

Power supply

The power supply (next page) is elementary - solid-state voltage rectification (voltage doubler), no filtering chokes, and very low energy storage due to low capacitance in the filtering circuit.

Since the voltage on the voltage doubler's capacitance (50µF) and the first elco (100µ) are almost equal, the energy storage for the output stage is $E=0.5CV^2 = 0.5*150*10^{-6}*308^2 = 7.1$ Joule for both channels, which is very low. The energy for the screens is $E=0.5CV^2 = 0.5*100*10^{-6}*285^2 = 4$ Joule and for the preamp stages $E=0.5CV^2 = 0.5*330*10^{-6}*265^2 = 11.6$ Joule, which is fine since their current draw is minute. So, the selection of capacitance values is wrong. Instead of 100µF + 100µF + 330µF, it should be 330µF + 330µF + 100µF!

To calculate current draws we start from point C (the feed for preamp stages) and work backwards. The current through 5k1 resistor is $I_{BC} = (285-265)/5.1 = 20/5.1 = 3.92mA$. We calculated 0.96mA per triode, so $4*0.96 = 3.84mA$, close enough.

The current through 1k resistor is $I_{AB} = (305-285)/1 = 20/1 = 20mA$. Once we subtract the $I_{BC} = 4mA$, that means the four screen grids are pulling 20-4=16mA or 4 mA each.

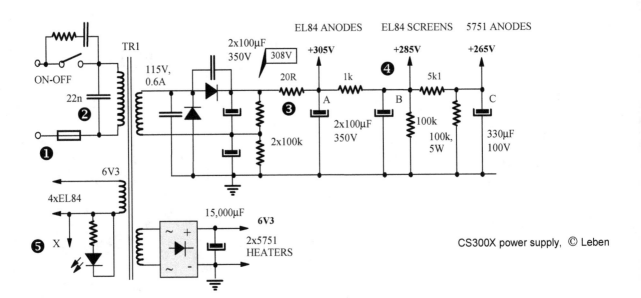

CS300X power supply, © Leben

Finally, the current through the 20R resistor is the total current of the amplifier $I_{TOT}=3/20 = 150$mA. However, the 3 Volts measured difference probably wasn't very precise, and everything hinges on its precision. The four anodes are pulling 150-20=130mA or $I_A=32.5$ mA each! We have already determined the cathode currents of 33.7mA, out of which 4mA is screen current, leaving 33.7-4 = 29.7mA anode current.

Starting from the mains side, notice that the circuit diagram does not show the ground or earth terminal (1). Does the mains transformer have an electrostatic shield connected to the ground? It does. By law, the fuse and the on-off switch must be in the same leg (L for live). Here they are not; the fuse is in the neutral leg.

Notice the 22n capacitor across the mains input (2). Again, by law, that must be a Class X2 capacitor. It is not clear from the photo if the capacitor used was approved for such duty or not.

The first filtering resistor (20R) should be replaced by a 200-300mA choke (3), and so should the 1k resistor. In pentode output stages, the screen voltage should be rock-steady; here, it isn't (4).

As the anodes are pulling more and more current (at higher volumes, class B operation), the voltage in point B (fed to the screens) will drop and thus reduce the output power, acting as a dynamic limiter. So, using an HV regulator for screens would be the best solution. Short of that, a large choke would improve things, but Leben uses neither.

Finally, some good news. Notice that one leg of the AC heater supply for all four EL84 is connected to the cathode of one of the four tubes, at $9V_{DC}$ approximately (5). Such elevated heater voltage helps minimize hum and increase the S/N ratio, with improvements of up to 10 dB reported.

The wiring is neat, reminiscent of the typical Japanese component placement and interconnection used in vintage amplifiers such as Luxman, Trio, Kenwood, Lafayette, Pioneer, Sansui, and many others of that era. The wire runs are short, so compliments to Leben's amp makers. Pity the design side is not as good as the build quality!

Test results

The -2dB frequency range is specified as 15Hz - 100 kHz, and while 15Hz seems accurate, the upper -2dB frequency is barely above 40kHz.

Notice also a pronounced peak at around +3dB at 35kHz. This is due to the very strong NFB used and ultrasonic resonances of the output transformer. Normally such resonant frequency is much higher, in the 60-80kHz region, far removed from the upper audio limits.

The output transformers seem to be quality units with a high Q-factor, resulting in a low damping factor of the resonant circuit formed by their leakage inductance and parasitic capacitances.

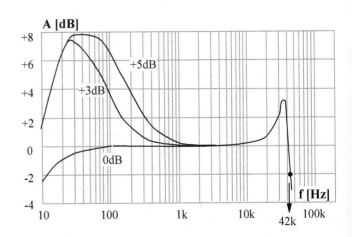

Leben CS300 amplitude-vs. frequency test results, redrawn with permission from Stereophile, http://www.stereophile.com

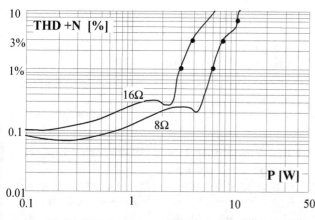 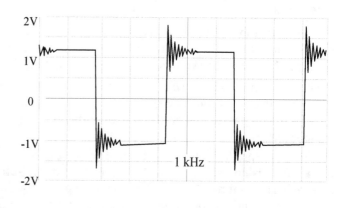

Leben CS300 THD+ noise versus output power (ABOVE LEFT), and 1kHz square wave reproduction on the 4Ω tap (ABOVE RIGHT), redrawn and reproduced here with permission from Stereophile, http://www.stereophile.com

A bass boost is achieved by switching additional RC combinations in series with the permanently connected one, although both provide a boost of around 8dB, peaking around 30-50Hz.

The manufacturer claimed a THD distortion of 0.7% at the 10 Watt output level. According to the measurements by *Stereophile* magazine using the 8Ω output, the distortion at 10W was 7-8%, ten times higher! The amplifier reached 1% distortion at 6Watts and 3% at 8Watts. At the 16Ω tap, the story was even worse.

The reproduction of 1kHz square wave is very poor, showing pronounced underdamped ringing, which will manifest itself as harsh and irritating treble.

This amplifier is very close to instability; it's a pity *Stereophile* didn't test it with a simulated electrostatic speaker, where such amplifiers easily become unstable driving such capacitive loads.

Although the 3k negative feedback resistor is bypassed by a 270 pF capacitor, whose purpose is to dampen and reduce such oscillations, the designer failed to achieve this aim. Regardless of the output tap, the output impedance is very high, 3–3.5 ohms. Even a strong NFB could not reduce the high distortion and high Z_{OUT} to acceptable levels!

Design project: EL84 PP amplifier

Let's see if we can build a better amp for a fraction of the cost. The answer is, of course, yes we can! For this simple design, we chose the concertina or split-load inverter. It provides no gain but is well balanced and not impacted by tube aging. Its maximum output voltage swing isn't as large as that of other phase inverters (using two triodes), but a wide voltage swing isn't needed for the sensitive EL84 power pentodes.

A simple input common cathode triode stage provides more than enough gain.

The input and phase splitter stages use 12AT7, a very common audio tube. Many NOS varieties are still available, plus half a dozen current manufacturers all make sure that tube roller's dreams come true at once! Notice DC-coupling between the two stages, so only a pair of coupling capacitors in the signal path.

You can also plug in 12AU7 tubes instead of 12AT7, but the input sensitivity will be reduced, so an active preamplifier may be needed in that case.

It's a pity we could not have an A-B shoot-out between Leben CS300 and this baby.

TUBE PROFILE: ECC81 (12AT7)
- Indirectly-heated duo-triode, Noval socket
- Heater: 6.3V/300mA or 12.6V/150mA
- V_A= 300V_{DC}, V_{HKMAX}=90V_{DC}
- P_{AMAX}=2.5W, I_{KMAX}=15 mA
- TYPICAL OPERATION:
- V_A=100V, V_G=-1.0V, I_0=3 mA
- gm = 3.75 mA/V, µ=62, r_I = 16.5 kΩ

MEASURED RESULTS (UL mode):
- BW: 6Hz - 69 kHz (-3dB, at 1W into 8Ω)
- BW: 18Hz - 43 kHz (-3dB, at 10W into 8Ω)
- V_{NOM}= 11V_{RMS}, P_{NOM}= 15W
- V_{MAX}= 14V_{RMS}, P_{MAX} = 24W

The audio section

The amplification of the first stage (after mild global NFB) is A=22, the second stage, cathodyne or split-load inverter, provides no voltage gain, while the output stage amplifies 23 times. The overall voltage amplification of the amplifier is A=2.5V/0.1V = 25 times of A= 28 dB! the feedback ratio is β=330/(22,330) = 1.48%.

Each output tube has its self-biasing RC network, so pairs don't have to be precisely matched; providing the imbalance in anode currents is below 5mA, the output transformer doesn't seem to mind. The ultra-linear mode with 33% of primary windings on the screen grids combines the best of both worlds, the pentode's power and triode's sonic signature.

We haven't reproduced the square wave test results, but they are much better than Leben CS300 seen a few pages back. There is no oscillation, and the split-load phase splitter does a better job of keeping out-of-phase grid drive signals balanced throughout the frequency range. Thus, there is no provision for any adjustments; if tweaking with oscilloscopes and function generators makes you happy, you can easily incorporate such a feature in the cathode leg of the phase splitter (instead of the 33kΩ resistor), as covered elsewhere in this volume.

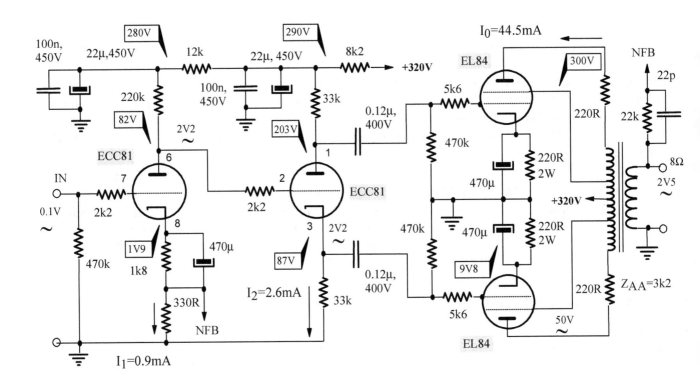

The power supply

The power supply cannot be simpler, EZ81 (6CA4) rectifier, CLC filtering, the total energy for output tubes is approx. $E=0.5CV^2$ = 22 Joule.

Leben CS300X uses 7J for both channels, so six times higher energy storage here!

The 10Ω resistor limits the inrush current upon power-up when filtering elcos are discharged. That reduces the stress on the EZ81 rectifier tube! An NTC resistor can also be used here, of course.

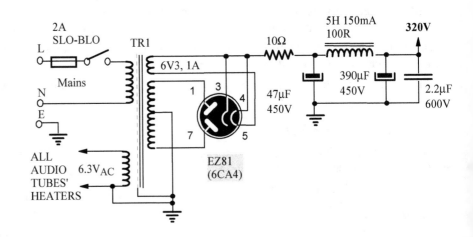

A variation on the same theme

A quirky approach to chassis sourcing and design was taken here. Two powder-coated steel wire drawers out of a wardrobe storage unit bolted together to form the outside frame for a monoblock amplifier. The internal steel chassis is entirely separate, covered with a much denser wire steel mesh on top to make it finger-proof.

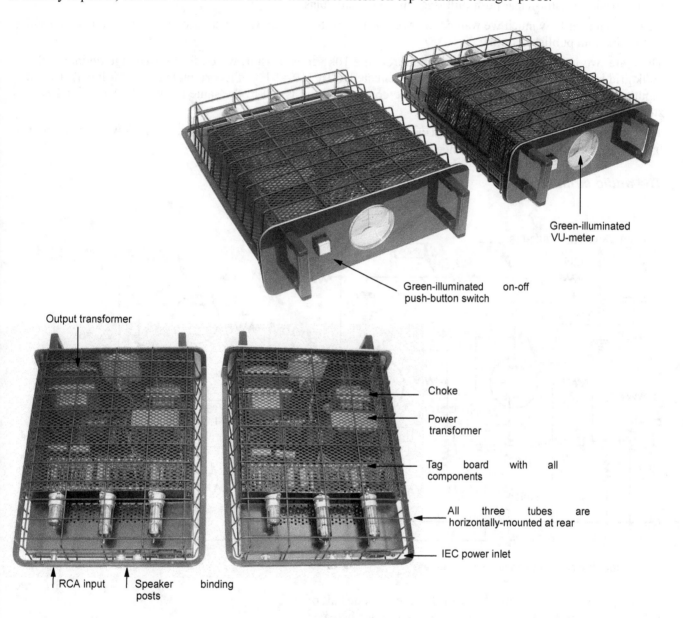

The front fascia was cut out of thick mirror-finish red Perspex. Add a stylish analog meter and surplus handles, and voila, a pair of very sleek and elegant monoblock amplifiers!

The 1 mA FSD (Full-Scale Deflection) analog meter is connected to the speaker output in a bridge configuration. Two silicon diodes rectify the music signal, so the meter's deflection changes with the signal's loudness, acting as a VU-meter! The 18k resistor in series with the meter adjusts the sensitivity so meaningful deflections are achieved. Depending on the meter you use, you will have to experiment to find the optimal value for that series resistor.

The meter bought on eBay did not have any backlighting, so we added green LEDs at the back, powered from the 6.3V tube heater supply, for a very discreet and classy illumination.

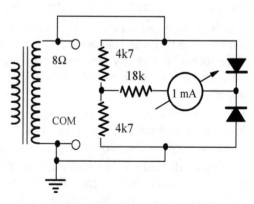

ABOVE: Ordinary analog mA-meter, connected as a VU-meter in parallel with the load

Shindo Sinhonia

Shindo Sinhonia was a monoblock tube amplifier released in 1986. There isn't much information available online, and the Shindo website gives very little detail on its specifications page. Even the photo is wrong, showing the Cortese amplifier on the Sinhonia page. The output power of 40 Watts in Class A is claimed, produced by two F2a beam tetrodes in ultra-linear push-pull stage using cathode bias.

Again, no matter how much we wanted to have one on our test bench, we must base our analysis on an unverified circuit diagram published online.

There are two inputs. The fixed input is fed through a 10k series resistor, which forms a voltage divider with the 100k grid leak resistor, meaning the signal is attenuated to 10/110 = 91% of its original value (9% down). The other input passes through to the 100k (log) volume control pot and is fed to the same point, followed by a 10k grid stopper resistor (quite a high value).

The triode of the triode-pentode 6BM8 (ECL82) tube is used in the first stage, with a very low anode current of 0.6mA and a very high anode resistance (100kΩ), to increase the gain of the stage.

The audio circuit

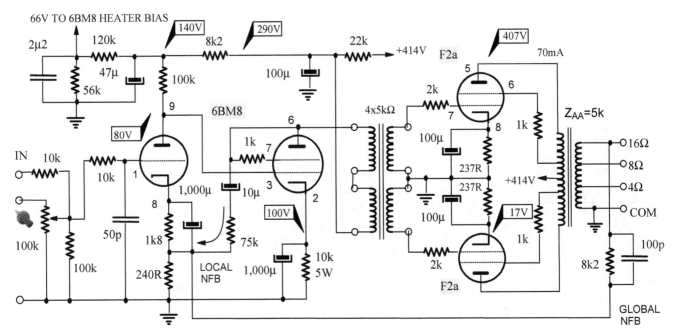

ABOVE: Shindo Sinhonia audio section, ©Shindo Labs

The output is directly coupled to the grid of the pentode half of 6BM8 tube, which is strapped as a triode (1k resistor between pin 7, the screen grid, and pin 6, the anode).

The two 5kΩ primaries of the interstage transformer are connected in series and are in the anode circuit of the 2nd stage. Since its grid is at +80V (the anode voltage of the first stage), the cathode had to be raised to +100V, thus the high-value cathode resistor (10k).

A 100V voltage drop on a 10k resistor means the DC current through the 2nd stage is 10mA, and that current flows through the primary of the IS transformer.

Judging by the online photos of the amp, the IS transformer is tiny; one wonders how it can take 10mA of primary current without taking the core into saturation. Unless, of course, it is made with an air gap.

It would be very illuminating to see the test results of this design, including its square wave reproduction!

TUBE PROFILE: ECL82 (6BM8)

- Indirectly-heated audio triode-pentode
- Noval socket, heater: 6.3V/780 mA
- P_{AMAX} (T): 1W, P_{AMAX} (P): 7W
- I_{AMAX} (T): 15mA, I_{AMAX} (P): 40mA
- Max. plate/screen voltage: 300V
- Typical operation triode:
- V_A=300V, V_G=-1.3V, I_0=1.1 mA
- gm = 2.5 mA/V, r_I=28kΩ, μ=70
- Pentode section triode-connected:
- gm=4.2 mA/V, r_I=1.6kΩ, μ=6.7

Going back to the input tube, its maximum heater-to-cathode DC voltage is 100V, which is why its heaters are powered from a separate 6.3V winding of the mains transformer.

Instead of one end being grounded, it is connected to the high voltage decoupling circuit. The 56k bleeder resistor was chosen to provide 66V to elevate the heater of the 6BM8 tube.

> TUBE PROFILE: 6AU4
>
> - Indirectly-heated single diode for TV horizontal damper service
> - Octal socket, heater: 6.3V/1.8A
> - $V_{AMAX}=300V_{DC}$, $V_{HKMAX}=$ -900V or +100V_{DC}
> - P_{AMAX}: 6.5W, I_{KMAX}=210 mA
> - Voltage drop at 140mA: 15V

Triple negative feedback used

The main reason for going through all that DC-coupling and interstage coupling trouble was the designer's aim of eliminating coupling capacitors from the signal path. However, from the point of view of audio puritans, there are two significant issues. The first one is the cathode biasing of the output stage. The quality of its elcos is critical. In this design, these 100μ elcos aren't even bypassed by a good quality film&foil capacitor.

Secondly, and more importantly, this amplifier uses three types of negative feedback. The first is obvious; it is the global NFB from the 16Ω output tap to the cathode of the first stage. The second is local NFB from the anode of the 2nd stage to the cathode of the 1st (the same point into which global NFB feeds)! Notice that the NFB signal passes through a 10μF electrolytic series capacitor.

Last but not least, the ultra-linear operation of the output stage involves local NFB since part of the audio signal from each tube's anode is fed back to its screen grid.

The power supply

Center-tapped HV secondary winding, two TV damper diodes feeding an RC filter, followed by the LC filter.

Shindo likes using TV damper diodes for their long heating time, which delays the appearance of high voltage on power tubes' anodes long enough to allow them to fully warm up!

With 147μF total capacitance, the energy storage of this amplifier is very low, only around $E=0.5CV^2 = 13$ Joule.

Unless the circuit diagram is wrong, the rectifier diodes which require 6.3V for their heaters are heated with only 5 Volts. While this prolongs the already long heater warm-up time, it also reduces the life of these tubes, so such practice is generally not recommended.

However, in this case, 6AU4 diodes are cheap and cheerful, so replacing them every year or so wouldn't be a major expense.

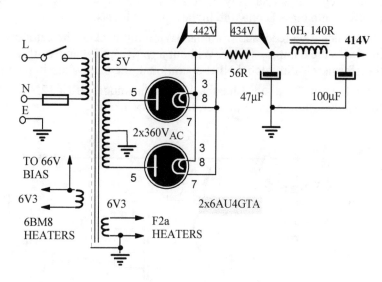

ABOVE: Shindo Sinhonia power supply, © Shindo Labs

Design project: E12L transformer-coupled PP amplifier

A completely symmetrical arrangement of all tubes and transformers on a spacious and beautiful Suntec (made in Taiwan and sold online) chassis. We bought one champagne gold and one black aluminum chassis. The gold one looked too flashy; the black one looked too morbid and depressing. However, once we swapped the front panels around, both the gold chassis with black fascia used here and the black chassis with gold front looked stunning.

We added a couple of large solid machined, gold-plated knobs for a double-contrast: gold-black-gold (next page). The rotary on-off switch is on the left (1) with volume control on the right (2).

The RCA inputs (6) are very close to the volume control pot and the input tubes. We even gave it a name, MSPL technology, which stands for "Minimal Signal Path Length"!

If commercial manufacturers can liberally use meaningless acronyms and catchy phrases, you and I can use our own terms for something that is very meaningful and makes lots of sense, can't we? Plus, why use somebody else's terms when you can invent your own?

The power transformer (3) is flanked by the output transformers (4), with interstage transformers in front (5), between the input and output tubes, to keep the MSPL approach throughout.

The vintage SAF audio transformers (made in France) had one primary and two 600Ω secondary windings, so we employed them here as phase splitters.

The first stage is a SRPP using ECC40 duo-triode with approx. 3mA of current and voltage gain of 15. The 600Ω interstage transformer could not take any DC current through its primary, so capacitive coupling had to be used.

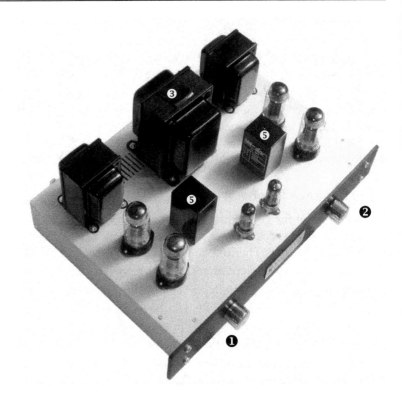

The interstage transformer acts as a phase splitter in a 1: 1+1 ratio, so with 1V at the input of the amplifier (pin 3 of ECC40), there is $15V_{AC}$ on its primary and on the grids of each output tube.

The voltage gain of the output stage is about ten times. Cathode biasing is used, separate for each tube. The idle cathode current is relatively high, around 72mA, meaning the output stage is biased to operate mainly in class A, crossing into class B only the highest power levels.

Global negative feedback from the 8W output back to the cathode of the first stage widens the frequency range and increases the damping factor. Without any NFB, the bandwidth was from 60 Hz to 13 kHz due to the unfortunate IS transformers. With global NFB operational, the bandwidth improved to 28Hz - 19 kHz.

Once the mild local NFB feedback from the anodes of EL12N tubes back to IS transformer's secondary was added, the bandwidth improved further, to 19Hz - 46 kHz.

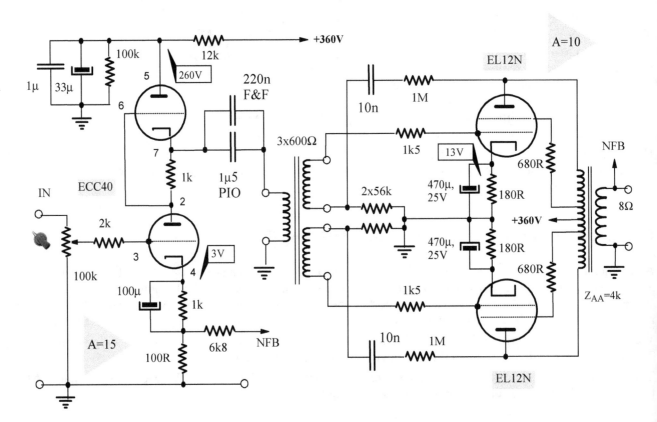

The power supply

Simplicity supreme, solid state bridge as a rectifier and two separate CLC filters (one for each channel). The energy reserve for each channel is approx. $E=0.5CV^2 = 78$ Joule per channel, six times more than Shindo Sinhonia.

The inrush current surge (charging two 390μF elcos) is significant; if it bothers you, install an NTC resistor after the rectifier.

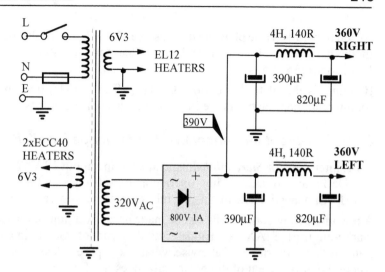

Grid stopper resistors

Grid stopper resistors should be mounted and soldered as close as possible to their grids. Since that is not obvious from a circuit diagram, where the line between the resistor symbol and the socket pin can be interpreted as a length of wire, sometimes a dot is placed on the grid to act as a reminder to the constructor to solder it *directly to the tube socket pin*.

TRADE TRICKS

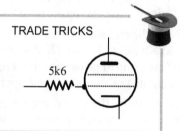

Listening impressions and possible improvements

A very detailed and clean sounding (but never sterile) amplifier due to various reasons: 1. Solid-state rectification. 2. Both ECC40 and EL12N tubes are relatively neutral-sounding. 3. SRPP input stage usually sounds cleaner than most common cathode stages. 4. Dual NFB cleans up the act even further. 5. There is only one coupling capacitor in the signal path, a synergetic combination of paper-in-oil warmth and film & foil speed and transparency. 6. Elna Cerafine cathode bypass capacitors in the first and output stage.

Since it is so revealing, the amp does not suit lesser quality audio systems or poor recordings. If you wish to warm its sonic signature a bit, replace SS rectification with tube rectifiers; mercury-vapor rectifier tubes are excellent sounding. You could also replace the SRPP input stage with two ECC40 or 6SN7 in parallel common cathode stage. Use high anode current, at least 10 mA, which is easy for such powerful double-triodes.

TWENTY PREAMPLIFIER TUBES IN A PUSH-PULL OUTPUT STAGE?

Humans tend to overcomplicate things, and some audio designers are true masters of overcomplicating, for instance, paralleling twenty voltage amplifier triodes in a push-pull output stage. I cannot think of one good reason for such an approach, can you?

COMMERCIAL BENCHMARK: E.A.R. Yoshino V20 amplifier

V20 must be the quirkiest tube amplifier ever, not just because of its unique looks resembling Mickey Mouse's ears. The name reminds us of an automotive engine with cylinders mounted in a V-shape. Since push-pull amps work in the electrically equivalent manner, it can be justified as an analogy.

V20 uses ten paralleled duo-triodes per channel, hence its name. The quirkiness comes from the choice of output tubes, ECC83 triodes.

There are duo-triodes with much higher anode dissipation, such as 12BH7, 6CG7, 5687, and many others, whose use in this manner could be understood if not justified. Yet, its designer Tim De Paravicini chose 12AX7, the most feeble triode of all commonly used preamp tubes, with a very low anode dissipation of only 1 Watt!

E.A.R. Yoshino V20 amplifier
MANUFACTURER'S SPECIFICATIONS:

- BW: 12Hz-80kHz (no power level specified?)
- $P_{OUT} = 24W$ per channel into 4, 8 and 16 Ω
- THD: 0.5% (at what output power?)
- Signal/Noise Ratio: 90dB (A-weighted).
- Weight: 49 lb.

So, ten in parallel (five physical tubes, since 12AX7 is a duo-triode) would be equivalent to using a triode of 10 Watts anode dissipation, slightly more than the flea-powered 45 triodes (8.5 Watts) and much less than the low powered 2A3 triodes (15 Watts)!

10 Watts is the plate dissipation of 10Y (VT25) triode, but since these haven't been in production for decades, commercial manufacturers are limited in their choice of tubes.

Confused about the "enhanced triode" mode? Well, it ain't what it used to be!

In his interview for *Stereophile* magazine, talking about 12AX7 triodes operating in "enhanced triode" mode in V20, Mr. De Paravicini said, "They operate as pure triodes and are always driven hard in pure Class-A," to which he added, "Enhanced triode means the tubes are never allowed to be running in what I call the cutoff mode."

A few interesting conclusions can be made here. Even luminaries such as Mr. De Paravicini are not immune to using marketing fluff. 12AX7 are triodes, and they cannot operate in any other way but as "pure" triodes! Is there such a thing as "impure triode"? Likewise, what does "pure" class A mean? There are two types of Class A operation, A_1, where the grid current of the power tube does not flow, and A_2, where the grid current flows. When marketers and designers use the term "pure Class A," do they mean Class A_1, as if Class A_2 is something less pure?

Now, liberal analogies and meaningless language aside, notice his explanation of the "enhanced triode mode." In his most famous design, De Paravicini used PL509 or Pl519 beam power tubes driven through their screen grids instead of the control grids, which were held at zero AC potential, but at a positive DC potential. This linearizes the output stage just as the ultralinear operation does. He called that mode "enhanced triode mode."

Apparently, the verbal analogy came from the solid-state world, as a comparison between JFETs, which operate with negative gates (equivalent to triode's grid), and MOSFETs, which can operate in an enhanced mode, with their gates at a positive DC potential. While MOSFETS can physically operate in depletion or enhanced mode, triodes cannot. Nothing is or can be enhanced; it is simply a verbal analogy. A more appropriate but less catchy sounding term would be "linearized triode" or "screen-driven pseudo-triode."

"Under these conditions the valve behaves effectively as a current-controlled device, rather than the more normal voltage-controlled mode,' said Mr. De Paravicini. That is another way of saying that the I_A-V_A curves resemble pentode's (more horizontal - higher internal resistance) more than a triode (more vertical curves, lower internal resistance). Pentodes are imperfect current sources, while triodes are closer to voltage sources.

In this case, Paravicini has completely changed his definition, and now it makes even less sense. It is a regular triode mode operation with anode current swings not entering the low current region, where distortion is high, and a tube is close to or into a cutoff, where anode currents stop flowing during signal peaks, resulting in high distortion.

While the avoidance of the cutoff region can always be justified, Paravicini takes things to another extreme. To avoid losing a large part of the voltage swing, he biases the output 12AX7 tubes positively, meaning a grid current is flowing. Since 12AX7 grids weren't designed to take any current (extremely thin wire used), this begs the question of reliability and tube life - how long will these tubes last in such a regime?

Talking about his V12 amplifier, a more sensible model that soon followed, which used paralleled EL84 output pentodes instead, De Paravicini provides us with a hint: 'I wanted more grunt, a good, honest 50W/ch, and more extended tube life.' Is "more extended tube life" a euphemism for "very short tube life" in the V20 amp?

Audio history: E.A.R. 509 amplifier

Since we could not get the circuit diagram for the V20 amplifier, to understand the "balanced bridge output stage," we need to have a quick look at de Paravicini's classic, E.A.R. 509 amplifier. From a few scant technical details revealed in the Stereophile review of V20, it seems to use a similar topology. While in E.A.R. 509 fixed bias is used, the V20 output tube bank is self-biased.

The output stage is equally loaded in the cathode and anode circuits, making its voltage gain unity. Since it provides no voltage gain, three balanced voltage amplification stages (differential amplifiers) are needed in V20 to provide the very high grid drive voltages required, followed by a cathode follower stage, needed for its low output impedance to feed the current drawing grids of output tubes.

E.A.R. 509 does not have that 4th cathode follower stage of V20 since the output grids are capacitively coupled and cannot cross into Class A_2 or AB_2!

What makes that stage a "bridge" is the capacitive cross-coupling of the output tubes.

The cathode of the upper tube is connected to the anode of the lower tube (point A in the diagram above), and the cathode of the lower tube is connected to the anode of the upper one (point B), both through capacitors of course, since anodes and cathodes are at different DC potentials.

NEW? IMPROVED? MODERN PUSH-PULL AMPLIFIER DESIGNS

This is done to ensure signal symmetry at higher levels, so when one output tube starts conducting (turns "on"), it forces the opposite power tube to turn off correctly. The cathode and screen grid primary windings of each tube are also capacitively coupled but not cross-coupled.

Since 12AX7 output tubes are triodes, there is no need for the dedicated screen winding (the one connected to +260V in the diagram above), as in E.A.R. 509. The screens of PL509 beam tubes have a much lower voltage rating than their anodes, thus the need for another winding and another lower voltage power supply.

In E.A.R. 509, the fourth primary winding provides NFB to the driver stage with ECC85, while in V20 NFB signal is taken from the output transformer's primary side back to the 1st and 2nd stages.

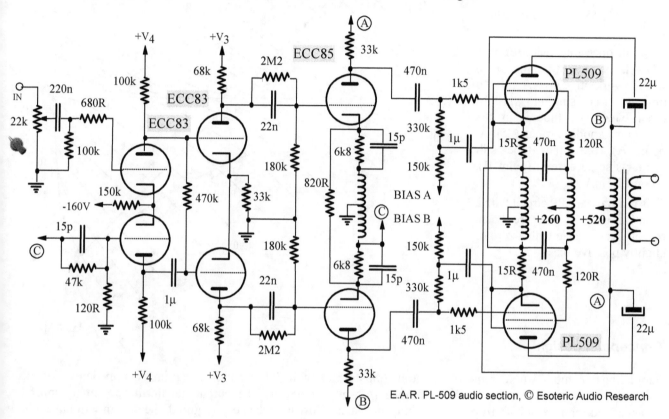

E.A.R. PL-509 audio section, © Esoteric Audio Research

The most sensible triode for a V20 PP circuit?

We could find only one small signal duo-triode whose published anode characteristics include positive grid bias curves, which is 5670, so let's use that great sounding and reliable tube to outline how one would proceed with this kind of conceptual design. First, we'll study a push-pull stage with only a pair of triodes (one physical duo-triode tube) and then extrapolate the conclusions to ten of them in parallel.

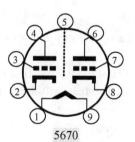

TUBE PROFILE: 5670 (6N3P)
- Indirectly-heated duo-triode
- Noval socket, heater: 6.3V/350 mA
- V_{HKMAX}= 100V_{DC}, P_{AMAX}=1.65 W, V_{AMAX}=330V, I_{AMAX}=18 mA
- TYPICAL OPERATION in PP:
- V_A=300V, Z_{AA}=27k, R_K=800Ω (V_G=-7.8V), I_0=4.9mA, I_{MAX}=6.3 mA
- V_{INPP}=20V, P_{OUT}=1 Watt

Class A push-pull 5670 stage

The 5670 datasheet includes basic parameters of a PP stage with a -7.8V bias and 5mA quiescent current. The anodes are at a very high voltage of 300V! This choice of operating point would take the tube deep into Class B operation. To get a feeling for the output power levels obtainable from a pair of 5670 triodes, let's have a look at the Class A triode push-pull stage.

The composite load line is $Z_{COMP} = Z_{AA}/4 = 5k\Omega$, and we read the current and voltage swings:

$I_{AMAX}=25mA$, $\Delta I_A = 2I_{AMAX} = 50mA$, $V_{AMAX}=275V$, $V_{AMIN}=25V$, $\Delta V_A=250V$

The output power is estimated as

$P_{OUT} = \Delta I_A \Delta V_A / 8 = 0.05 * 250/8 = 1.6W$.

From the operating curves of both tubes it is clear that the stage would work in Class A all the time, never crossing into Class B. Even when one tube is at the maximum anode current (25mA, point A), the other tube still conducts some current (point B), around 2.5mA.

This is a standard PP circuit with a negative bias. If you are keen, repeat this exercise with a zero or slightly positive bias!

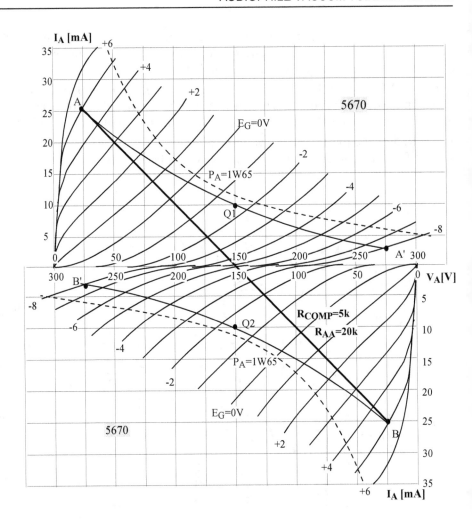

Conceptual design

Paralleling ten tubes would require an output transformer with 500Ω composite primary impedance (or 2kΩ anode-to-anode impedance) and would produce 16 watts of power. The output transformer's primary must be dimensioned for at least 250mA of current, but since such a low impedance is required, the design will have a low number of primary turns but wound with a large diameter wire.

Another demanding requirement is imposed on the driver stage. Since the output stage would cross into Class A_2 for a considerable portion of its operation, the driver stage must be a cathode follower using a powerful triode with a low internal impedance.

To produce 16 Watts on the 8Ω output tap, the secondary signal voltage must be $V_S=\sqrt{(P*R)} = \sqrt{(16*8)} = 11.3V$. Since the voltage ratio of the output transformer is 15.8:1 (attenuation of 0.0633), the anode-to anode primary voltage must be $V_P=15.8*11.3 = 178V$!

That normally wouldn't be a problem, but due to equal cathode and anode primary windings (distributed load), a strong NFB is introduced in the output stage, so that stage does not provide any voltage gain, and neither does the cathode follower driver.

Both have voltage "gains" of slightly less than one (say 0.9) for a cumulative loss of 0.8. That means the first two voltage amplification stages must provide all the voltage gain of $A_1*A_2 = 178/0.8 = 222$! This assumes input sensitivity of 1 Volt.

To get a feel for gain distribution, we can initially assume equal amplification in the first two stages, so $A_1=A_2=\sqrt{222} = 14.9$! Thus, the minimum gain we need is around 15 for both preamp stages.

On page 46 of Volume 2 of this book, we described a fully balanced Class A2-capable 2A3 push-pull amplifier circuit. The first three stages of this design could be identical to that design. The only difference is that the 2A3 output stage provided some gain, while this output stage provided none, so the first two stages must provide all the voltage gain.

The first stage used 6BQ7A duo-triode, the second stage was designed around 5614 duo-triode (12AU7), while the CF driver was 5687 duo-triode. The gains of the input stages are about 10 (each), and since a total gain of 100 wouldn't be enough here, we need to use higher gain tubes such as 12AX7 for the first stage at least, or, if NFB is used, for both stages.

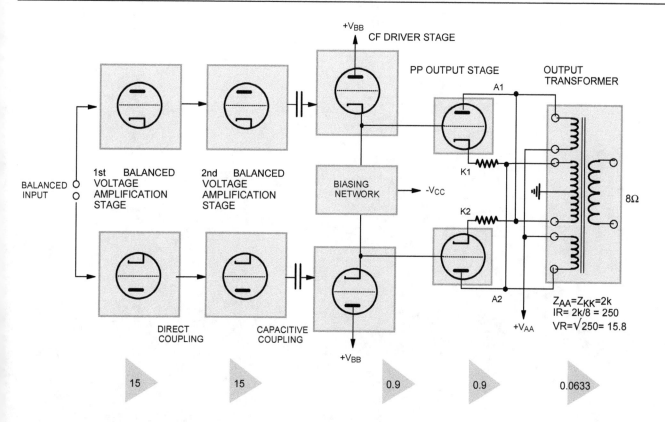

The input capacitance of the output stage with 10x 5670

Another issue remains to be assessed, which is the power required to drive the output stage. We need to consider the very low grid resistance in Class A_2 (with grid current flowing), which is generally under $2k\Omega$, and the capacitive loading of the driver stage. Only then would we be able to choose the driver tube.

5670 has a grid-to-cathode capacitance of 2.2pF, while its plate-to-grid capacitance is 1.1pF. Since the output stage operates with a gain of slightly less than one, the Miller effect is not an issue, which is good news. So, to get the total input capacitance, we simply add the two tube capacitances together and multiply the figure by ten, the number of paralleled triodes in each leg of the PP stage: $C_T = 10*(C_{GK}+C_{AG}) = 10*(2.2+1.1) = 33$ pF.

Unless you have a deep feeling for tube electronics and an elephant memory, this figure probably does not mean much to you, so to put things into perspective, let's compare it to the input capacitance of 300B. The Queen has $C_{GK}=8.5pF$ and $C_{AG}=15pF$. Ignoring the Miller's effect (assuming operation in the same type of design with no voltage gain), the total input capacitance driving one 300B tube instead of ten 5670 midgets in parallel would be $C_T = 23.5pF$.

Since 5687 can drive 300B even when the output stage has a gain (around A=3, so the dynamic capacitance is increased due to the Miller effect from $C_{AG} = 15$ pF by the factor of 4 to 60 pF), we can conclude that one 5687 will drive a bank of ten 5679 triodes without any problems.

However, we have disregarded the increase in input capacitance from the wiring in these deliberations. In the case of ten paralleled Noval triodes, the cabling is much longer (three wiring loops, anode, cathode, and grid) than with a pair of 300B. Thus, the total input capacitance will easily exceed 100pF, possibly even exceeding 200pF, while (with careful wiring) the input capacitance of a 300B triode stage would stay around 50pF.

Since a high-power cathode follower driver stage and a special output transformer with distributed loading are needed, this kind of design clearly fails *The Simplicity Test*.

Keep in mind that the anode power rating of 300B is 40 Watts, while ten 5670 in parallel have a power rating of only 16.5 Watts. So, if we don't gain output power, what do we get, improved sonics?

The sonic benefits are dubious. Why would ten tiny triodes per side sound better than one larger triode? There is always a mismatch in their characteristics, and as they age (unevenly!), that mismatch gets larger and larger, smearing the sound. Paralleling tubes cannot be justified from the purely sonic point-of-view and is usually done primarily to increase output power!

Three balanced stages plus a push-pull output stage, three preamp, and ten output duo-triodes (per channel!) and all that trouble for only 16 watts of output power? If anyone can find a better example of audio extravagance, I'd really like to hear about it!

CUTE, CLEVER OR CONTROVERSIAL? INTERESTING IDEAS FROM TUBE AUDIO'S PAST AND PRESENT

- Brook 10C amplifier and its Automatic Bias Control (ABC) circuit
- Modern auto-biasing schemes - Prima Luna's and Tenta Labs' approaches
- H.H. Scott's Dynamic Power Monitor circuit
- Moth Audio's "positive ground amplifier"
- The cascode differential input stage for stereo & bridged mono operation
- Bel Canto's way of converting a stereo SET amplifier into a monoblock with doubled power
- Marantz T-1: 300B triodes as preamp and driver tubes

What audio equipment did you like or find well engineered?
"As far as equipment, it was home made amplifiers. Also, I liked an early amplifier of 1948 vintage made by Brook. The Brook amplifier used 2A3 tubes and put out 10 watts with very low distortion."
Paul Klipsch, in an interview for Vacuum Tube Valley magazine, issue 13, 2000

Brook 10C amplifier and its Automatic Bias Control (ABC) circuit

Brook 10C amplifier was a direct competitor of Bell 2145-A. 10C also featured a combined capacitive and inductive (transformer) coupling, a 1:1 interstage transformer between the 3rd stage (balanced driver) and the CF stage, and a pair of 30nF film caps between the IS transformer's primary and secondary windings. At low frequencies, transformer coupling is predominant since the reactance of the coupling caps is high ($X_C=1/\omega C$), and at higher frequencies, as the X_C drops, the capacitive coupling takes over.

The topology is also almost identical to Bell 2145-A. Even the tubes used were electrically identical; instead of 6SN7 duo-triodes, Brook used 6J5 or 7N7, 1/2 of 6SN7!

If you thought that automatic bias control in push-pull amplifiers is a recent idea, made possible by fast, low drift, and high gain operational amplifiers, you are in for a surprise. The Automatic Bias Circuit (ABC) used in this Brook amplifier was patented in Oct. 1944 (the application was filed in Nov. 1941, the Pearl Harbor days)!

The inventor, Lincoln Walsh, and his offsider J.R. Edinger were the names behind Brook Electronics, Inc. of New Jersey. The US patent no. 2,361,889 is available for download as a pdf online, so we will cover only the basic principle behind the ABC circuit.

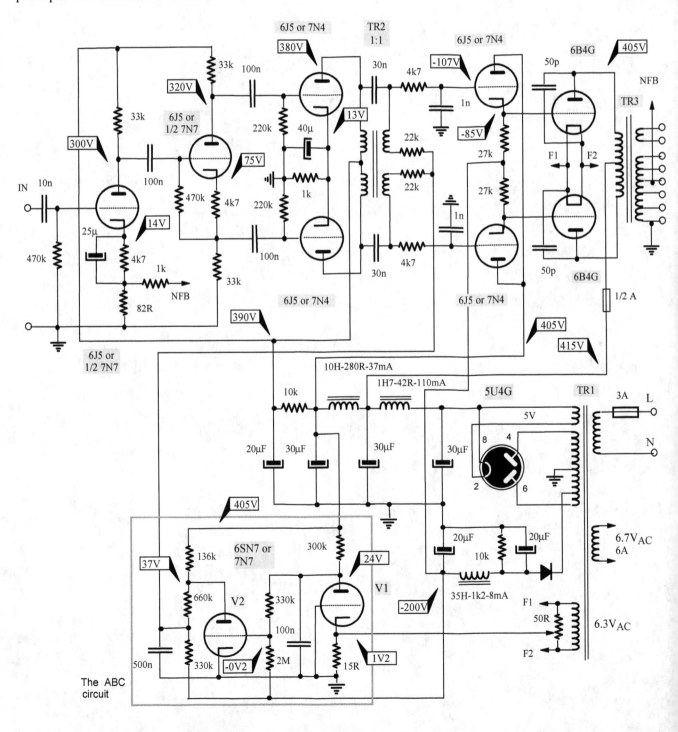

Datasheets specify power output of two 6B4 in Class AB₁ push-pull as 10 watts with 5% distortion if cathode (self) bias is used, but 15 watts (a 50% increase!) at 2.5% distortion (halved distortion levels) with fixed bias.

While Brook 12A3 was a 2A3 PP amplifier that produced 10 watts, model 10C3 produced a whopping 30 watts from a pair of 6B4G tubes (an octal and 6.3V heater version of 2A3). This was made possible by two design factors, the aforementioned automatic or, more precisely, *adaptive* biasing, and the fact that the output stage was driven by a balanced cathode follower stage, DC coupled to its grids, meaning it could easily cross over into Class AB₂, with grid current flowing.

The gain of the 1st stage is 3 (9.5dB), the 2nd stage (split-load inverter) reduces the signal by 30% (gain of 0.7), so most of the overall voltage gain comes from the 3rd stage, 25 times or 28dB. The IS transformer and the CF driver stage provide no gain (A=1 in both cases), and the output stage has a voltage gain of around 3!

Notice the relatively high bias of around -85 Volts in this design, which means that the output stage is biased very close to pure Class B operation. It operates in Class A₁ only at very low power outputs, under 5 Watts, when it crosses into Class B₁ mode, and then into Class B₂, with grid currents flowing.

The Automatic Bias Control circuit

The ABC circuit is essentially a two-stage DC amplifier. The cathode current of both output tubes flows through the 15Ω pilot or sensing resistor, connected between the cathode of the first triode and the ground. In the actual Brook amplifier, a current of 80mA flows through the pilot resistor, so the cathode voltage of V1 is $1.2V_{DC}$.

The grid of V1 is grounded, while the output from its anode feeds a high impedance resistive voltage divider (330k-2M) connected between two DC sources, a positive voltage of +405V and a negative source of -200V. The grid of the 2nd stage (V2) is at a low negative potential (-200mV in this case).

Tube V2 and its own resistive divider (136k-660k-330k) form a similar DC amplifier, the output of which is the bias voltage applied to the secondary of the interstage transformer.

In static conditions, without any signal, if the anode current through output triodes raises for any reason, the cathode of V1 becomes more positive, its bias becomes more negative, its current drops, and the voltage on its anode increases.

The grid of the following tube, V2, then becomes more positive, causing its anode voltage to become more negative (to drop) which means the voltage at the grid bias line also becomes more negative, and increased negative bias on the output tubes reduces the output current to the previous equilibrium value.

With small amplitude signals, the anode currents of the two output tubes are equal, as per the waveforms on the right.

The positive change in current is larger than the negative, but the change of the average anode current is negligible for small signals.

A large signal causes the total output current to increase as in waveform d), increasing the cathode voltage of V1 and making its bias more negative, just as in the previous discussion. This happens twice per signal cycle (period). However, the anode voltage of V1 cannot increase immediately; its rate of rise is limited by the RC circuit, the 300k anode resistor, and the 100n capacitor to the ground.. Their time constant is $\tau = 0.1*10^{-6}*0.3*10^{+6} = 0.03$ seconds, so the -3dB frequency is $f = 1/2\pi\tau = 5.3Hz$.

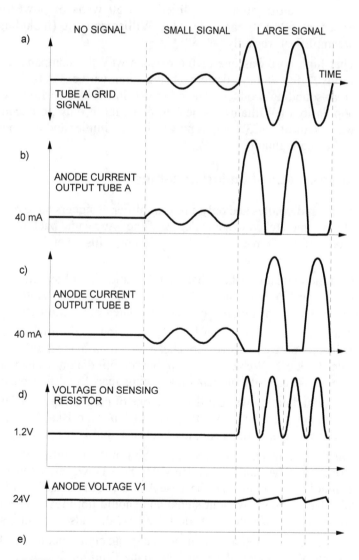

ABOVE: Current and voltage waveforms at significant points in Brook amplifier and its ABC circuit

This time constant is relatively long compared to the periods of audio frequency signals (above 20Hz), so before the anode voltage of V1 raises appreciably, the output tubes' currents drop to their zero-signal value, as does the grid bias of tube V1. The anode of V1 conducts and discharges the timing capacitor down to its original voltage (idle or zero-signal state), in this case, +24V.

So, the average DC voltage on the timing capacitor remains virtually constant, just a small ripple as in graph e), as does the bias voltage of the output tubes. In a standard self- or cathode-biased output stage, as the signal increases, so does the positive voltage on the tube's cathodes, making the bias more negative, which tends to oppose the rise in output currents, thus acting as a dynamic limiter and reducing the maximum output power attainable.

This servo system keeps the bias practically constant and thus maximizes the output power of the amplifier and its overall efficiency. As a bonus, it also reduces harmonic and intermodulation distortion of the output stage.

The competition

Brook amplifiers were not cheap by any standard. In 1952, model 10C3, without the interstage transformer or Automatic Bias Control, sold for almost $200, while the flagship model 10C3 cost $315. McIntosh 50W-2 amplifier with 50 watts of power was priced at $249.50, and Heathkit Williamson kit (including a preamplifier) was only $69.50!

That high cost was one of the reasons why the company didn't survive for long, despite being innovative and making great-sounding products. The other nail in Brook's coffin was the popularity of Williamson's design. Although it only produced 15 watts, half of 10C3's output power, it was simpler and cheaper to make and maintain.

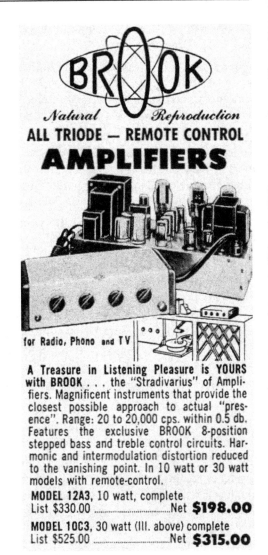

Brook advertisement, 1952, © Audio magazine

Modern auto-biasing schemes

Many audiophiles and only a few amplifier designers have raised the issue of unmatched output tubes (and they are always unmatched to some degree). The issue came back to the fore in the 1990s when Joe Curcio started offering mods for Dynaco ST70, including his version of the auto-bias feature. Check his website http://www.curcioaudio.com.

Commercial amp makers that include some kind of automatic bias adjustment can be counted using fingers on one hand only. Perhaps the best known is the Adaptive AutoBias™ by PrimaLuna. The company released its first products, the ProLogue One, and Two integrated amplifiers, in 2003 in Europe.

These amplifiers are of push-pull persuasion and sell in the US$2-3,000 range (stereo designs), while monoblock pairs are more expensive. Not cheap for China-made amplifiers, but the company developed a good reputation, and sales took off. We've never seen their circuit diagrams, had one of their amps on the test bench or even listened to one, so we cannot comment on anything apart from their biasing approach.

A circuit diagram was published on one of their competitors' websites (Doge Audio), which also make their gear in China and tout their own PAB (Precision Auto Bias™) approach, which is microprocessor controlled and user initiated (by pressing a couple of buttons) and is not permanently operating in the audio circuit like PrimaLuna's.

On their website, Doge says this: 'You may argue us that there are already auto-biases on the market, like the PrimaLuna "Adaptive AutoBias™". Yes. However, even this bias is also friendly to use, it is not a proper technology to set the bias, because Adaptive AutoBias™ works in real time and all tube experts will tell you that the bias has to be set without signal and should not move depending the signal. Doing a real time bias creates more distortion and absolutely doesn't decrease it. It also can limit significantly the power in low frequencies.'

Poor English aside, the claims in that quote contradict those of PrimaLuna and of another prominent tube designer, Menno Vanderveen, the brain behind the TentLabs Negative Bias Supply module, which sells for 179.- euro (2015). More on that product soon.

PrimaLuna's approach

According to the online diagram, the common power supply feeds a TL084, an IC containing four high-speed, JFET input operational amplifiers, each configured as a DC servo system. The circuit compares the DC voltage on the cathode (K) to the referent DC voltage in point R, measured as 0.47V, and changes the negative voltage in point G (control grid bias) to maintain it constant and equal to the voltage at R.

The +/-11V$_{DC}$ voltages are fed to the integrated circuit, while the +11V$_{DC}$ regulated voltage is also used to derive the referent bias voltage of 0.47V. This is equivalent to the voltage drop a 47mA cathode current of one output tube would make on a 10-ohm current sensing resistor between its cathode (K) and ground.

Making a cathode 0.47V positive relative to the control grid CG is the same as making the CG 0.47V negative with respect to the cathode, and this negative voltage, as we should have learned by now, is called the bias voltage.

The referent voltage (point R) is adjusted by the 10k trimmer pot, which changes the current through two forward-biased Schottky diodes. Their voltage drop depends on the magnitude of the current, as illustrated below for the BAT85 Schottky barrier diode. Keeping in mind that a LOG vertical scale was used, assuming both diodes are identical (may or may not be the actual case), with a 0.235V drop on each, the current is around 0.5 mA.

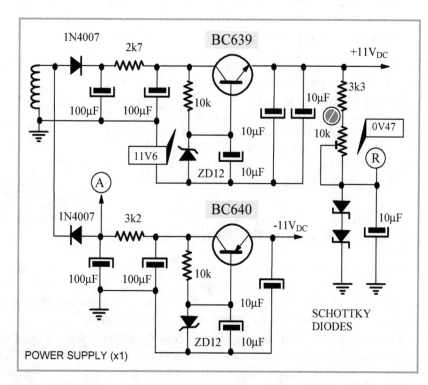

ABOVE and BELOW: The alleged (unconfirmed) circuit diagram of PrimaLuna's Adaptive AutoBias™
Source: http://www.doge-audio.com/doge%20PAB%20versus%20adaptive%20autobias.htm

BELOW: The relationship between the forward current and forward voltage drop for BAT85 Schottky diode at two temperatures, 25 degC and 85 degC

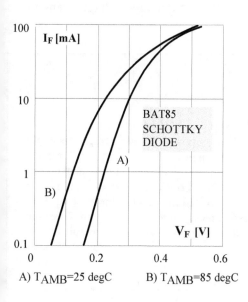

A) T$_{AMB}$=25 degC   B) T$_{AMB}$=85 degC

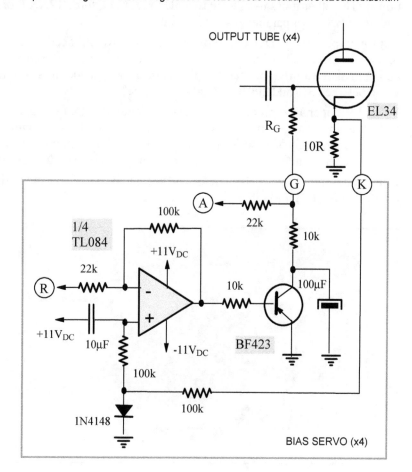

The dual +/-11V_{DC} power supply starts from the common secondary winding of the power transformer, which is half-wave rectified (1N4007 diodes) and CRC filtered and then fed into series-pass transistors (BC639 for the positive supply and BC640 for its negative equivalent). 12V Zener diodes act as voltage references.

Vanderveen - TentLabs approach

The Negative Bias Supply module does not control the existing bias but provides its own (integrated) source of negative bias voltage in two ranges, -85V to -2V and -160V to -2V. The ranges are selected by moving a jumper on the 100*100*40 mm PCB. The board controls two channels and is actually two boards that can be split; one contains the transformer and the power supply, and the other has four identical control circuits.

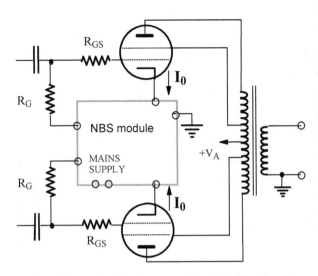

ABOVE: The in-principle diagram of TentLabs Negative Bias Supply module's connection to a typical ultra-linear PP output stage. Connection for one channel is shown; everything is duplicated for the other channel on the same PCB.

H.H. Scott's Dynamic Power Monitor circuit

Established in 1947, H.H. Scott was a vintage audio equipment manufacturer named after its founder Hermon Hosmer Scott (1909-1975). While their later (early-mid 1960s) amplifier models are well regarded by today's audiophiles and vintage amp restorers, the earlier models aren't that common or well known. Models 265-A and 280 were both mono amplifiers marketed as "laboratory standard".

265A produced 70 watts of music power (65 watts RMS) from a quad of 1617 tubes (military version of 6L6) in parallel push-pull, run as pentodes with fixed bias. There were two 12AX7 duo-triodes in preamp and phase splitter stages, one 6080 series pass regulator triode in the power supply, one 6AM8A in the dynamic power monitor circuit, and two 5U4 rectifiers.

Model 280 used a quad of EL34 pentodes and produced 80 watts short-term RMS or 65 Watts continuous RMS power. The other tubes used were two 5U4 rectifiers, three 12AX7, one 6AL5 as a bias and output signal rectifier, and one 6BX7 as a paralleled voltage regulator.

Both incorporated a "Dynamic Power Monitor", which, according to H.H. Scott, "reduces possibility of speaker burnout on overload by limiting maximum continuous output to any value desired between full power and 10 watts", and both also featured speaker damping control, which "permits continuous adjustment of output impedance to any value between 3% and 200% of load impedance".

V1 and V2 operate as an ordinary push-pull pentode output stage, but the DPM circuit can also be applied to triode and ultra-linear stages. In this case, the anode voltage is fixed, while the DPM circuit varies or limits the screen supply voltage only. The series-pass tube V5 can be a smaller triode such as 6BX7. In pentode operation, the anode current depends much more on the screen voltage and very little on the anode voltage, anyway.

In a triode output stage, the screen and the anode are connected, so their common supply voltage needs to be controlled by the DPM circuit.

265-A 70-watt POWER AMPLIFIER

A distinguished amplifier for the perfectionist. Exclusive adjustable "Dynamic Power Monitor" control allows full output on music, with maximum speaker protection. Damping factor continuously adjustable from 30/1 to 0.5/1. Class A circuitry throughout. Flat from 12 to 80,000 cps. Intermodulation distortion less than 0.1%; harmonic distortion less than 0.5% at full output.
$200.00 net*.

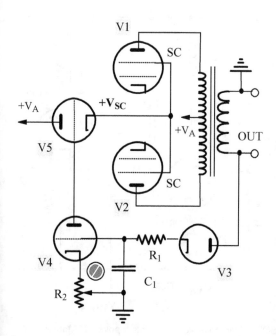
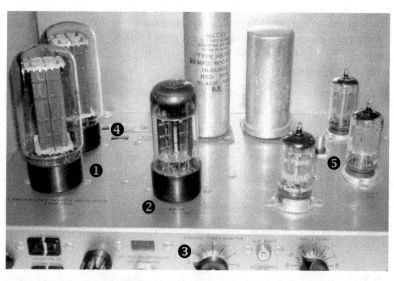

ABOVE: Partial rear view of H.H. Scott 280 amplifier: 1. dual 5U4 rectifiers 2. 6BX7 duo-triode 3. DPM adjustment for maximum power 4. Screen overload level adjustment and neon indicator, and 5. cathode bias adjustment
LEFT: The in-principle diagram of Scott's DPM

The total current through the series regulator tube is much higher in that case, and the series-pass tube must have a much higher anode dissipation rating, even higher than the output tubes!

The purpose of the DPM circuit is to protect the loudspeaker from being overdriven by the amplifier by limiting the amplifier's long-term output power level without affecting its short-term power peaks, as in music crescendos. This time delay is determined by the time constant of the R_1C_1 passive integrator circuit and is in the order of 0.25 seconds.

The music signal across the load (loudspeaker) is half-wave rectified by diode V3, integrated by the already mentioned R_1C_1 circuit, and brought to the control grid of the control tube V4. A triode is shown, but usually, a pentode is used for higher gain and better response, just as with series voltage regulators in power supplies. In a sense, that is exactly what a DPM circuit is, a series voltage regulator, which reduces its output when a large output peak is detected instead of keeping the output voltage as steady as possible!

R_2 sets the cathode bias of V4 and ultimately determines the maximum permissible output level since that tube drives the grid of the series pass tube V5. The $+V_A$ supply voltage at the input of the series-pass tube in H.H. Scott 280 is +520V, while the screen voltage at its output in quiescent conditions (no signal) is around +400V.

V5 is 6BX7 (both triodes in parallel), and V3 is half of the 6AL5 duo-diode (the other diode is used as a rectifier for a bias voltage of the output tubes), while V4 is 12AX7.

A detailed circuit diagram of model 280 is available for download from the web; it would take us too long to redraw it and discuss it here, but by all means, study its variable damping and special DPM circuitry. After mastering the material from Volume 1 and 2 of this book, you should have enough knowledge to fully understand its design and implement such a circuit in your own design if you wish to go that route.

Moth Audio's "positive ground amplifier"

Issue 17 (2001) of the now-defunct magazine "Vacuum Tube Valley" featured a two-page review of Moth Audio's 300B stereo amplifier. The power output was specified as 10 watts per channel with THD of under 1% (40Kz - 20 kHz @ 1W). The -3dB frequency range was 15Hz - 40 kHz, and the input sensitivity of 300 mV for 1W output.

This design is interesting for two reasons. Firstly, instead of the usual output audio transformer with galvanically isolated but magnetically-coupled primary and secondary windings, an autotransformer is used. The 6Ω output is simply a tap on its single winding.

Since the supply voltage to a SE stage with 300B triodes is in the order of 350-400V_{DC}, depending on the particular design, one speaker terminal would have say 400V_{DC} on it, the other a few volts less, which would be dangerous and illegal. So, what the designer (Craig Uthus) had done, is reverse the polarity of the amplifier's power supply. That is our second point of interest here.

The other end of the OPT's primary winding, normally at +400V_{DC}, is now grounded, and that is the speaker's COM terminal. The cathodes of the input (driver) stage and the output stage must be at a much lower voltage than their anodes, so a negative high voltage (350-400V_{DC}) is brought to where the ground bus would typically be.

The single driver stage uses a high mju triode (6AN4) with anode choke to maximize the voltage gain, which is almost as high as the triode's mju of 70! The amp uses external bias, which is adjustable by the user.

The use of autotransformers was chosen to reduce the significant cost of sectionalized windings in output transformers since the construction of autotransformers is much simpler. All that sectionalizing and insulation between the primary and secondary layers are not needed anymore.

More importantly, Moth Audio claims a wider bandwidth and a reduced phase shift.

However, notice that unless a coupling capacitor is used at the speaker output, nothing will prevent a DC current from flowing through the loudspeaker coil, which is never a good thing.

Likewise, since the input signal must be brought in between the cathode and the grid of the input tube, one end of the volume control potentiometer is at a very high negative DC voltage $-V_{AA}$, so a series capacitor must be used. This will prevent a DC current from flowing through the pot and the output stage of the previous preamp or signal source.

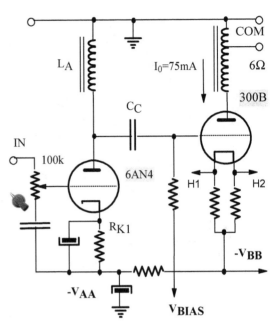

ABOVE: One possible in-principle topology of Moth Audio's design

6AN4 driver stage

6AN4 is a single triode designed for UHF applications. It is in the 12AT7 class regarding its anode power rating (both at 2.5Watts) and mju (70). However, it has a higher mutual conductance (10mA/V compared to 5.5mA/V for 12AT7) and slightly lower internal resistance, 7kΩ versus 10kΩ for 12AT7.

Let's see how 6AN4 would perform with a resistive and inductive load. We don't know the voltages and component values used in the Moth Audio's design, but this estimation will give us an idea of the tube's potential.

Even without this estimation, one look at these nonlinear curves tells you that the distortion will be unbearable. Indeed, in this case, with a resistive anode load, it is above 10%.

TUBE PROFILE: 6AN4

- High μ UHF triode
- Noval socket, heater: 6.3V/225
- V_{AMAX}=300V, P_{AMAX}= 4 W
- I_{AMAX}= 30mA
- gm=10 mA/V, μ=70, r_I = 7 kΩ

GRAPHIC ESTIMATION

A-Q1-B:
R_L=24k, V_{BB}=300V, V_{bias}=-2V
V_0=180V, I_0=5mA
ANODE VOLTAGE SWING:
$\Delta V = V_B - V_A$ = 240-88 = 152V
$A_V = -\Delta V_A/V_G$ = -152/4 = -38
ΔV_+= 92V, ΔV_-= 60V
$D_2 = (\Delta V_+ - \Delta V_-)/2\Delta V$ *100 [%]
= 32/(2*152)*100 = 10.5%

X-Q2-Y:
V_{bias}=-1.5V, V_0=197V, I_0=10mA
ANODE VOLTAGE SWING:
$\Delta V = V_X - V_Y$ = 285-98 = 187V
$A_V = -\Delta V_A/V_G$ = -187/3 = -62.3
ΔV_+= 99V, ΔV_-= 88V
$D_2 = (\Delta V_+ - \Delta V_-)/2\Delta V$ *100 [%]
= 11/(2*187)*100 = 2.9%

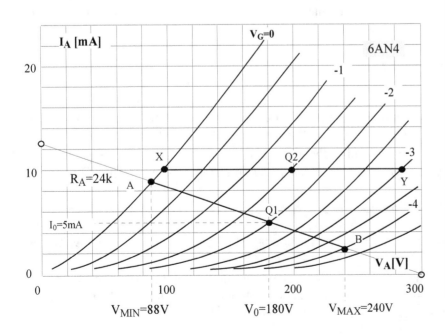

The review in VTV magazine mentions a "reactance drive circuit" in the Moth amplifier, so assuming that means the use of anode choke, let's see what would such a stage produce (load line X-Q2-Y).

A much better result, but the 2nd harmonic is still hovering around 3%, which is often taken as a borderline case. Now we can understand why the designer chose inductive loading.

The cascode differential input stage for stereo & bridged mono operation

Lazarus H1-A amplifier hails from 1990. This 50 Watts per channel hybrid features a 6DJ8 input cascode that provides all the voltage gain, feeding the MOSFET drivers and the MOSFET output stage, both configured as source followers, thus providing no voltage gain at all. The gain of a cascode stage is approximately $A \approx gm_1 * R_A$, in this case $A \approx 12.5*39 = 488$, or 53.7 dB!

The circuit is featured here to illustrate a clever way these amplifiers can be bridged for mono operation. With the stereo-mono switch in the stereo mode, as illustrated, the two preamp cascodes operate independently, the joint of their 475Ω cathode resistors is grounded by one pole of the switch (point X to GND).

With the mono mode selected, the switch disconnects point X from the ground and connects both anode resistors together. The two identical but separate anode power supplies V_{BB1} and V_{BB2} are then connected in parallel.

The 10k resistor between point X and -90V cathode power supply becomes a common cathode resistor for both cascodes, turning the two separate cascodes into a differential amplifier!

Regardless of what input is used, the balanced XLR or the two unbalanced RCA inputs, the two outputs are out-of-phase with regard to one another, as in every differential amplifier.

Thus, in bridged mono operation, one channel will be propagating a positive signal, the other a negative, and the speaker would be connected between two + (positive) speaker binding posts.

The negative speaker binding posts would be left floating, unused. The two out-of-phase output voltages would effectively be connected in series, driving one speaker, thus doubling the output power.

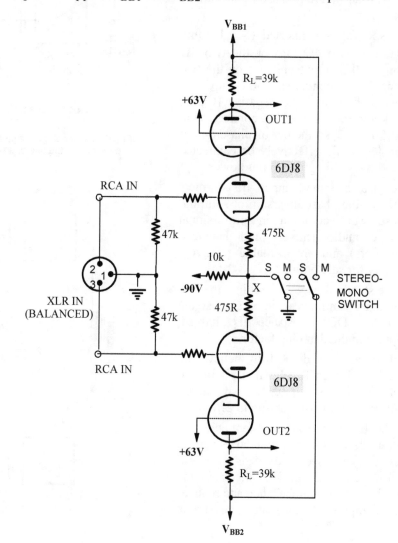

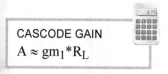

CASCODE GAIN
$A \approx gm_1 * R_L$

ABOVE RIGHT: The input stage of Lazarus H-1A hybrid amplifier

Bel Canto's way of turning a stereo SET amplifier into a monoblock with doubled power

After their well-received SET40 stereo amplifier, Bel Canto released SET80 monoblocks, which were SET40 amps cleverly bridged for mono operation.

It may pay to outline the topology of a typical SET stereo amplifier first, with signal polarities indicated by voltage pulses in main points. Both channels share the common ground (-) as a signal return and the same HV (high voltage) power supply.

The usual way to bridge two channels for mono operation is to connect the + inputs together and parallel the output transformer taps (4Ω and 4Ω, 8Ω and 8Ω, and so on.

However, the channels are never identical, neither in terms of their gains and output voltages nor their frequency range, so their parallel connection "smudges" the sound and degrades the sonics.

Another option is to invert the phase of one channel so that one channel is propagating a positive signal. At the same time, the other amplifies the same signal but with an inverted phase (a "negative" signal).

The speaker is connected to + binding posts (the two 4Ω taps or the two 8Ω taps), so the output voltages on the two channel's secondaries add up, as illustrated below. The negative (COM) speaker terminals would be connected but not grounded, so the out-of-phase output voltages would effectively be connected in series, doubling the output power.

The use of small input transformers achieves the phase inversion or splitting. These were common in professional vintage audio (mixers and balanced amplifiers) and are available relatively cheaply for sale online.

In the standard way of bridging, the two secondary windings are in parallel, so the resultant DC resistance is halved, improving the damping factor.

Just in case you didn't fully believe my "No Free Lunch Rule" when I said that there is always a price to pay, here's another example of that rule in action!

In the Bel Canto arrangement, the two output transformers' secondaries are in series, so their secondary resistances and impedances add up. This increases the amplifier's output impedance and reduces the damping factor, an undesirable effect.

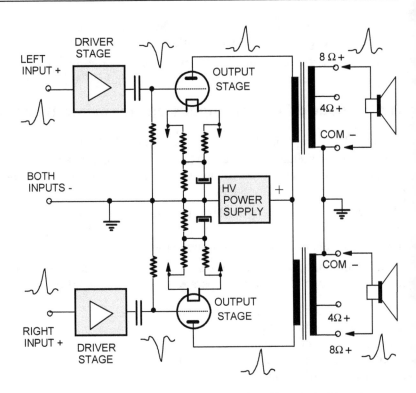

ABOVE: A topology of a typical stereo SET amplifier with a single driver stage

BELOW: The topology of a stereo amplifier with a single driver stage, converted into a monoblock by the use of an input transformer, as in Bel Canto SET80

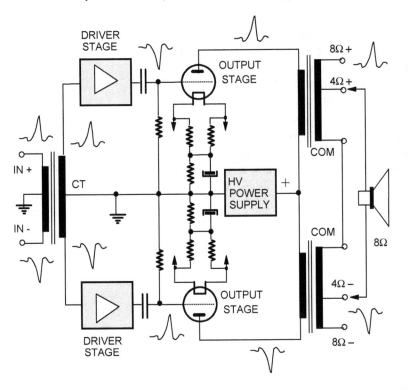

Marantz T-1: 300B triodes as preamp and driver tubes

Project T-1 started in 1989, and the final product was released six years later. It is claimed that only 50 were produced, but it is unclear is that 50 pairs or 50 individual monoblocks. Each monoblock weighs 65kg and cost 2,500,000 yen in 1995, or around US$25,000!

No test results or listening evaluations were ever published; however, a partial circuit diagram is available online, so we will use it here to deduce a few snippets about its operation and raise a few rhetorical questions about the design approach behind it.

Topology and the physical layout

The tall chassis dwarfs the four large 845 power tubes. A symmetrical topology is used - two power transformers and two chokes at the back, centrally located 5U4G rectifier tube, and all driver, output, and rectifier tubes spaced out in two neat rows. The volume control pot is on the left, and the on-off switch is on the right.

Each of the three selector switch positions connects the bias-check meter to the cathode resistors of a different amplification stage.

Separating sources of electromagnetic radiation such as power transformers and chokes on top of the chassis from the sensitive audio circuitry below (electromagnetically & electrostatically screened by the individual transformer enclosures and the chassis itself) is always the best option.

Unfortunately, most commercially available chassis are not tall enough for such an arrangement. If you have your chassis laser-cut and fabricated, you'd be wise to follow this approach. Plus, the wow factor will make your chassis stand out, just as Marantz T1 does.

While its modular chassis comprises die-cast aluminium panels, the internal supports are copper-plated steel.

The inside view (left) shows high voltage film capacitors on the far left (1 and 2) and far right (3), with a single large PCB carrying most of the power supply components (4).

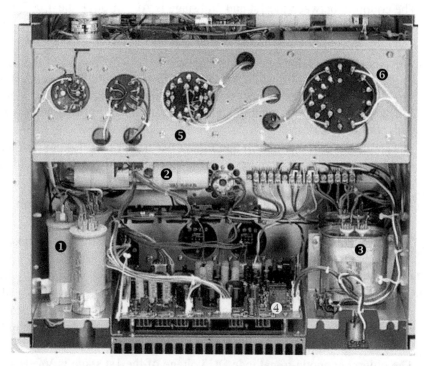

ABOVE: The inside view Marantz T1, with bottom cover removed.

KEY FEATURES:
- Fully balanced, zero-feedback design
- Very low input impedance (1 kΩ)
- Four transformers & one capacitor in the signal path
- The same audio triodes (845) were used as both output tubes and HV rectifiers
- Completely separate power supply with tube rectification (5U4) for the preamp and driver stages.
- The same power tube (300B) was used as a preamp tube in the first two stages.

All four audio transformers are mounted on the same horizontal plate underneath the front two rows of tubes (5), and follow the signal path (left to right on the photo). The input transformer and the first interstage transformer are of the same physical size, followed by the larger driver transformer and an even larger output iron (6).

Most specifications were not given (distortion, damping factor, frequency range), only input sensitivity of 0.75V for 50W output was claimed.

The audio circuit

50 Watts from a pair of 125 Watt-rated triodes is a very low power (one 845 in Class A2 SET circuit could deliver it), so it seems that the output tubes are run very conservatively. We will soon see if that is the case. The input impedance of the T-1 amplifier is very low, 1kW, so most tube preamplifiers will be unable to properly drive these monoblocks due to their high output impedance. This is a serious error of judgment by Marantz designers!

All three stages feature an almost identical topology. Cathode biasing is used, with two cathode resistors in series and a bias test point between them.

The values of the much higher upper resistance (closer to the cathode) determine the cathode and bias voltage (since grids are at an AC potential of zero). The resistance values for the lower resistors (to ground) were selected so that when connected to the common analog bias meter, the same reading is obtained in all three cases, despite different cathode voltages and currents.

The only differences between the three stages are:

The coupling between the 1st and 2nd stage is AC, through the 4µ7 250V coupling capacitor. Thus, there is no DC current through the interstage transformer TR2, just as with the input transformer TR1. None of the audio transformers used here can take any unbalanced DC current since they have no air gap.

Why such a high value of coupling capacitance (4µ7), instead of the usual 0.1 to 0.22µF? Also, notice that only a single capacitor is used, while the anode of the lower 300B triode is DC-coupled to the primary of TR2. Why isn't another coupling capacitor necessary here? After reading Vol. 1 and Vol. 2 of this book and fully understanding the concepts of AC and DC coupling and time constants of grid circuits, you should be able to answer both questions with confidence.

The anode DC current of 300B tubes in the 1st stage flows through 15k load resistors and is supplied by the 372 V_{DC} power supply, derived from the 2nd stage power supply of 390V_{DC}, after a decoupling circuit of 30H series choke and 40µF 240V_{AC} motor-start film capacitor (not shown here).

The voltage drop on the choke is 18 Volts, and since the 1st stage draws 2 x 12.5 mA (we will calculate that soon), its DC resistance is R_{CH}=18V/25mA = 720Ω.

The two identical primaries of TR2 are connected in parallel, while those of TR3 are in series, actually in a push-pull arrangement.

Let's calculate the DC currents and determine the quiescent operating points of the first two stages. The cathode voltage of the 1st stage is 38V, and since the total cathode resistance is 3k+120R, we have I_1= 38/3,120 = 12.2 mA. As a check, we also know that the DC voltage drop on 15k anode resistors is 372V-184V = 188V, so I_1= 188/15,000 = 12.5 mA. Close enough, although I expected an identical result from the otherwise pedantic Japanese! Ah, how standards are slipping everywhere ...

The quiescent anode-to-cathode DC voltage of the 1st stage is V_0=184-38 = 146V, so the power dissipated on the anode is P_0=I_0*V_0 = 0.0125*146 = 1.8 Watts.

Here we must raise the first philosophical question about this kind of design: Why should we use an expensive and low µ power tube with an anode power rating of 40 Watts in a voltage amplification stage with anode power dissipation of 1.8 Watts?

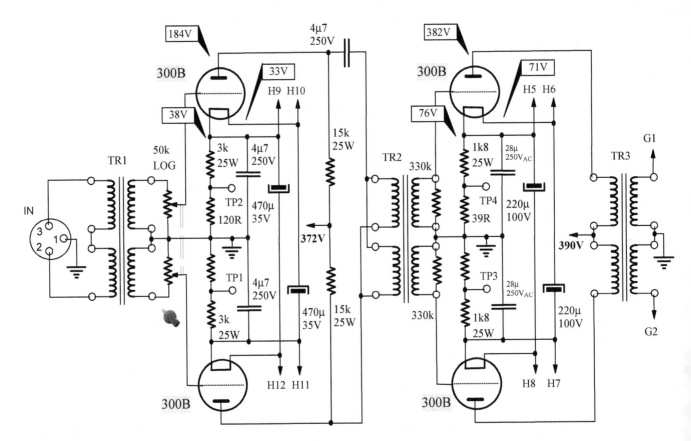

The first two audio stages of Marantz T1, © Marantz

> **CRITICAL QUESTIONS**
>
> 1. Can the use of expensive output tubes in preamplifier (gain) stages be justified?
> 2. Is the use of the same preamp/driver tube in a string of two or three identical stages wise, or should dissimilar tubes be used, so their sonic strengths and weaknesses compensate in a synergetic way?
> 3. Why are multiple coupling capacitors in signal path considered sonically undesirable while numerous input, coupling, and driver transformers aren't?

This choice makes no sense from the technical perspective, so there are only two possible answers. The first is that 300B's use is purely for marketing purposes. The perception of potential buyers is that one of the most expensive amps in the world should use as many 300B tubes (also one of the most expensive per watt of output) power) as possible.

The second reason could lie in the voicing of the amplifier. While some audiophiles don't like the warm sonic signature of 300B triodes, claiming they are "artificially colored," most do find it appealing. The first stage leaves a dominant imprint on the overall sound of any amplifier since its distortion and sonic character propagate through the rest of the stages. Thus, it makes sense to use a tube with superior sonics. However, the same tube is used here in the 2nd stage as well, and in the same type of circuit.

This raises another important question, the answer to which audio designers and audiophiles will never agree: Should we string identical or similar amplification stages using the same tube, or is it better (sonically) to aim for synergy and use different tubes in different stages, so their sonic strengths and weaknesses compensate?

Finally, if using four capacitors in the signal path is viewed as undesirable (distortion, phase shift, etc.), why is the use of four transformers (which also distort and cause phase shift) considered "hi-end, no holds barred pinnacle of audiophilia"? Is "inductive" or "magnetic" distortion of transformers and chokes better sounding than the "capacitive" or "dielectric" distortion of capacitors?

A few more points to conclude our numerical analysis of the preamp stages. The cathode voltage of the 2nd stage is 76V, and since the total cathode resistance is 1k8+39R, we have $I_1 = 76/1,839 = 41.3$ mA!

Since the CT of the TR3's primary is at the supply voltage of 390V, but the anodes are at 382V, the voltage drop on each half of the primary is 8V, meaning its DC resistance is around $R_{DC} = 8/0.0413 = 194\Omega$.

The anode-to-cathode DC voltage of the 2nd stage is $V_0 = 382-76 = 306V$, so the power dissipated on the anode of each 300B is $P_0 = I_0 * V_0 = 0.0413*306 = 12.6$ Watts.

The time constant of the 2nd stage is $\tau_2 = R_K C_K = 1,839*28*10^{-6} = 45.9$ milliseconds.

The -3dB frequency of its cathode circuit is $f_L = 1/(2\pi\tau_2) = 3.1$ Hz. We didn't check the time constant of the 1st stage's cathode circuit and the associated lower -3dB frequency. As a practice exercise repeat this calculation for the 1st and the output stage.

> **TUBE PROFILE: 300B**
> - Directly-heated power triode
> - Heater: 5V, 1.2 A
> - Maximum anode voltage: 450 volts
> - Maximum anode dissipation: 36W (40W)
> - TYPICAL OPERATION:
> - $V_A=350V$, $V_G=-74V$, $V_{SIG}=74V_P = 52V_{RMS}$
> - $I_0=60$ mA, $I_{MAX}=77$ mA, $R_L=4k\Omega$
> - $gm=5.0$ mA/V, $\mu=3.85$, $r_I = 790\Omega$
> - $P_{OUT} = 7$ W @ 5% THD

Estimating the interstage transformer voltage ratios

The voltage amplification of the first stage is $A_1 = v_{OUT}/v_{IN} = -\mu*R_L/(r_I + R_L)$. For 300B with $R_L=15k$, $r_I=800\Omega$ and $\mu=3.8$, the voltage gain is $A= -3.8*15/(0.8+15) = -3.6$. Since A_2 is roughly of the same value, $A_1*A_2 = 13$.

Since the cathode of the output stage is at $120V_{DC}$, the peak signal voltage at its input should not exceed that value in Class A_1 operation. The effective value of the $120V_{PEAK}$ signal is $120/1.41 = 85V_{RMS}$.

The nominal sensitivity of this amplifier is $0.75V_{RMS}$ at its input for full power, so we can estimate that the total required amplification from its input to the grids of the output tubes is $85V/0.75V = 113$ times.

The two 300B stages (without the three coupling transformers) provide a gain of 13, so the voltage gain of the three transformers needs to be around $113/13 = 8.7$! One possibility is that each of the three trafos provides a gain of two (step-up ratio of 1:2), for an overall gain of $2*2*2 = 8$, close enough to the estimated 8.7!

Driver transformer insights

Japanese "Zuoshi Audio" website (http://www.zuoshi.com/product-2550.html) lists the specs for their driver transformers ZS-3345 and ZS-3345B (priced at US$934 each in Sept. 2015), which are claimed to be equivalent or at least a suitable replacement for those used in Marantz T-1.

They are of 1+1 to 2+2 voltage and turns ratio with 6kΩ to 20kΩ specified impedance ratio (half-to-half). However, 1:2 voltage ratio means the impedance ratio is IR= VR^2= 2^2 = 4, so with 6kΩ impedance of each primary half, the impedance of each secondary half should be 24kΩ and not 20kΩ! Minor details aside, the DC resistance of each primary is 130Ω, while the secondaries measure at 189Ω each. The frequency bandwidth is claimed to be from 3Hz to 117kHz (-3dB), an excellent result.

The maximum primary current is 170mA, while the secondary wire is sized for 30 mA. Since there is no air gap, the maximum allowable DC imbalance current is only 5mA, so well-matched 300B tubes are mandatory!

The output stage

The cathode voltage of the output stage is 120V, and since the total cathode resistance is 1,522Ω, we have I_3= 120/1,522 = 79 mA.

The anode-to-cathode DC voltage is V_0= 970-120 = 850V, so the power dissipated on the anode of each 845 triode is $P_0=I_0*V_0$ = 0.079*850 = 67 Watts, which is very low for a tube rated at 125 Watts.

Since we know the operating points Q1 and Q2 of the two output tubes, let's slap them onto the composite anode curves and estimate the primary anode-to-anode or PP (plate-to-plate) impedance of the output transformers used.

This graphical technique was covered in detail through quite a few examples in Volume 1 of this book, so the logic behind its construction and associated details will not be repeated here.

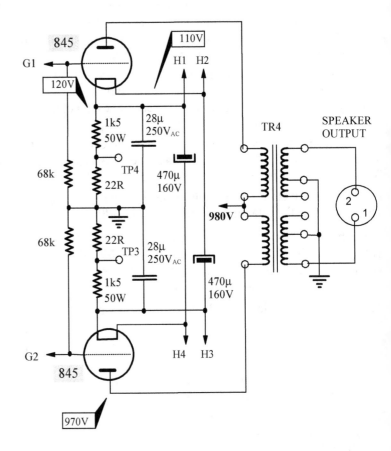

TUBE PROFILE: 845

- Directly-heated power triode
- Socket: 4-pin jumbo
- Anode power rating: 100W
- Thoriated tungsten filament/cathode
- Heater: 10V, 3.25 A
- Max. anode voltage: 1,250V
- gm=3.1 mA/V, µ=5.3, r_I = 1,700 Ω

ABOVE: The output stage of Marantz T1, © Marantz

The graphic analysis of Marantz T1 output stage

The two anode characteristics are aligned along the V_0=850V vertical line. Once its intersections with two I0 = 80mA lines are found, these are the quiescent points Q1 and Q2. The bias voltage from the curves seems to be closer to 110V than 120V from the circuit diagram, but that is within the margin of error of such an imprecise graphical method.

Points A and C must lie at the E_G=0V lines (zero grid voltage - maximum signal), while points B and D must lie vertically opposite, on the curves for double the bias voltage, around -220V.

With a bit of "wiggling around," the operating path for each tube (must be identical) will be curved as pictured. The composite load line RL connects points of zero bias (A and C). Let's calculate its slope from the graph. The voltage swing (read from the upper graph, but the same result must be gained if the bottom curves are used) is ΔV=1,350 -350V = 1,000V. The current swing is approx. twice the maximum current swing of one tube (155 mA), so ΔI=2*155mA = 310mA. Finally R_L= ΔV/ΔI= 3,226 Ω.

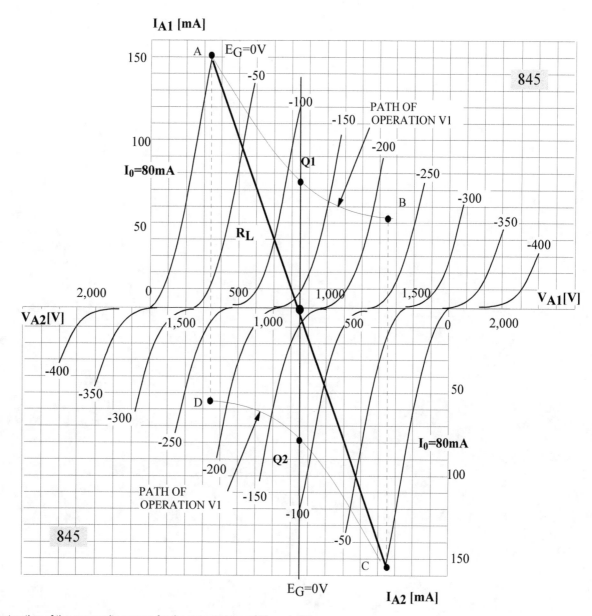

ABOVE: The construction of the composite curves for the output stage of Marantz T1

Now, if you remember the discussion from Vol. 1, the actual anode-to-anode or primary impedance is four times higher than the composite load R_L, or $Z_{AA} = 4R_L = 12,900\Omega$ or 12.9 kΩ!

Another reason for the relatively modest 50Watt output is the fact that this amplifier operates entirely in Class A and never crosses into class B. How do we know that?

Look what happens to I_2, the anode current of V2, when the anode current I_1 of V1 is at its maximum (point A). The current I_2 in point D is not zero but at a relatively high level of 55 mA. That means that the other tube (V2) is also conducting even at the lowest point. The definition of Class A push-pull operation is both tubes conducting the audio signal at all times without ever entering a cutoff region, which is the case here!

THIS PAGE WAS DELIBERATELY LEFT BLANK

THRIFTY TIPS & TRICKS: TIME & MONEY SAVING IDEAS

- Quality components from vintage tube test equipment
- Saving on power supply parts
- DIY project: High voltage test bench power supply for amplifier development work
- Professional quality, custom made chassis in two weeks or less
- How cheap and nasty tube sockets can end up costing you big $$$!

"I was brought up in an era when thrift was still considered a virtue."
John Paul Getty, American industrialist, world's richest private citizen in 1966 (Guinness Book of Records)

Quality components from vintage tube test equipment

Everyone's after vintage tube amps and preamps these days, and even budget units of dubious quality are fetching incredibly high prices, way above their actual value. Shrewd recyclers know better and look elsewhere for quality components that can be reused. There is no better option than the vintage tube test and measurement instruments. Compared to most audio gear from the 50s and 60s, which was built to a strict budget, tube instrumentation was expensive and built to a much higher standard, including parts of superior quality and precise values.

Example #1: Heathkit and Eico electronic switches

In the 1950s and 60s, CROs (Cathode Ray Oscilloscopes) were expensive instruments, especially professional instruments such as Tektronix and HP. Even the more affordable versions aimed at DIY and hobby users, such as those made by Heathkit and Eico, were not cheap at all. Most were of a single-channel type to save on parts and reduce complexity. Soon, their designers realized that users wanted two channels to compare waveforms of two amplifier channels and measure the phase shift between two voltages or between the input and output of an amplifier, for instance.

Electronic switches were a low-budget alternative to expensive dual-trace scopes. Two inputs are brought in, the unit operates as a free-running (astable) multivibrator, alternatively passing one and the other input signal to the output, fed to a single-channel scope. Due to the inertia of the human eye, two traces would be observed on the scope's screen. Nowadays, these switches are obsolete; nobody would use them for their intended purpose, so they are cheap, below $50, often even under $20! The most common models are Heathkit ID-22 and Eico 488.

First of all, you get a power transformer for a preamplifier, with a center-tapped HV winding of 20-80mA capacity and LV secondary supplying heater voltage(s). There is even a filtering choke that you can also use in preamplifiers. These two components alone will recoup the cost of this entire investment, and then some!

BELOW: One humble and unwanted Heathkit electronic switch such as ID-22 gives you enough tubes and power supply parts to build a phono stage and a line preamplifier!

You also get high-quality tubes. Mullard often supplied Eico with tubes, and Heathkit used Telefunken as their supplier, so a feast of quality brands there. Most of these vintage units had not been used often, or not at all, so the tubes inside them are practically NOS.

Eico 488 has a 6X5 rectifier, two 12AU7, and a pair of 6AU6 pentodes, which also make very linear and good-sounding triodes.

Heathkit switches such as ID-22 or S-3 are even better for salvage purposes. Apart from the mains transformer and choke, you get a 6X4 rectifier, two 12AX7, three 12AU7, and a 6C4 single triode (half of 12AU7), all made by Telefunken in Germany!

Example #2: HP 425A microvolt-amperemeter

Hewlett-Packard tube voltmeters such as the 400 series (versions D, H, and L) and many other models are basically single-channel preamplifiers of the highest quality. They can measure AC signals from 1 mV to 300V, ranging in frequency from below 10 Hertz to 4 Megahertz! How many audio preamps do you know with such a wide frequency bandwidth? Their input impedance is 10 MW, and for monitoring purposes, there is even a very low impedance (50Ω) cathode follower output.

So, don't use these and other vintage beauties such as Tektronix tube scopes just as an Aladdin's cave of quality parts. Pay them, and their designers respect and study their circuit diagrams and user manuals. You will learn more than in most books on tube amplifiers (except this one, of course)!

HP 425A microvoltmeter is the best model for salvaging purposes. It will give you a large, quality-made, fully shielded power transformer (1) with dual primaries, suitable for most preamplifiers of up to six audio tubes!

While DIY constructors living in the USA have a great choice of salvaged mains transformers, those of us from 230 or 240V mains countries are much more limited. Luckily, electronic giants such as Tektronix, HP, and even Heathkit (later models) were selling their equipment worldwide, so most of the mains transformers inside their gear have dual voltage primaries. These can be connected in parallel for 115V operation or series for 230V, another reason to get them whenever you can.

THRIFTY TIPS & TRICKS: TIME & MONEY SAVING IDEAS

You will also get a 6X4 rectifier (2), OA2 and OB2 voltage regulator tubes (3), and three lovely duo-triodes, 5751 (low-noise version of 12AX7, perfect for phono stages), ordinary 12AX7 (as if there was anything ordinary about that legendary tube?) and 12AT7 (4). There were two Mullard UK 12AX7 tubes in one unit and a Telefunken 12AT7! These three could fetch more on eBay than what you pay for the whole 425A, including international shipping, meaning you get everything else for free.

You also get quality tube sockets, binding posts, fixed and trimmer resistors, switches, terminal strips, and handy aluminium mounting plates of all shapes and sizes. Best of all, all multi-strand hookup wire is of the highest quality, in all kinds of colors, and, wait for it, made of pure silver!

Models 400D, 400H, and 400L are less desirable since they use tubes of no interest to DIY audio constructors. Apart from the 6AX5 rectifier and 5651 VR tube, there are 5x 6CB6 pentodes, 1x 6U8 triode-pentode and a 12B4 triode.

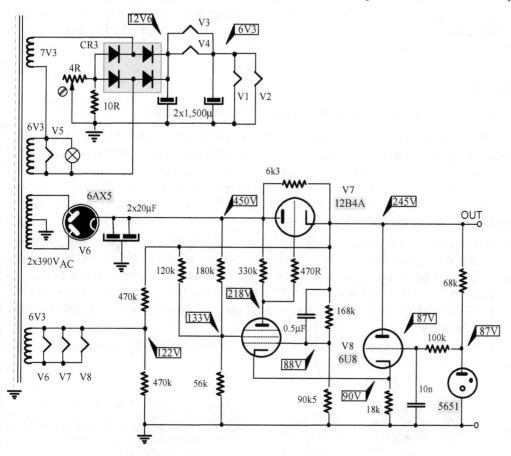

LEFT: HP 400 series power supply. To save space, the primary connection is not shown.

BELOW: HP 425A and with its slip-on aluminium cover removed, to reveal its internal construction and major parts.

While 425A uses a simple passive shunt regulator with OA2 and OB2 tubes, the 400 series features a sophisticated voltage regulator with 12B4A as a series-pass triode, 5651 voltage reference tube, and 6U8 as a two-stage error amplifier. Very few modern high-cost audiophile line stages use such an intricate voltage-regulated power supply.

The three tubes in the voltage regulator circuit (V6, V7 & V8) have their own secondary winding providing 6.3V for their heaters, elevated to half of the output voltage (245V/2=122V) through the 470k+470k voltage divider. Since 6AX5 is indirectly heated, it does not need its own, dedicated secondary winding for heater supply.

Tubes in the measuring circuit (V1-V4) are heated by a DC voltage. Also, look at the clever way that 12.6V voltage can be adjusted by simply connecting a 4R rheostat and a fixed parallel resistor (10R) between one end of the rectifier bridge and ground.

The triode section of the 6U8 triode-pentode tube works as a cathode follower, keeping the cathode of its pentode section (error amplifier) at a constant potential. The rest is a fairly standard series regulator, with a 0.5µF startup capacitor to the control grid, while the screen grid is kept at a fixed 133V, referenced to both the input voltage (via 180k resistor) and output (via the 120k resistor).

Saving on power supply parts

Motor start and audiophile metallized polypropylene high voltage capacitors

We have already talked about the cheapest upgrade you can make to any vintage or new design, and that is to replace electrolytic power supply filtering capacitors with film caps. If you like brand names, the 47µF "Solen Fast Cap" is of metalized polypropylene construction, rated at 630V_{DC}, and sells for US$22.95 each. Since it has exposed axial leads, it cannot be mounted above the chassis, but there are polypropylene film caps that can - motor start (MS) capacitors. Some have plastic cases; others are in metal housings. The same metal clamps for securing electrolytic capacitors can be used, bolted to the chassis from underneath.

Most of the currently produced units come from China, and higher capacitance units (30-47µF) generally cost under US$10. However, beware, not all MS capacitors are film type; many are bipolar electrolytics, and these aren't sonically any better than unipolar ones! Since the sellers do not emphasize this fact, two methods can be used to identify them. If the item description does not mention either "polypropylene" or "film," it is likely a bipolar elco. Secondly, they pack higher capacitances into the same size cases, as in the photo below, typically three times higher, 150µF versus 30µF for polypropylene film.

Motor start caps are rated in AC volts, but to convert that to DC volts; simply multiply the AC rating by 2.82. So, a 450V_{AC} rated motor start cap is suitable for up to 1,270V_{DC}!

Four typical made-in-China motor start capacitors. L-R: polypropylene film with snap-on terminals (450 V_{AC}), polypropylene film with integral leads, bipolar electrolytic with snap-on terminals, and bipolar electrolytic with screw-type terminals

DC VOLTAGE RATING OF MOTOR-START & OTHER AC-RATED FILM CAPACITORS

$V_{DC} = 2.82 * V_{AC}$

Two low voltage power transformers back-to-back

One of the most expensive tube amplifier parts is the power transformer. Since mains transformers with low voltage secondaries are much more common and much cheaper (due to their widespread use in solid-state equipment), simply buy two such mains transformers and connect them back-to-back. TR1's primary winding is your primary, while TR2's primary winding is your secondary (output) winding.

The transformers don't have to be identical; their voltage taps and even power ratings can be different. The main thing is that the smaller transformer's power rating can handle the load. They can be R-core, C-core, EI-core, or toroids; it does not matter at all.

Only three of many possible connections are indicated. To get the same V_{OUT} as the primary mains voltage, simply connect 15V tap to 15V tap, or 12V tap to 12V tap, and so on, as per option a).

THRIFTY TIPS & TRICKS: TIME & MONEY SAVING IDEAS

To get a lower V_{OUT}, connect a lower tap on TR1 to a higher tap on TR2, as per the example b). For a higher output (as you would need in $115V_{AC}$ mains countries such as the USA), connect a higher tap on TR1 to a lower tap on TR2, as per the example c).

$V_{OUT} = V_{IN} * TAP_1 / TAP_2$

The higher the TAP_1 volts and the lower the TAP_2 volts, the higher the output voltage V_{OUT}! The heating voltage is taken from the appropriate secondary of TR1, in our case, 6V. You can take it from the 9V tap, rectify it and regulate it down to 6.3 Volts.

So, the second benefit (apart from saving money) is the adjustability (in steps) of the output voltage.

Case #1: 230V mains transformers

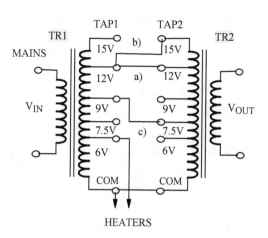

Say you need an AC voltage of around $320V_{AC}$, which will be rectified and filtered into a DC anode supply voltage of $400V_{DC}$.

Since $V_{OUT}/V_{IN} = TAP_1/TAP_2$, you have $TAP_1/TAP_2 = 320/230 = 1.39$! You need to find two taps that will give you that ratio. With 15V and 12V, 15V/12V=1.25, which is too low, or 15V/9V=1.66, which is too high. Let's try 12V/9V=1.33; that is as close as we can hope for - the output voltage will be 230V*1.33 = 306V!

If 306V is too low (better to have a higher voltage than a lower one, you can always kill a few volts, but you can never add any!), try 9V/6V=1.5, which would give you $345V_{AC}$ at the output.

Case #2: 115V mains transformers

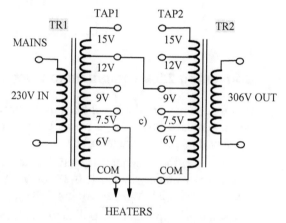

Since the difference between typical HV levels in a tube amp and 115-120V mains voltage in the USA is much greater than in the 230-240V mains countries, the secondary ratios are much greater! Say you need an AC voltage of around $320V_{AC}$, which will be rectified and filtered into a DC anode supply voltage of $400V_{DC}$.

Since $V_{OUT}/V_{IN}=TAP_1/TAP_2$, you have $TAP_1/TAP_2 = 320/115 = 2.78$ or close to 3! The highest ratio obtainable with two transformers from case #1 is 15V/6V = 2.5, which would result in an output of only 288V.

You'd need to find two mains transformers whose low voltage secondary ratios would be close to 3, for instance, TR2 with V_S=6.3V (say a vintage tube heater transformer) and TR1 with V_S=18V (say a 2 x 9V transformer from a small solid-state amplifier), which would give you a ratio of 2.86 and HV of 2.86*115V=329V, as illustrated on the right!

ABOVE: Case #1 (230V mains)
BELOW: Case #2 (115V mains)

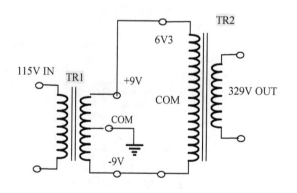

Power rating considerations

So far we have only been considering voltages, but these back-to-back transformers must have adequate current ratings as well. The power rating of a transformer and its secondary current ratings are always specified by the manufacturers, although primary current draws aren't, but that can be calculated quickly and easily.

As an example, say that each channel of the stereo amp will draw 100 mA_{DC} or 200 mA_{DC} in total. Once we convert that into AC current and add a safety margin let's say the primary of TR2 must be rated for 300mA. The voltage on its primary is now 306V (case #1 above right), so its power rating must be at least P=0.3*306= 92VA.

TR1 is a bit different, since we are assuming that it must supply the 6V heater supply for the amplifier as well. Let's say we have two preamp tubes drawing 600mA and two power tubes drawing 2A each. The additional dissipation on the 6V winding of TR1 is PH= 3.2A*6V = 19.2 VA. So, the total power rating of T1 must be 92VA (to supply T2) plus 20VA for the heaters or 112VA total.

We've neglected losses in both transformers, and they are typically 5-10%. Adding 10% to TR2 power draw we get a minimum of 100VA and for TR1 we get P_{MIN}=1.1*(100+20) = 130VA!

Practical example: BTB universal power supply for tube preamplifiers

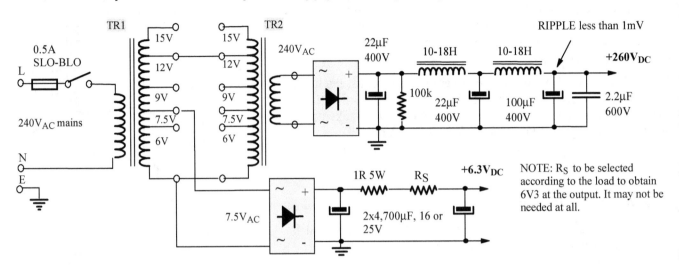

Converting 220V equipment to work on 240V mains or 100V equipment on 115V mains

If you don't have or don't feel like buying or winding a 240 to 220V transformer or an auto-transformer yourself, just buy a standard low voltage mains transformer of a suitable VA (apparent power) rating (minimum 30VA), and connect it as per the diagram below.

In this case, the secondary voltages are symmetrical (15-0-15V), but they don't have to be; we only need one secondary voltage, say 0-24, 0-28, or 0-30V. Instead of the nominal $240V_{AC}$, our workshop's mains voltage is usually around $248V_{AC}$, so we got $218V_{AC}$ out, perfect!

The same principle can be applied to amplifiers designed for $100V_{AC}$ mains (Japan), so they can be used in $115V_{AC}$ countries such as the USA. In summary, you are dropping part of your mains voltage across the transformer's secondary winding and taking the reduced AC voltage off its primary winding for the actual amplifier (load).

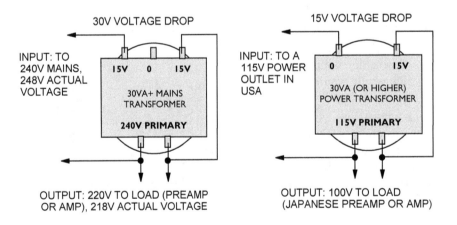

FAR LEFT: How to use a small low voltage transformer with a 30V or 2x15V secondary to run 220V amplifiers on 240V mains.

LEFT: How to use a small power transformer with a 15V secondary to run 100V Japanese amplifiers on 115V mains voltage.

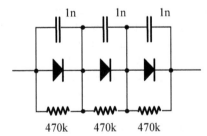

Connecting solid state diodes in series

Just as capacitors of lower voltage rating can be wired in series to work on a higher voltage, so can identical silicon diodes be wired in series to increase their overall voltage rating. Remember, we are not talking about the operating or forward voltage, but about the PIV, Peak Inverse Voltage. The PIV rating of a diode or a string of identical diodes should be at least double the actual PIV the diode or diode string is subjected to.

So, if you have a bunch of low voltage diodes and want to avoid the expense of buying high voltage ones, simply string them along!

To equalize the voltages, just as with capacitors, bypass each diode in a string with a 330-470k resistor. The power rating is not critical; they can be 1/2 Watt or 1 Watt resistors. Also, bypass each diode with a 1nF ceramic capacitor, as per the diagram.

Of course, the current rating of all diodes in the string must be adequate since the same current flows through all of them. Talking about the current rating, unfortunately, you cannot parallel two or more silicon diodes to increase the overall current rating.

THRIFTY TIPS & TRICKS: TIME & MONEY SAVING IDEAS 241

DIY project: High voltage test bench power supply for amplifier development work

Commercial alternatives

Vintage high voltage regulated power supplies are available for sale online but are limited to 125 mA of current and 400V_{DC}, with maximum power delivered to a load of only P = 0.125*400 = 50 VA. That could be enough for low-powered single-ended designs with 2A3 or 300B triodes but is inadequate for larger SE and many high-power push-pull circuits.

All these vintage HV power supplies use the same series regulated design with two triode-strapped 6L6 tubes in parallel, so it doesn't matter which one you get, Eico, Precision (PACO), Knight, or Heathkit. By far, the most common are various reincarnations of Heathkit's "Regulated H.V. Power Supply."

Later units used solid-state diodes instead of tube rectifiers and Zener diodes instead of VR (Voltage Regulator) tubes and were sold under the name Heath Zenith and Heath Schlumberger. Those later models feature transformers with dual primaries (115-230V), while the other three brands mentioned were only made for 115V mains voltage.

If you plan to do lots of circuit experimentation and testing, you should build your own better and more powerful HV power supply. You can go as high in voltage and current capabilities and other functionality as you need or want.

Ours can go up to 600V and 300 mA or 180 VA, three and a half times more powerful than those vintage units designed by accountants and financial controllers. Before we look at its design, let's refresh our knowledge of how these regulated power supplies work from Volume 1 of this book.

ABOVE: Vintage high voltage regulated power supplies such as Heathkit IP32, Heath Zenith IP2717 (pictured) or Eico 1030 are useful but limited vintage workshop instruments, not just in amplifier design but can be are heart of your own tube testing and matching setup.

The series pass tube as a variable resistor

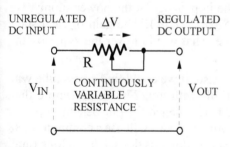

ABOVE: The operating principle behind series voltage regulators

BELOW: Basic series voltage regulator with a triode error amplifier and ZD reference.

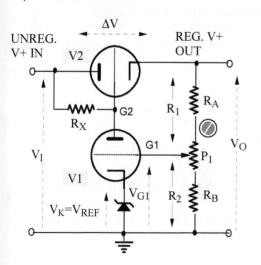

A variable resistor connected between the input and output terminals would neutralize any changes in input voltage or output load current by adjusting its resistance, thus changing the voltage drop across its terminals. A higher input voltage means a larger voltage drop on that resistor and a constant output voltage. A larger load current means lower resistance, lower voltage drop, and a constant output voltage.

However, a resistor cannot do that by itself, so a servo-circuit employing an error detector and a negative feedback design needs to be designed. Series regulators have a power tube in series with the output, and since it must pass the full load current, it is also called a "pass tube."

The whole regulator is a negative feedback DC amplifier that keeps the output voltage steady by opening and closing the pass tube as needed, just like the cruise control on your car keeps the speed constant by opening or closing the throttle.

Series regulator with a triode error amplifier

Triode V1 is a DC error amplifier. Part of the output voltage is taken from the voltage divider network of two fixed resistors and one trimpot and fed to the triode's grid. The voltage is $V_{G1} = V_{R2} = KV_O = R_2/(R_1+R_2)V_O$! The gain of this amplifier whose load is R_X is $A_1 = -\mu_1 R_X/(R_X+r_I)$.

The change of the anode voltage of the error amplifier drives the grid of V2, the series pass tube, so $V_{G2} = A_1 V_{G1} = -\mu_1 K R_X/(R_X+r_I) V_O$.

The change of the plate current of V2 is $\Delta I_{P2} = KA_1 Gm_2 \Delta V_O$.

The ratio of the change in output voltage to the change of output current is the output impedance of the voltage regulator: $Z_O = \Delta V_O/\Delta I_{P2} = 1/(KA_1 Gm_2)$, where A_1 = amplification factor of the error amplifier, gm_2 = transconductance of V2 and $K = R_2/(R_1+R_2) = 50/147 = 0.34$.

For the 200 mA regulator with 6080 as a series pass tube and 12AX7 (both triodes in parallel) as error amplifier, illustrated on the right, we have $A_1 = -\mu_1 R_X/(R_X+r_I) = -100*470/(470+33) = -100*0.93 = -93$. Since $gm_2=13$ mA/V, and $K=R_2/(R_1+R_2)$, the output impedance is $Z_O = 1/(KA_1 Gm_2) = 1/(0.1875*93*0.013) = 4.4\ \Omega$!

The 33nF capacitor between the grid of the error amplifier and the output terminal is called a speedup capacitor. It makes factor K=1 for ripple and sudden surges, increasing the amplification at higher frequencies and thus improving the regulating speed of the regulator.

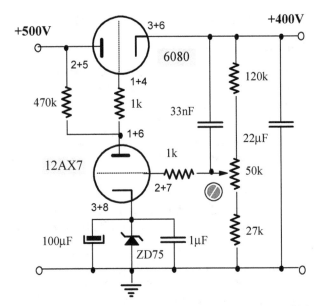

ABOVE: Practical implementation of a simple series high voltage regulator with paralleled 6080 duo-triode as a pass tube and 12AX7 error amplifier

The 470kΩ anode resistor of the error amplifier (R_X) is connected to the unregulated side, which is not ideal. The performance would be improved if it was connected to the other (regulated) side of the pass tube, but then there is a danger that the whole circuit would hang up and not start regulating.

Zener diodes are noisy and should be bypassed by a film capacitor. The elco in parallel stabilizes the voltage even further. We used 13mA/V for gm_2, although 6080 triode has gm of 6.5 mA/V since V2 is actually two 6080 triodes connected in parallel (one physical tube).

Our design

It is now obvious how improvements in performance could be made. Although we used the triode with the highest amplification factor for our error amplifier (12AX7 has μ=100), to reduce the impedance of the power supply, we could use its two triodes in cascade, for an overall amplification factor of 88*88 =7,744! The gain of each stage would be 88 and not 93 due to the higher internal resistance of a single triode. That would lower the Z_{OUT} by the same factor (88 times), bringing it under 54 mΩ!

We could also use a pentode since their m is much higher than that of triodes. Also, if we could find a pass tube with a higher gm, that would lower the output impedance. One such tube is E130L with gm=27.5 mA/V, more than double that of two 6080 in parallel. However, we don't need such superior performance for our benchtop power supply and experimentation purposes (as we would desire in a hi-end tube amplifier), so we will use a common KT88 for the pass tube and a 6AU6 pentode for the error amplifier. Both perform very well in this design, last a long time, and when they do fail, are cheap to replace.

Should you need a higher current capacity, you could use one of the recently developed power tubes, KT120 or even KT150, but then again, why not parallel two cheaper KT88 tubes and double the current output? That would also double the mutual conductance of the pass tube and lower the output impedance of the power supply.

TUBE PROFILE: 6AU6

- Sharp cutoff pentode
- 7-pin mini socket (B7G base), heater: 6.3V, 300 mA
- Controlled warm-up time (11 seconds)
- Maximum ratings:
- V_{AMAX}=330 V_{DC}, V_{SMAX}=330 V_{DC}, P_{AMAX}=3.5 W, V_{HKMAX}=200V
- As a triode:
- V_A=250V, V_S=250V, V_G=-4.0V, I_A=12.2mA, gm =4.8 mA/V, μ=36, r_I=7.5kΩ
- As a pentode:
- V_A=250V, V_S=150V, V_G=-1.0V, I_A=10.6mA, I_S=4.3mA, gm =5.2 mA/V, μ=5,200

The unregulated power supply

The voltage doubler was used simply because we had a transformer with a 240V secondary available. Feel free to use a 2-diode rectification (with a center-tapped secondary of 2 x 480V) or a solid-state bridge rectifier (480V secondary without a CT).

Likewise, the dual power supplies producing +260V and -260 also use voltage doublers and CRC filtering.

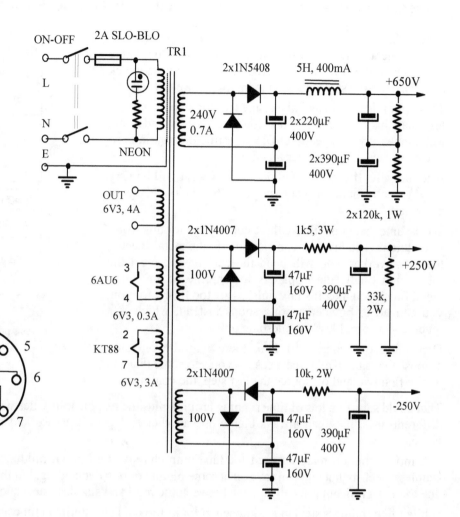

The practical regulator design

Three Zener diodes are used to keep voltages on the 6AU6 electrodes at fixed levels. The cathode is at $-250+150 = -100V$ and the screen grid is at $-100+44V = -66V$! The very high anode resistance of $470k\Omega$ ensures a high amplification factor; the 6AU6 anode output drives the grid of the series tube via a 1k grid stopper resistor. The 6AU6 pentode is not critical; EF86 or almost any similar pentode can be used.

Switching between the bias and the anode voltage metering is annoying and often inconvenient, so separate voltmeters are used for the two DC voltage outputs.

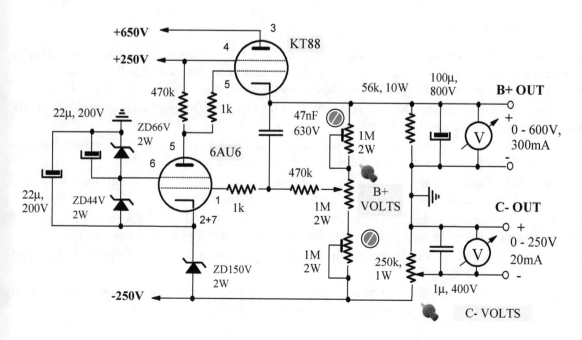

Professional quality, custom made chassis in two weeks or less

If you plan to make more than two or three amplifiers and dislike metalwork, this option is for you. The same approach applies to both the stereo designs and monoblock chassis. The five steps are outlined in the flowchart (right).

This option is not financially viable (it's too expensive) for less than six or so chassis, but don't let that dissuade you. Remember, you can get other DIY constructors to join in so you can place a larger order together.

Alternatively, if you are of a lone wolf type, it will be very easy to sell some of your surplus chassis on eBay or elsewhere.

We assume that you will use three contractors for this work: laser cutters (usually they can also supply the metal sheets), the metalworker who will bend the cut chassis, weld the joints, and finally electroplaters for chrome or any other metal finish you like. If shiny finishes are too flashy for you, you can go to powder coating shops instead, which is a much cheaper and less troublesome option.

Depending on where you live, there may be one-stop "shops" that can do all three steps for you and some that can do the first two and outsource the last step, the coating.

①	DETERMINE THE MOST LIKELY TOPOLOGY (NUMBER OF PREAMP & POWER TUBES) AND THE REQUIRED SIZE
②	DRAW THE CHASSIS
③	SEND THE DRAWING TO LASER CUTTERS TO GET ITEMIZED QUOTES FOR PROGRAMMING AND CUTTING (INCLUDING QUANTITY DISCOUNTS, IF ANY)
④	SEND THE DRAWING TO METALWORKERS TO GET ITEMIZED QUOTES FOR FOLDING, WELDING AND CLEANING
⑤	TAKE ONE FINISHED CHASSIS TO METAL PLATERS AND GET QUOTES FOR CHROME, BLACK CHROME OR COPPER PLATING

ABOVE: The 5 steps involved in getting your tube amp chassis professionally made

That could save you lots of time running around, plus the responsibility lies with only one supplier. If you use three different ones, they can always blame one another for specific issues and problems and play the buck-passing game.

The most common material used for tube amp chassis is a 1.6mm mild steel. Alternatively, you could opt for stainless steel so that you can save on chrome plating or powder coating, but the final cost will most likely be higher due to the significant price difference between the mild and stainless steel sheets!

While I like stainless steel in the kitchen, it looks too cold and sterile in tube amps. Plus, many amps made in China use it, and you don't want your pride and joy to look mass-produced in Shenzen, do you?

The most striking (and most expensive) finish is black chrome. I am surprised Chinese amplifier makers are not using it more often since it is much cheaper there, as everything is.

I still remember my telephone conversation with a local chrome plating business owner in Perth, Australia. I got quotes from a few platers in eastern Australian states (Sydney and Melbourne) to make us one chassis at $100 (in 1998, so not exactly cheap), and he quoted $260. When I asked him how he could justify such an exorbitant price, he laconically replied, "Mate, I'm the only one in Perth who does black chrome!"

Tapered-front monoblock chassis

Although this monoblock chassis (next page) assumes a single-ended amplifier with one output tube, it can easily be modified to accommodate two power tubes. However, it may be a tad too small for two large bottles such as 845, 813, GM70, or 211. Of course, you can always scale its dimensions to enlarge it.

Likewise, there are cutouts for two preamp tubes and no cutout for a tube rectifier (we used solid-state rectification), but these can also be changed to suit your topology.

The RCA input hole is at the rear, together with holes for speaker binding posts, fuse, and IEC power input. A rectangular cutout for the on-off power switch is at the rear, on the top surface, for easier access.

Although a bit flashy, these chassis looked stunning in chrome, and when powder coated, the finish was more low-key. It is easy to add timber trim onto the two short parallel sides if you wish.

The holes for mounting transformers, chokes, and transformer covers are not marked on the drawing above; you can add them to suit the exact dimensions of the magnetic components and the covers you choose to use.

There are only six welds (vertical joints that determine the chassis height, which is 70mm), a total weld length of 420mm.

The front hole on the bottom plate (*) is for outside access to a bias adjusting potentiometer inside the chassis. If your design uses cathode- or self-bias, you can delete that hole from the drawing.

THRIFTY TIPS & TRICKS: TIME & MONEY SAVING IDEAS

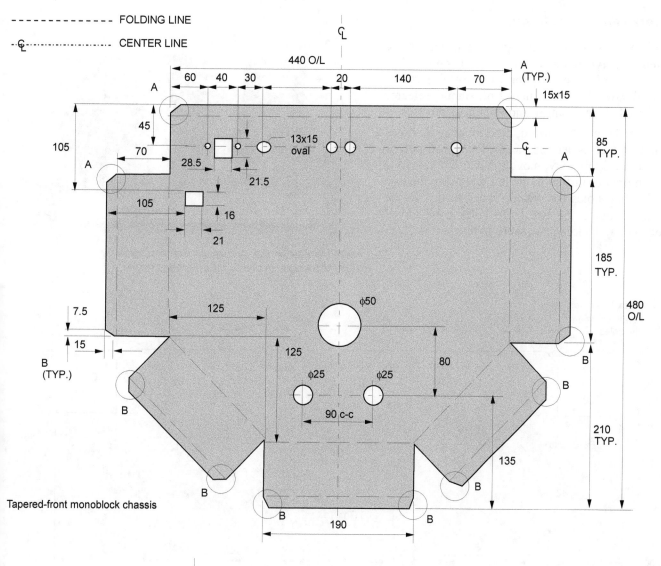

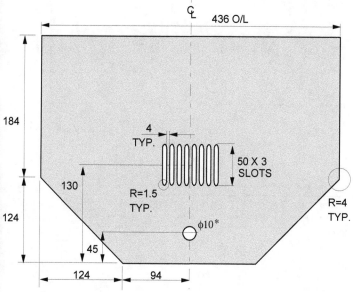

As for fixing the bottom plate to the finished chassis, the best option is to ask the metalworker to insert stud nuts (also called rivet nuts). Alternatively, insert U-nuts yourself.

There should be at least one stud along each of the three shorter sides (towards the front), three studs along the back (the longest side), and one or two along the 184 mm sides. These aren't marked on the drawing; the best is to consult the metalworker before finalizing the drawing and laser cutting.

ABOVE LEFT: Tapered front monoblock chassis - bottom cover (all dimensions in mm)
BELOW LEFT: U-nut, side view (left) and stud nuts (right)

Stereo amplifier chassis

This chassis is for an integrated amp with three inputs and two pairs of speaker binding posts. The holes at the front are for (left-to-right) rotary on-off switch, neon indicator, input selector switch, balance control, and volume control.

Most holes can be moved, deleted, or more holes can be added as per your needs. For a push-pull or parallel SE topology, move the two left holes for preamp tubes further to the left and move the two right-hand holes for preamp tubes further to the right, bringing the two large holes a bit closer together and adding two more in line.

ABOVE: Integrated SET amp using the chassis below
BELOW: Stereo amp chassis (all dimensions in mm)

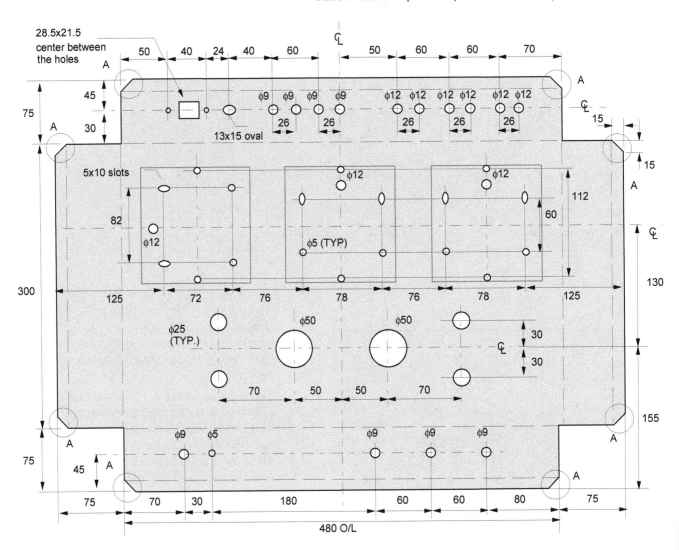

Universal transformer cover

Twenty-five years ago, there were no China-made transformer covers for sale on eBay, so we had to get them made locally. Again, if you are making a few amps, this may still be a better and even cheaper option since Chinese covers are typically priced at around US$20-30 plus at least US$20 postage each. US$50 at the time of writing is AU$70. If you make more than 10 or 20, you should get a better price locally, plus you can have any size, shape, material, or finish you like.

Powder coating is usually the fastest and cheapest finishing option. However, if it fits your overall visual scheme, chrome-plated transformer covers make any amp look striking.

THRIFTY TIPS & TRICKS: TIME & MONEY SAVING IDEAS

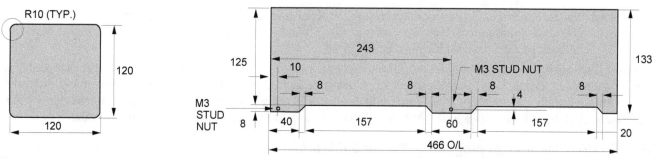

ABOVE RIGHT: transformer cover - sides (NTS)
ABOVE LEFT: transformer cover - top (NTS)

A black chassis with black transformer covers looks like a coffin for your dearly departed pet; a bit of black-chrome contrast makes a huge difference, especially if you position them to get multiple reflections of the tubes, an effect that is even more beautiful at night time.

Black chrome is sometimes used for solid machined knobs (for instance, in MingDa preamplifier), and it looks very smart and high-end. If you don't want to go to the expense of having the whole chassis coated in it, you can do only the transformer covers, so have a chrome chassis with black chrome covers and knobs, a true visual feast!

Cash register drawers as cheap chassis

Cash drawers of the standard size (430x430mm) are perfect for use as chassis for stereo amps but are too big for monoblocks. These made-in-China smaller drawers (380 mm deep x 335 mm wide), illustrated on the right, have a 32% smaller footprint but would still be too large for monoblocks with only one power tube.

However, with two large transformer covers at the back, an interstage transformer case, a large choke, a rectifier, and two power tubes (in push-pull or SE parallel), they are perfect for more complex monoblock projects.

Best-of-all, they cost only AU$45 each (including postage), which was around US$30 in Nov. 2015.

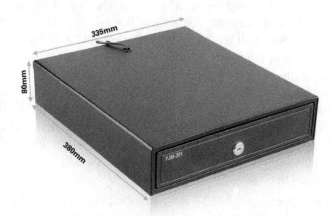

Medium-sized cash drawers (380x335mm) are close to optimal for larger monoblock and smaller stereo amplifiers.

How cheap and nasty tube sockets can end up costing you big $$$!

CASE STUDY: "RimLock" sockets from China

RimLock system is a typical example of how a better product does not always win or even survive in the marketplace. Compared to the de facto standard Noval sockets, RimLock sockets held tubes much tighter (a nipple in the glass base of a tube would lock into the metal socket), so tubes could not pop out in transport for instance, due to vibration and shock, as they can with Noval sockets.

However, RimLock sockets were more complex and thus more expensive to manufacture, and RimLock tubes needed to have a protruding nipple in their glass base, another cost-increasing factor. The bean counters usually win battles between engineers and accountants, which is what happened here - the superior product lost!

Chinese manufacturers are very good at identifying a gap in the market.

Vintage RimLock sockets and current Chinese production "equivalent". Notice the total absence of any locking or positioning capability in Chinese-made sockets.

They started making modern and good-looking versions of RimLock sockets, using quality ceramics and often even gold-plated metal rings around them. However, their lack of critical (or any) thinking is astonishing! In this case, the supposedly RimLock sockets have no locking facility at all.

Moreover, the locking pin serves another purpose: to act as a locator pin, making sure the RimLock tube can only be inserted one way, the right way! In this case, a RimLock tube can be inserted into these sockets in many different ways, with potentially disastrous results! You could end up with a high anode supply voltage on the heater, which would instantly burn out, or with a high positive voltage on the grid, which would also destroy the tube.

CASE STUDY: Incorrectly plugged-in 2A3 or 300B tubes

Large mechanical tolerances are a common issue with some of the currently produced UX4 sockets (for tubes such as 45, 2A3, and 300B). There are two thinner and two thicker pins, but if the diameter of smaller holes on the socket is slightly larger and the diameter of large pins on a tube is slightly smaller, a tube could be plugged-in the wrong way, making a tube "roller" blissfully unaware of an impending disaster!

This moment of distraction happened to us recently; luckily, the four 2A3 tubes were not damaged due to a fortunate probability outcome. Assuming the possibility just described that a 4-pin tube could be inserted the wrong way, there are three ways to do so. Let's analyze them. Remember, pins 1 and 4 are large diameter (heater + cathode), and pins 2 and 3 are of smaller diameter, as illustrated.

If we rotate the tube 90o clockwise, we get the outcome illustrated in b) The grid is now connected to where the anode is supposed to go, which is at one end of the output transformer's primary winding, at high voltage with respect to the cathode. Since there is a grid-leak resistor of 220-470kW between the grid and the ground, assuming 350V anode voltage, a current of 0.75-1.6mA will flow through it, which is fine.

One end of the heater is in the right place, at a low potential, which could be at ground potential, at 2.5V, or at 5V, depending on how the heater is grounded, so let's call that "heater voltage." However, the anode is now connected to the other end of the heater supply, meaning there is 350V anode voltage across the heater, so the heater will act as a fuse and burn out almost instantaneously, destroying the tube.

Rotating a tube 180o as in c), the heater is now connected between the HV transformer (where the anode is supposed to go) and the coupling capacitor and grid resistor, where the grid is supposed to be. The high-value grid-leak resistor is now in series with the heater filament and will limit the current to 0.75-1.6mA, which is fine, and the filament will not get damaged.

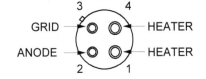

a) correct orientation (bottom view)

The heater voltage of 5V is now applied between the grid and the anode, so nothing will happen. In this scenario, the tube will be fine.

However, depending on the amplifier design, if there is no coupling capacitor, and the grid of the output tube is biased from a negative bias supply (typically -70V), the voltage across the heater will be -70V bias voltage plus the anode voltage of say 350V, and with 420V across it, the heater will definitely burn out!

If we rotate the tube 90o anti-clockwise as in d), one end of the heater is in the right place, the other is connected to the grid point, where the coupling capacitor and grid resistor are joined.

There is no DC voltage at that point, so the heater is effectively in series with the grid-leak resistor, and even with full 5V at the other heater end, a minuscule current will flow through it, and the tube will not light up! The anode is connected to the other side of the heater supply (0V, 2.5V, or 5V), but in any case, that is such a low voltage that it can be ignored. Thus, the tube will not be affected.

However, if fixed bias is used, the -70V or so will now appear across the heater, enough to destroy it! However, most bias supplies don't have a high current capability, so this may save the heater from burning out.

In conclusion, it all depends on the topology of the output stage. With cathode bias and a coupling capacitor, one of the three wrong ways to plug a tube in will result in its destruction, and two will not affect it at all, so you have a 33.3% chance to lose your precious investment! In amplifiers with fixed bias, the tube(s) will most likely be destroyed in all three situations!

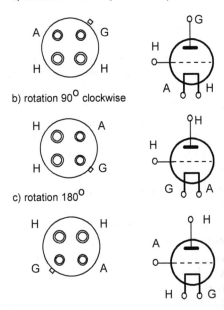

b) rotation 90° clockwise

c) rotation 180°

d) rotation 90° anti-clockwise

OUTPUT AND INTERSTAGE TRANSFORMERS: FROM COMMERCIAL BENCHMARKS TO DIY DESIGNS

- Design: 15 Watts, 2k7 primary, universal single-ended output transformer
- Design: 2k3:6Ω single-ended output transformer
- Design: 1k8 primary SE output transformer for two 300B in parallel
- Design: Preamp output transformer (line driver) 5kΩ :150Ω
- Design: The universal interstage and phase-splitter transformer project
- TVC (Transformer Volume Control)
- The way it used to be: Williamson output transformer (10kΩ PP)
- Ertoco toroidal output transformers
- Vintage commercial example: Universal 10W Hi-Fi PP transformers
- Comparing the parameters and fidelity of audio transformers
- Lessons learned from the vintage German power transformer
- A few important lessons from power supply filtering chokes

15

"When you come to a fork in road, take it."
Yogi Berra

Design: 15 Watts, 2k7 primary universal single-ended output transformer

A SE transformer of 2.5-3.5kΩ primary impedance and 15 Watts audio power can be used for many output triodes and triode-strapped pentodes, such as 2A3, 300B, EL34, KT88, F2a, EL153, and many others. We have covered the design of audio transformers in Volume 2, so we will not explain too much here but go through the mechanics.

Core sizing

EI96 laminations are a reasonable starting point for most output transformers. EI114 size (assuming the same or similar stack thickness) gives you a much larger cross-sectional area, resulting in a lower -3dB limit and improved bass performance. However, the parasitic capacitances and leakage inductance also increase due to the larger dimensions, which negatively impacts the high-frequency performance, so a compromise is inevitable.

Now we need to determine the stack thickness via the cross-sectional area of the center leg, using the empirical formula $A_{EF}=k\sqrt{(P/f_L)}$. Since k is usually around 20-25, let's be modest and choose k=20 and the lower -3dB frequency of 20Hz. Now the effective cross-sectional area is $A_{EF}=k\sqrt{(P/f_L)} = 20*\sqrt{(15/5)} = 17.3$ cm².

The effective area A_{EF} is 3-6% smaller than the gross cross-sectional area A. We have $A = A_{EF}/0.96 = 17.3/0.96 = 18$ cm².

Since the width of the center leg for EI96 lams is 3.2cm, the stack thickness we need is $S=A/3.2 = 18/3.2 = 5.6$ cm. The standard bobbin widths are 50, 55, and 60 mm; any of them could be used here, but 55mm is the closest.

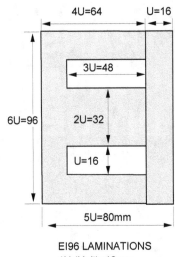

EI96 LAMINATIONS
1U (Unit)=16mm

Primary voltage, primary & secondary number of turns

To get 15 Watts of output power, the primary voltage needs to be $V_1=\sqrt{(PZ_P)} = \sqrt{(15*2,700)} = 201V$.

Now we need to choose B_{MAX}. The lower the B_{MAX}, the higher N_1 (and, consequently, N_2) and the larger the winding area needed. If we select 0.8 Tesla as AC magnetic flux density, once the DC flux density is added, we should still be under the saturation level of around 1.5T for grain-oriented stainless steel, so plenty of headroom.

$N_1=V_1 10^4/(4.44 f_L B_{MAX} A_{EF}) = 201*10^4/(4.44*12*0.8*17.6) = 2,680$ turns. With $Z_P=2,700$ Ω and the load $Z_S=8$ Ω we get the impedance ratio of IR= 2,700/8 = 337.5 and turns or voltage ratio of $TR=\sqrt{(IR)} = \sqrt{(337.5)}=18.37$.

Finally, we can calculate the required number of secondary turns as $N_2=2,680/18.37 = 145.9$ or 146 turns.

Sectionalizing

The leakage inductance of 1:2, 2:1, 3:2, and 2:3 designs is too high, the magnetic coupling is inferior, and the upper -3dB frequency would be under 10 kHz, which is fine for a guitar tube amp, but not for an audiophile amp. The smallest number of sections you should ever use is 3:4, but in this case, we want a superior design, so let's choose a 4:5 arrangement, four primary and five secondary sections.

Wire sizing and window fitting

PRIMARY: The primary must be able to take $I_0 = 80$mA, and up to $I_{MAX} = 160$mA of current at full power. Primary wire diameter: $d_1=0.71\sqrt{(I_{MAX})} = 0.71\sqrt{(0.16)} = 0.28$ mm, but since no amplifier works at full power all the time, we can safely use 0.23mm wire.

SECONDARY: With an 8 Ω load and 15 Watts of power, the secondary AC current will be $I_2=\sqrt{(P/R)} = \sqrt{(15/8)} = 1.4$A so the smallest wire diameter we should use is $d_2=0.71\sqrt{(I_2)} =0.71*0.615= 0.84$ mm

However, we will have five secondaries in parallel so that the total current will be equally divided, 1.5A/5 = 0.3A through each secondary section, so $d_2=0.71\sqrt{(0.3)} =0.39$ mm!

The window length of EI96 laminations is 48 mm. The walls of a typical plastic bobbin are 1.5 mm thick (on each side, so 3mm total).

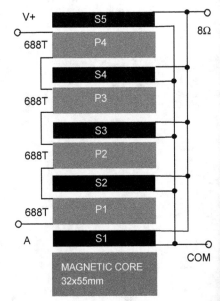

ABOVE: The winding diagram

We should also allow 1 mm of clearance for the insulation (cut slightly wider than the CL, to ensure the last turn of the next section does not drop down onto the previous layer), a total of 4mm. So, the winding length or Coil Length (CL) is CL= 48-4 = 44 mm!

We need 146 turns in the secondary and 44mm/146 turns = 0.30 mm diameter wire. So, if we use 0.3 mm wire, with perfect winding (100% horizontal fill factor - HFF) we could fit all 146 turns in a single row.

Since we are not professional winders, let's assume 80% HFF and two layers per secondary section and see what size wire we could fit in: 44*0.8/146*2 = 0.48mm!

So, let's use 0.45mm wire (if you have 0.5mm wire, you could also use that). Remember, the larger the diameter of the secondary wire, the lower the secondary resistance and the higher the damping factor of the amplifier.

PRIMARY: Maximal number of turns per layer TPL_{MAX}= CL/d = 44mm/0.23mm = 191! Allowing for a horizontal fill factor of 0.9, we can fit 191*0.9 = 172 turns and 2,680/172 = 15.6 layers, meaning we need 16 layers. With 4 layers in each of the four sections, 4x4=16, perfect!

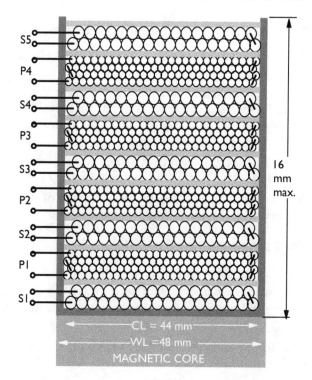

ABOVE: Cross-section of the coil (not to scale)

Checking the vertical fit

The window of EI96 laminations is 16mm tall, but we lose 2mm on bobbin thickness and another 1mm on top for clearance, so our total winding height must not exceed 13 mm! The height of the primary winding is 16 layers*0.23 mm = 3.7 mm, the secondary is 10 layers*0.45 mm = 4.5 mm high.

Using 0.35mm Mylar insulation, the eight insulation layers between primary and secondary sections are 2.8 mm thick, so our total winding height is 3.7+4.5+2.8= 11 mm.

We need to allow for the "bulging factor" or "bowing." The winding is always taller in the middle of the window; it bows or bulges out. It depends on the tensioning skills of the person winding the transformer but is usually 5-15%, so assuming the worst case of 15%, 11mm*1.15 = 12.65mm, just under the 13mm maximum.

Sizing the air-gap

Some designers determine the air gap's width based on the core's cross-sectional area (A=a*S), using the formula $g=0.2*\sqrt{A}$.

The main factor determining the required air gap size is the total DC magnetizing force, which may be expressed in AT (Ampere-Turns). The type of the magnetic material used (its permeability) and the size of the laminations (the length of the magnetic path LMP) also impact the size of the gap, but their effect is so small compared to that of the magnetizing force that it can be neglected, at least in the first approximation.

Perplexingly, this rule-of-thumb formula does not involve ampere-turns in any way!

Secondly, from our experience, the values obtained by this method are too high; if you are keen to use a formula, $g=0.05\sqrt{A}$ would be more accurate.

Design: 2k3:6Ω single-ended output transformer

This winding diagram, found on the web, is claimed to be for output transformers used in Shindo Cortese amplifiers, which use one F2a power tube per channel, working as a pentode. Regardless of the accuracy of this claim, let's analyze it and see what kind of performance can be expected and if its design can be improved. The sectionalizing is very simple, three primary and only two secondary sections and a single output impedance.

The gross cross sectional area is A=a*S = 3.8*6 = 22.8 cm². Assuming A_{EF}=0.96A = 21.9 cm², our rule of thumb formula $A_{EF}=20\sqrt{(P/f_L)}$ tells us that at 10 Watt output level, the stack of that size can go down to $f_L=P*(k/A_{EF})^2$ = 10*(20/21.9)² = 8.3Hz, a great result.

Allegedly, a primary impedance of 9.5H was measured, while the frequency response was specified as 8Hz-110 kHz, but the referent power level was not specified.

Primary & secondary number of turns

The primary signal voltage is $V_1=\sqrt{(PZ_P)} = \sqrt{(10*2{,}300)} = 151V$.

Now we need to choose the induction level B_{MAX}. The lower the B_{MAX}, the higher N_1 (and, consequently, N_2!) and the larger the winding area needed. If we select 0.8 Tesla as AC magnetic flux density, once the DC flux is added, we should still be under the saturation level of around 1.6T for GOSS, so plenty of headroom.

Since $N_1=V_1 10^4/(4.44 f_L B_{MAX} A_{EF})$, we can express the product $f_L B_{MAX}$ as $f_L B_{MAX}=10^4 V_1/(4.44 N_1 A_{EF}) = 6.75$ [HzT].

Again, assuming f_L of 10Hz, the induction of $B_{MAX}=0.675$ Tesla seems conservative, leaving lots of headroom.

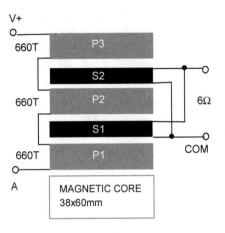

The winding diagram

Wire sizing and window fitting

There are 3x660 = 1,980 primary turns and 100 secondary turns. Thus, turns and voltage ratio is T=1,980/100 = 19.8, while the impedance ratio is TR^2 = $19.8^2 = 392$. A 6Ω speaker impedance would reflect onto the primary side as 6*389= 2,334 Ω.

The primary wire is specified as AWG30 size, meaning its diameter is 0.255mm, while the secondary wire is AWG 22 with a 0.65mm diameter.

The window length of EI114 laminations is 57 mm. The winding length or Coil Length (CL) is CL= 57 - (2 x 2) mm = 57-4 = 53 mm

The walls of a typical bobbin are 2mm thick, plus we could allow 1 mm of clearance for the insulation (cut slightly wider than the CL, to ensure the last turn of the next section does not drop down onto the previous layer)! So, CL=52mm!

PRIMARY: Maximal number of turns per layer TPL_{MAX}=CL/d = 52mm/0.26mm = 200

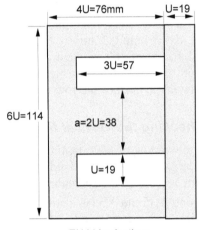

EI114 laminations

Since 660/200 = 3.3, let's assume 4 layers in each primary section. That would mean 660/4 = 165 turns in each layer and a Horizontal Fill Factor of HFF= 165/200 = 82.5%. That is reasonable.

SECONDARY: TPLMAX= 52mm/0.65 mm = 80 turns, so to fit 100 turns, two layers are needed, with 50 turns in each layer. HFF= 50/80 = 62.55, which is very low. The designer/transformer maker should have used a much larger diameter wire to fill both layers. That would also reduce the secondary winding resistance and increase the amplifier's damping factor.

The secondary layers should always be fully filled since such a large diameter wire gives the coil rigidity and mechanical strength. This means that the size of the secondary wire should be chosen that with the given number of turns per layer (50 in this case), the whole width of the bobbin is filled. So, 52/50 = 1.04, so 1mm diameter wire should be used.

The window is 19mm tall, but we lose 2mm on bobbin thickness and another 1mm on top for clearance, so our total height of the winding must not exceed 16mm!

Primary height: 12 layers*0.26 mm = 3.12 mm, secondary height: 4 layers*0.65 mm = 2.6 mm. Assuming 0.1mm insulation between layers and 0.2 mm between sections, the insulation height is 11*0.1 + 5*0.2 = 2.1mm.

The total winding height is 7.8 mm. Around 8 mm coil height in a window that is 19mm tall (16 mm useable height) means that Vertical Fill Factor or VFF is less than 50%, which is too low, so we'll use larger diameter wires.

2nd iteration: Coil height with larger diameter primary and secondary wire

A larger diameter primary wire could also be used, say 0.35mm instead of 0.255mm.

For the primary, TPL_{MAX}=CL/d = 52mm/0.35mm = 148 and 660/148 = 4.6, so we can comfortably fit that into 5 layers in each primary section. For the secondary, TPL_{MAX}=CL/d = 52mm/1mm =52 so with careful & tight winding we could fit 100 turns into 2 layers.

Primary height: 15 layers*0.35 mm = 5.25 mm, secondary height: 4 layers*1 mm = 4 mm. With an insulation height of 14*0.1 + 5*0.2 = 2.4 mm, the total winding height is 11.65 mm. Even with a 15% bulging factor, the coil height will not exceed 13.4mm, which is better, much closer to the 16 mm useable height.

Design: 1k8 SE output transformer for two 300B triodes in parallel

We bought two of these transformer sets (4 C-cores in each) from an eBay seller in Bulgaria. The cost was US$30 each plus US$40 airmail (US$100 total). The seller specified sheet thickness of 0.3mm, maximum flux density B_{MAX}=2.2T and permeability µ=8,000 at B=1.1T.

The C-cores had "strip width" D=50mm and "build up" E=12.5mm. The other dimensions were F=25mm, C=2E+F=50mm, G=50mm and B=90mm. The mean magnetic path length, also called "the length of flux", was 25.1 cm, gross cross-sectional area was A = 2*E*D = 2*1.25*5=12.5 cm², so the core's power rating was approx. P=$(2*A_{EF})^2$ =$4*A_{EF}^2$ = 600VA.

Determining TPV and number of turns

Assuming 0.98 stacking factor, the effective cross-sectional area is A_{EF} = 0.98A =12.25cm². The transformer equation says TPV = N/V = 10^4/(4.44BA_{EF}f), so now we need to decide on B and f. We want these transformers to transmit the full power down to 20Hz, so let's use f=20 and B=1.1 and see what we get. TPV=N_1/V=10^4/(4.44*1.1*12*20) = 10,000/1,172= 8.5

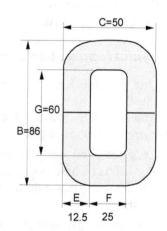

It is possible to get 12W from a single 300B, so one would expect 24 Watts from two in parallel. The primary voltage will be V_1=√(PZ_P) = √(24*1,800) = 208V, which means N_1 = TPV*V = 8.5*208 = 1,768 turns.

Since the impedance ratio is IR=1,768/8 = 221, the turns (voltage) ratio is TR= √IR = √221 = 14.87, and finally, N_2=N_1/TR = 1,768/14.87 = 119 turns.

Window fitting & layering

The width of the winding window is G=60mm, and the window height is F=25mm. We lose 4mm on bobbin's walls thickness (horizontally) and 2 mm vertically, so the coil or winding length is CL=60-4=56 mm and the maximum coil or winding height is CH= 25-2=23 mm.

Maximum primary turns per layer is TPL_{1MAX}= 56mm/0.5mm =112. Since 1,768 turns/112 TPL=15.8, use 16 layers; the actual TPL=1,768/16 = 110!

Maximum secondary turns per layer is TPL_{2MAX}= 56 mm/0.9 mm = 62. We need 119 turns, 119/62 = 1.92, so we need 2 layers. Actual TPL_2 = 118/2 = 59.

Checking window height

With 4 primary & 3 secondary sections, there are seven insulation layers; with 0.2 mm thick Mylar sheets, the total thickness of insulation is than 1.4mm.

The primary height is 16 layers*0.5 mm = 8 mm, secondary height is 6 layers*0.9 mm = 5.4 mm. We have seven insulation layers between primary and secondary sections plus one final layer on top, with 0.35mm Mylar insulation that is 2.8 mm, meaning our total winding height is 8+ 5.4 + 2.8 = 16.2 mm. We need to allow for the "bulging factor" or "bowing effect of 15%, so 16.2 mm*1.15 = 18.6 mm. We have 23 mm, so a very comfortable fit.

Measured results

MEASURED RESULTS:
- R_P= 48 Ω, R_S= 0.6 Ω
- L_P= 4.8 H, L_L=4.1 mH
- C_{PS1000} = 2.6 nF
- BW @1W: 17Hz-27kHz (-3dB)

Due to the large diameter of the primary wire, the primary's DC resistance is low, under 50 ohms, which is a great result. The leakage inductance is only around 4 mH, another outstanding achievement, as is the primary-to-secondary capacitance of 2.6 nF.

However, the primary inductance is relatively low, less than 5H, which indicates that the magnetic material used in these cores has low permeability.

The primary inductance L_P can be increased by increasing the number of primary turns N_1 (and N_2 as a consequence). We have an extra height of 4mm, so as an exercise, recalculate everything to completely fill the winding window vertically!

Design: Preamp output transformer (line driver) 5kΩ :150Ω

The way it used to be: vintage benchmark by Tamura

Tamradio A-4714 are vintage audio transformers by Tamura, Japan, with a 5kΩ primary and 600Ω CT secondary. The frequency range was specified as 30Hz - 20kHz (-1 dB), the insertion loss was 0.2 dB, and the maximum primary DC current was 60mA.

Just as with all vintage transformers, they rarely come up for sale due to their limited supply and desirability and are priced accordingly (read: extremely expensive).

ABOVE: The side view of a Tamura A-4714 output transformer, clearly showing its winding arrangement

Typical application circuit and the required specifications

One possible line-level preamplifier stage with Tamura A-4714 is illustrated. The secondaries are in series for single-ended operation, but balanced outputs are possible by grounding the CT instead. The DC current through the primary is around 18mA.

Let's settle on the following specs: I_1=20mA, f_L=15 Hz, Z_P=5kΩ, Z_S=150 Ω ! There is no need to allow 60mA in the primary as Tamura did; 20mA is just right for driver tubes such as 5687, 6SN7, 6CG7 or 12BH7.

The smaller gap will give us higher effective permeability and higher primary inductance. Since the DC current is relatively small, the required air gap is only around 0.05 mm.

We will reuse the laminations and the bobbin from surplus Taiwanese low voltage mains transformers (photo below) since they use GOSS laminations.

The bobbin is vertically split in two (primary on one and low voltage secondaries on the other half), but that divider can be cut off.

The EI66 laminations with 50mm stack give us A_{EF}=0.95*A= 0.95*2.2*5.0 = 10.45 cm^2.

With Z_P=5,000 Ω and Z_S=150 Ω, we get the impedance ratio IR= 5,000/150 = 33.3 so the turns ratio is TR=√IR =√33.3 = 5.77!

Window sizing and maximum coil height

We have chosen 4:3 sectionalizing, four primaries in series, and three secondaries in series. Now we have to determine the maximum number of turns we can fit into the available window, which is 33x11mm.

The window length is WL=33mm, so the coil length is CL=33-(2*2)-1= 28mm (B=2mm for the bobbin thickness on each side plus 1 mm clearance for insulation).

Window height: WH = 11mm

Coil height: CH= WH-B-1= 11-3= 8mm

Normally, with power and output transformers, we would calculate the minimum required wire diameter based on the primary and secondary current levels, but in this case, the current is only up to 20 mA, so 0.1 mm wire would be sufficiently large. However, we have chosen a larger diameter wire, 0.135 mm, to reduce copper losses and make the winding job easier. Thin wires are fiddly to wind and can break.

We will use a reverse process in this calculation, similar to that used for chokes. The thickness of insulation will be determined first, and the leftover height will then be used to determine the maximum total number of turns that can physically fit into the rest of the window height.

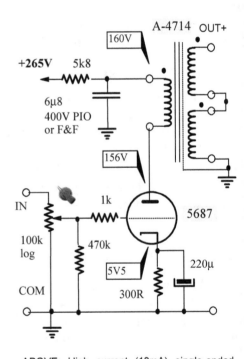

ABOVE: High current (18mA) single-ended preamplifier (line stage) with 5687 triode and Tamura A4714 output transformers

ABOVE: Some mains transformers use quality GOSS laminations and can be rewound as audio transformers

INSULATION THICKNESS DI:

There are 7 layers of insulation, including the final (outside) layer, total insulation thickness is DI=7*0.25 = 1.75 mm

GROSS COPPER HEIGHT (GCH):

GCH= CH-DI= 8-1.75= 6.25 mm

Factoring in 15% bulging factor we get the Net Copper Height of NCH=GCH/1.15=6.25/1.15 = 5.4 mm

INITIAL LAYERING:

Maximum number of layers: NL=NCH/d=5.4/0.135= 40

Maximum TPL (turns per layer): TPL_{MAX}= CL/d = 28/0.135 = 207

Actual TPL=0.9*207 = 186 (assuming 0.9 horizontal winding factor)

The total number of turns we can fit is TT= NL*TPL = 40*186 = 7,440

Since TT = N_1+N_2 =N_1+N_1/5.77 = N_1*(1+0.173), we can determine N_1 = TT/1.173 = 7,440/1.173 = 6,343 and N_2= 7,440-6,343 = 1,097

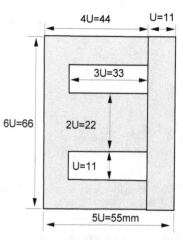

ABOVE: EI66 laminations

Final layering and final winding data (after adjustments)

Number of primary layers: 6,343/186 = 34.1, so we will use 34 layers. Since 34/4 = 8.5, two of the sections will have eight layers, and two will have nine layers (2*8+2*9=34).

Number of secondary layers: 1,097/186 = 5.9, so we will use six layers in total, or two layers per section.

FINAL DATA:

PRIMARY: 2*8*186 + 2*9*186 = 2*1,488 + 2*1,674 = 6,324 turns

SECONDARY: 3 * (2*186) = 3*372 turns = 1,116 turns

Turns ratio: TR = 6,324/1,116 = 5.667, so IR= TR^2 = 32.11

The initial goal was an IR of 33.3; due to our fitting adjustments, we got 32.1, which is close enough. Instead of the 5k : 150Ω impedance ratio, we will have 5k: 156Ω, which makes no practical difference whatsoever!

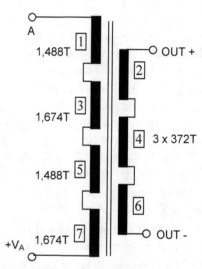

ABOVE: Winding diagram with 4:3 sectionalizing. The framed numbers indicate the winding order of transformer's sections.

Design: The universal interstage and phase-splitter transformer project

Commercial benchmark: Hashimoto A-105

Hashimoto A-105 is a versatile interstage transformer with the possibility of using both the primary and the secondary in series or parallel connection. The secondary windings can be connected in series, parallel, or configured in center tap push-pull connections, with 1:1, 1:0.5, or 1:0.5+0.5 turns and voltage ratio, respectably.

Judging by the photos on the manufacturer's website and by the dimensions of its metal case, it seems to be wound on a pair of tiny C-cores, confirmed by only 0.8 kg of total weight (including the steel case).

MANUFACTURER'S SPECIFICATIONS:

- Hashimoto A-105, made in Japan
- Dual primary & dual secondary driver transformer
- US$596.00 for a pair + $42 transport on ebay (in 2015)
- Three voltage ratios are possible, 1:1, 1:2 and 1:0.5
- Z_P (series) 5 kΩ, parallel 1.2 kΩ
- L_P = 30H each half, 60H (series), 15H (parallel)
- Maximum primary current: 15mA in series or 30mA in parallel
- Primary resistance: 350 Ω, secondary resistance 450 Ω
- Frequency range: 25-35,000Hz (+/-2dB)
- Dimensions: 52 (W) x 58 (D) x 85 (H) mm, weight 0.8 kg

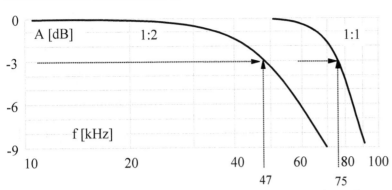

ABOVE: High frequency response of Hashimoto A-105 interstage transformer for two different voltage ratios

The manufacturer's amplitude-frequency graphs show the high-frequency response for 1:2 ratio (secondaries in series) and 1:1 ratio (secondaries in a phase splitting arrangement).

Notice how f_U increased from 47 kHz for the 1:2 ratio to 75 kHz for the 1:1 connection. As we have seen in Volume 2 of this book, the higher the step-up ratio, the lower the bandwidth of an interstage transformer!

Other commercial examples - Sowter and Audio Note

Sowter type 8423 is a 1:1+1 transformer on EI78 laminations with a 5k nominal impedance of each of its three windings. The primary inductance is 25H (frequency not specified?), the maximum primary current is 50mA, and the bandwidth is modest, 30Hz-30kHz (-3dB). In 2016 it sold for £133.57 (approx. US$190.-) each.

Audio Note UK offers 33 (!) different interstage transformers, grouped by the maximum primary current (10, 20, and 30mA), so 11 models in each group. The cheapest model in each group uses EI laminations (M4 material) and costs £155.40 each. The following three models also use copper wire and leads, but C-cores made of better grade magnetic materials - "Improved HiB" (£353.60), "Super HiB" (£424.00) and "ultra HiB" (£635.20 each).

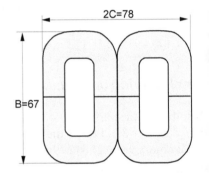

ABOVE: The specified overall dimensions of 78 and 67mm, indicate this configuration of Audio Note IS transformer cores.

If you thought those prices were steep, the next group uses copper on the primary side and Audio Note silver lead-out wires on the secondary side.

It is unclear if that includes silver secondary windings or only the lead-outs? Depending on the magnetic material used, they range in price from £789.60 to £2,438.00 each!

The last three models are fully wound with silver wire and cost from £2,109.80 to £3,392.76 each, excluding the VAT ("Value-Added Tax"). The most expensive units use "Super Perma" 50% or 55% nickel cores. The impedance and voltage ratios are the same for all models, 1:1, and all are of the same physical size.

Hammond 126A is a 1:1 interstage transformer with a 5k impedance. The maximum primary current is 15 mA, and the bandwidth is specified as 20Hz-20kHz (-1dB). The laminations are 76x63 mm. The magnetic material is specified as 29M6. The stack thickness is 38 mm, which means the cross-sectional area of its center leg is $A = a*S = 2.6*3.8 = 9.9$ cm^2. Each transformer weighs 3.125 lb. (around 1.4 kg). In 2015 these sold for US$79.- each, much cheaper than the competing Sowter units.

While the claimed primary inductance was 59H (frequency not specified), we measured only 11-12H with a digital LCR meter. There is a significant difference in the inductance between the two transformers; however, inductance varies with airgap, and simply tightening one transformer's bolts harder can cause such differences.

Hammond claims that "all models use bifilar wound windings for exceptional coupling and bandwidth," so one would expect no DCR differences between primary and secondary windings, such as 190Ω and 186Ω, the difference in test results between our two samples. One transformer had (190-177)/177 =0.073 or 7.3 % higher DCR.

MEASURED RESULTS:

- TRANSFORMER #1:
- R_P= 177 Ω, R_S= 176 Ω
- L_{P120}= 12.4H, L_{P1000}= 11.3 H
- L_L=0.4 mH, C_{PS1} = 64 nF, C_{PS2} = 2.2 nF
- TRANSFORMER #1:
- R_P= 190 Ω, R_S= 186 Ω
- L_{P120}= 19.5H, L_{P1000}= 13.6 H
- L_L=0.42 mH, C_{PS1} = 64 nF, C_{PS2} = 2.4 nF

OUTPUT AND INTERSTAGE TRANSFORMERS: FROM COMMERCIAL BENCHMARKS TO DIY DESIGNS

In their reply to that question, Hammond attributed that difference to magnet wire, which has "a resistance tolerance of +/-15%" and "then, during the winding operation, with winding tension, there may be stretching that adds an additional 5%", meaning that variations of up to 20% are normal!

Hammond 126A transformers don't have dual secondary windings, so they cannot be used as phase splitters in push-pull amplifiers. We can do better in our design. Our universal IS transformer will have a split primary and a split secondary winding for both SE and PP applications!

Our design

The transformers described here were installed into PSET monoblocks with 300B tubes (featured in this Volume, pages 114-118) and sounded very good indeed. When Hammond 126A transformers were temporarily substituted, the lower -3dB frequency of the amplifier worsened from 22Hz with our transformers to 29Hz with Hammond ones, resulting in weaker bass.

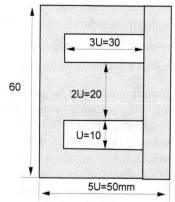

ABOVE: EI60 laminations

In this case, we adopted a pragmatic approach by taking both the laminations and the winding wire we had at hand and designing an IS transformer around those two givens. The wire was 0.11mm diameter, and the laminations were made of GOSS, reused from a pair of smallish but good quality vintage push-pull output transformers.

The window length is 30 mm. The winding length or Coil Length (CL) is CL= 30 - (2 x 2) mm = 26 mm.

Since the same wire is used for the primary and secondary, the maximal number of turns per layer for all windings is TPL_{MAX}=CL/d = 26mm/0.11mm = 236. Assuming a Horizontal Fill Factor of HFF= 80%, we can fit 236*0.8 = 188 turns in each layer.

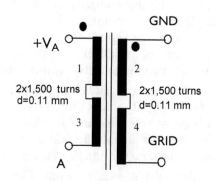

ABOVE: The winding diagram

The window is 10 mm tall, but we lose 2mm on bobbin thickness and another 1 mm on top for clearance, so our total winding height must not exceed 7 mm!

Since we have four identical windings, there will be 4 insulation layers of 0.2mm, or 0.8mm in total. After subtracting 0.8mm for insulation, we are left with a net winding height (copper height) of 6.2mm.

We have to take into account the bulging (bowing) effect, which increases the height of the winding in the middle of the bobbin by up to 15%, so 6.2/1.15 = 5.4mm.

The 0.11 mm wire is actually 0.14mm thick (with insulation), so we can fit a maximum of 5.4/0.14 = 38 layers. Since we have four windings, the total number of layers must be divisible by 4, so let's settle on 32 layers.

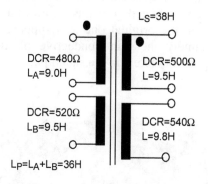

32 layers x 188 turns = 6,016 turns in total or around 1,500 turns in each winding, perfect!

The primary and secondary inductances are around 36-38 Henry, a good result for such a small core. The DC resistance of the two primary and two secondary halves are not identical since we didn't use a bifilar winding nor sectionalized the winding vertically. Still, the difference is fairly small, (520-480)/480 = 40/480 = 0.083 or 8.3%, which is acceptable.

ABOVE: The measured parameters of our universal interstage/phase splitter transformer on EI60 laminations

TVC (Transformer Volume Control)

The volume control transformer pictured (one per channel) is from a Malaysian-made Promitheous Audio preamplifier reviewed in Volume 1 of this book. The preamp can work as a passive TVC control unit or an active line stage with a 6DJ8 duo-triode per channel. The TVC is operational in both options.

The magnetic material is Z11/M6, EI-66 laminations are of 0.35mm thickness, the stack is 35mm thick.

The transformer is actually an auto-transformer, as per the diagram (next page).

The measured values of DC resistance, secondary (output) inductance at 1 kHz, the impedance at 1 kHz (all measured between the common and output terminals), and the previous step's attenuation are tabulated on the next page.

The attenuation figure for each position is referenced to the previous step and was calculated from the measured input and output signal RMS voltages. The attenuation is not uniform or regular, at least not according to our measurements; it ranges from -2.51dB to -6.22 dB.

The total attenuation from position 23 (maximum signal) to position 1 (minimum) is around -96 dB (the sum of the rightmost table column), making the average attenuation -4.17 dB per step.

The inductance of the whole winding is around 2.2H due to the low-grade magnetic material used, which equates to the input impedance of around 37kΩ at 1 kHz.

You may have already identified the significant weaknesses inherent in transformer volume control.

First of all, due to its magnetic core's highly nonlinear hysteresis curve, TVC distorts significantly.

Secondly, its inductance forms an LC circuit with its parasitic capacitances. The input capacitance of the following triode, so resonant peaks are present in its amplitude vs. frequency characteristics, resulting in an oscillatory nature in its response to step and square-wave signals. The Promitheous Audio's line stage oscillogram tells you all you need to know (right).

Thirdly, the input impedance of a TVC-based amplifier or preamplifier is not constant; it varies with the signal's frequency since it is a vector sum of a relatively constant resistive and a widely varying inductive component.

There is also a capacitive component, of course, but let's simplify things a bit; the aim here is not to impress you with my knowledge of vector algebra or to complicate things unnecessarily, but to illustrate the pitfalls in using TVC.

In fact, that's exactly what TVC is - an unnecessary overcomplication, a "solution" that creates at least three new problems. All the newly-created problems are much more serious than the original one it is supposed to solve, namely the thin conductive plastic film of potentiometers in the signal path.

MEASURED RESULTS:

SP	DCR [Ω]	L	Z [Ω]	ΔA [dB]
1	0.0	0.9 µH	0.58	
2	0.3	0.3 µH	0.98	-4.56
3	0.7	14.8 µH	1.40	-3.10
4	1.1	30.9 µH	1.87	-2.51
5	1.5	81.3 µH	2.67	-3.09
6	2.0	152 µH	3.74	-2.93
7	2.6	293 µH	5.88	-3.93
8	3.1	512 µH	9.76	-4.40
9	3.8	863 µH	16.14	-4.37
10	4.6	1.33 mH	26.02	-4.15
11	5.5	2.01 mH	40.79	-3.91
12	6.6	3.2 mH	66.26	-4.21
13	7.9	5.14 mH	107	-4.16
14	9.5	8.17 mH	172	-4.12
15	12.3	16.7 mH	341	-5.94
16	16.2	36.7 mH	683	-6.03
17	21.7	82.8 mH	1,398	-6.22
18	29.2	202 mH	2,764	-5.92
19	40.1	409 mH	5,392	-5.80
20	55.7	782 mH	10,640	-5.90
21	78.3	1.41 H	18,110	-4.62
22	111.4	2.01 H	27,610	-3.66
23	146.2	2.22 H	37,300	-2.61

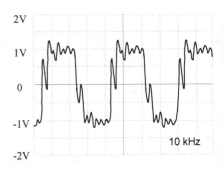

ABOVE:
SP = switch position
L = inductance (at 1kHz)
Z = impedance (at 1kHz)
ΔA = attenuation form the previous position
RIGHT: The fidelity (or rather "lack of") of a 10kHz square wave signal reproduced by the Promitheous Audio line stage with TVC
BELOW: The inside view of one transformer.

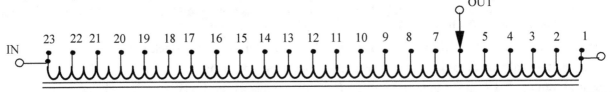

ABOVE: The auto-transformer TVC is the simplest configuration with only one winding.

ns, the audio field is bursting full of complex solutions to minor or nonexistent problems.

Let's look at a couple of frequencies. The reactance (the inductive component of the impedance) is $X_L=\omega L=2\pi fL$, meaning it increases with rising frequency f. Since L isn't constant but varies with f, that raise isn't even linear. The modulus of the impedance is $Z=\sqrt{(R^2 + X_L^2)}$. With 2.2H of input inductance and 146Ω of input resistance (tabulated test results on the previous page), the input impedance at 20 Hz is $Z=\sqrt{[146^2 + (2*3.14*20*2.2)^2]} =\sqrt{[146^2 + 276^2]}$ = 312Ω!

Notice that the order of magnitude of the resistive and inductive component of the input impedance at such low frequency is roughly the same, 146Ω versus 276Ω!

If such a TVC were used at the input of an integrated power amplifier, the load on a preamplifier would be below 300Ω at 20Hz, and not even solid-state amplifiers present such a low impedance load. As a result, most preamplifiers will struggle to drive such a low impedance load, and the distortion will increase significantly.

At say 10kHz, the situation is different. $Z=\sqrt{[146^2 + (2*3.14*10,000*2.2)^2]} = \sqrt{[146^2 + 138,230^2]} = 138,230$Ω! The resistive component is now negligible compared to 138kΩ of inductive impedance, which is three orders of magnitude or almost 1,000 times higher.

The way it used to be: Williamson output transformer (10kΩ PP)

Having analyzed the Williamson amplifier in the chapter on classic amplifier designs, let's look at how its output transformers were designed and wound. Firstly, a look at the description from the original article published in The Wireless World:

"Core: 1.75 inch stack of 28A Super Silcor laminations. The winding consists of two identical interleaved coils each 1.5 inch wide on paxolin formers 1.25 x 1.75 inch inside dimensions. On each former is wound 5 primary sections, each consisting of 440 turns (5 layers, 88 turns per layer) of 30 SWG enameled copper wire interleaved with 2 mil. paper, alternating with 4 secondary sections, each consisting of 84 turns (2 layers, 42 turns per layer) of 22 SWG enameled copper wire interleaved with 2 mil. paper.

Each section is insulated from its neighbors by 3 layers of 5 mil. Empire tape. All connections are brought out on one side of the winding, but the primary sections may be connected in series when winding, two primary connections only per bobbin being brought out. Windings to be assembled on core with one bobbin reversed, and with insulating cheeks and a center spacer."

First, the conversion of dimensions and SWG (Standard Wire Gage) sizes. One mil is one-thousandth (0.001) of an inch or 0.0254 millimeters, so two mil paper is 0.05mm. The thickness of 3 layers of 5 mil tape is 15 mil or 0.38mm. 1.75" stack thickness is 4.45cm, the coil width is 1.5" or 38mm. The primary is wound with wire diameter ɸ=0.315mm, the secondary with wire f=0.71mm.

Secondly, we don't know what 28A size means, but we do know that the winding window is 3" or 7.62 cm wide. A look at the current EI size table reveals that EI152.4 lamination has the same window width, so we will base this transformer's analysis on that size.

The laminations in the original Williamson transformers were made of ordinary 4% silicon steel, so to get a high primary inductance of around 100H, Williamson had to use a large core. The cross-section of the center leg had an area A=a*S = 5.08*4.45 = 22.6 cm².

With much better GOSS and amorphous materials nowadays, you should be able to achieve the same or even better specs with smaller laminations such as EI114. Remember, the larger the core, the higher the primary inductance, but also the higher the leakage inductance and the distributed capacitance, and the lower the upper -3dB frequency limit of the transformer. That is why the bobbin was split vertically in half - to reduce the distributed capacitance and improve the upper-frequency limit of the transformer.

SWG is an abbreviation for Standard Wire Gauge, also known as Imperial Wire Gauge or British Standard Gauge. 30 SWG wire has a diameter of 0.315 mm, while 22 SWG is d=0.711mm. To add to the confusion, there is also AWG, an acronym for American Wire Gauge, which is different from SWG. For instance, 30 AWG would be 0.255 mm, while 22 AWG would be 0.644mm.

The winding arrangement

The primary impedance is specified as 10kΩ and the turns ratio as 76:1. That means the impedance ratio is IR=76² = 5,776! The article explains that the impedance of each secondary section is 10,000/IR = 10,000/5,776 = 1.73Ω.

Since there are five primary sections, all connected in series, each half has 5x440T=2,200 turns, but since there are two vertical halves, the total (plate-to-plate or anode-to-anode) number of primary turns is 4,400.

The number of turns in each secondary section should then be $N_2 = N_1/TR$ = 4,400/76 = 58, exactly as specified.

Once the bobbin walls are subtracted from half of the widow width of 38mm, each coil has a width of around 35mm. With 88 turns in each layer of ϕ=0.315 wire, that is 88*0.315 = 28mm.

With 42 turns in each secondary layer of ϕ=0.711 wire, that is 29*0.711 = 30 mm.

Of course, the physical diameters of magnet wires are always larger than their effective areas specified (copper area), so ϕ=0.315 wire may have a diameter between 0.33 and 0.36, depending on the number of insulation varnishing coats, one, two or three!

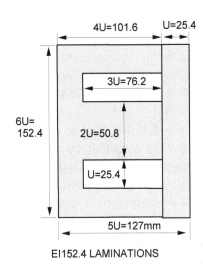

EI152.4 LAMINATIONS

RIGHT: The winding diagram of the original Williamson's output transformer

BELOW: Assuming the vertical secondary sections are connected in parallel (S1a and S1b, S2a and S2b, etc.), there are four possible ways of doing so, resulting in low output impedances of between 1.73 and 6.92Ω. Higher output impedances are possible without paralleling vertical sections, but the coupling would be unbalanced and the performance diminished!

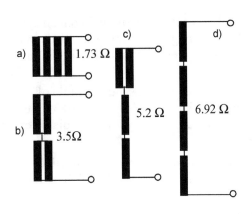

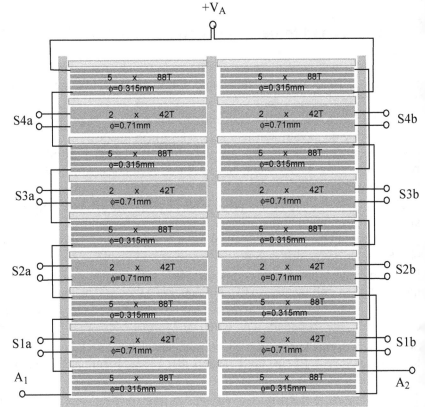

Ertoco toroidal output transformers

Ertoco is a transformer-making business in Vilnius, Lithuania. They started selling on eBay late in 2015, so we got a pair of their toroidal SE output transformers. They are rated at 10 watts with a 2k5 primary, suitable for 2A3, 300B, F2A, and EL153 in triode-connection and many other triodes and pseudo-triodes of around 30-40 Watts anode dissipation.

The square wave reproduction is quite good, but the primary inductance is low, only about 6H. With a relatively high leakage inductance of around 26mH, the quality factor is only 6/0.026 =250.

If you are on a tight budget or want to mount output transformers under a relatively thin chassis, these should be on your shortlist!

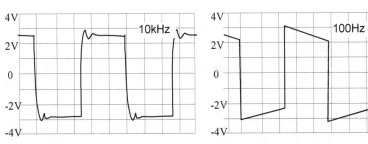

ABOVE: The test waveforms of Ertoco transformers at 100 Hz and 10 kHz.

OUTPUT AND INTERSTAGE TRANSFORMERS: FROM COMMERCIAL BENCHMARKS TO DIY DESIGNS 261

MANUFACTURER'S SPECIFICATIONS:
- Input / output impedance: 2500 Ω / 8 Ω
- BW: 10Hz - 130 kHz (-3dB)
- Primary / secondary DC resistance: R_1=50 Ω, R_2=0.3 Ω
- Maximum primary DC current: 150 mA
- L_P=6.5 H @ 100 mA, L_L= 26mH, QF= 250
- Rated power: 10W
- Diameter 105mm, height 40mm, weight : 1.6 kg

Vintage commercial example: Universal 10W Hi-Fi PP transformers

This vintage transformer by Tomoe Electro Eng. Works, Tokyo, is a good educational example, plus, if you wish to experiment with various primary and secondary impedances in your amplifier design, a pair of "universal" output transformers is a great tool!

These days most audiophiles would not go for a 10 Watt PP amplifier, when you can get more power from most single-ended designs. However, the same design and winding principles apply to larger transformers, so there are some valuable lessons to be learned.

The largest scrapless EI lamination that would fit in the 73mm tall case would be EI66, with the other dimension of 55mm to fit the 63mm width of the case.

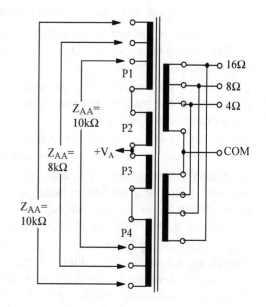

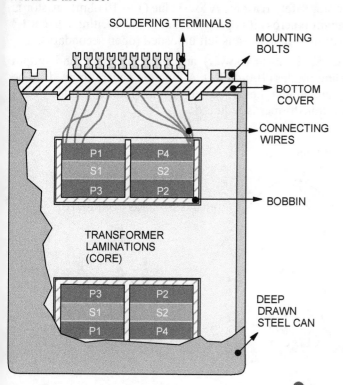

Vertically, one secondary section is sandwiched between two primary sections (2:1 sectionalizing), but 2:1 horizontal sectionalizing is also used, resulting in 4 primary sections and two secondary sections. Vertically splitting the winding in half reduces the shunt capacitance and increases the upper -3dB frequency f_U.

Notice the clever choice of the winding order. The two center sections of the primary (closest to the CT), P2 and P3, are wound first since there are no taps on them. Then, the two secondaries are wound side-by-side, followed by the topmost windings, primary sections P1 and P4, with three taps each.

The main benefit is the ability to connect all the taps on the two halves of the secondary windings in parallel (4-, 8- and 16Ω). Since they were wound side-by-side, they have the same number of turns, equal DC resistances, and, most importantly, equal voltages. Only bifilar winding would achieve the same result.

MANUFACTURER'S SPECIFICATIONS:
- Primary impedances: 5, 8 & 10 kΩ
- Secondary impedances: 4, 8 & 16 Ω
- Power rating: 10 Watts @ 50 Hz
- Insertion loss: 1 dB (efficiency of 80%)
- Frequency response: 40 Hz - 50 kHz (+/- 1dB)
- Size: 80 x 63 x 58 mm, weight: 1.0 kg
- For use with: 45, 2A3, 6V6, 6F6, 6AR5, 6AQ5

The same benefit is obtained on the primary side; P2 and P3 windings are identical, as are P1 and P4. Without vertical split, the windings further from the core would have a higher MLT (mean length of a turn) and higher DC resistance, so the two primary windings would be unbalanced!

Comparing the parameters and fidelity of audio transformers

In Vol. 2, we mentioned a quick way to assess the quality of a transformer's core by measuring its primary inductance at 120 Hz and 1 kHz, as all digital LCR meters do. The smaller the difference between the measured values at those two frequencies, the better suited the laminations are for audio use.

The oscillographic methods outlined below are more precise and do what the LCR method cannot do, and that is to evaluate the linearity and losses of the transformer's laminations, which can be deduced from the shape of its hysteresis curve and the magnitude and waveform of the magnetizing current. The test setup is very similar. While identical on the primary side, the main difference is the way an oscilloscope is hooked up to the secondary.

Observing transformer's magnetizing current on an oscilloscope

Once the primary winding is energized, a sinusoidal primary voltage (the mains voltage in power transformers or a sine signal for audio transformers) establishes a sinusoidal magnetic flux through the magnetic core. The flux lags behind the voltage by 90 degrees, so the flux peaks when the primary voltage drops to zero (point A).

A waveform of the primary (magnetizing) current can be drawn if point-by-point construction for a specific hysteresis loop is carried out. Due to the nonlinear dependence between the flux and the current, its waveform will be significantly distorted.

If displaying the hysteresis curve (the next experiment) is too complex or time-consuming, this simpler test setup that displays the magnetizing current can also be used as a proxy so meaningful comparisons between different transformers can still be made.

The broader and more tilted the hysteresis curve, the more distorted the magnetizing current and the higher the insertion losses and harmonic distortion of the audio transformer.

A variable autotransformer (popularly called Variac®) is used to adjust the amplitude of the mains signal used for this test. Its output is fed into an isolation (1:1) transformer for safety reasons. A low value (1 to 10 ohm) resistor is connected in series with the primary winding. The voltage drop across it is proportional to the exciting current I_1, and this voltage is observed on the oscilloscope. The transformer under test is left unloaded (open secondary).

The mains voltage is used as a signal source in the case of mains (power) and tube interstage and output transformers, which operate with similar primary voltage amplitudes (100-400V). A function generator needs to be used for low-level audio transformers such as MC step-up, and input types since very low amplitudes (under 1Volt) are needed. In that case, neither the variac nor the isolation transformer is required.

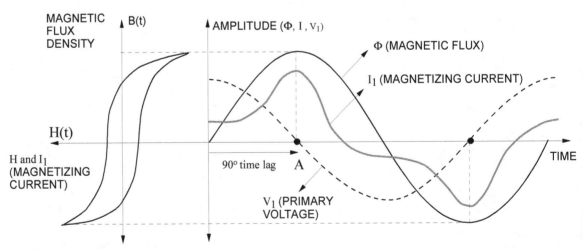

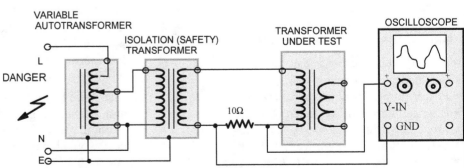

ABOVE: While the primary voltage and the magnetic flux it produces are undistorted sine waves, the primary magnetizing current isn't. Due to the hysteresis curve, it has the waveform illustrated.

RIGHT: The test setup for displaying the waveform of the magnetizing current on an oscilloscope

OUTPUT AND INTERSTAGE TRANSFORMERS: FROM COMMERCIAL BENCHMARKS TO DIY DESIGNS

This vintage advert from "Audio" magazine by Peerless shows basically identical magnetizing current of their competitors' output transformers (just different amplitudes, proportional to loss) and the (enlarged) waveform of their own push-pull output transformer, model S-240-Q, which looks like an almost undistorted sine wave, which seems too good to be true.

Talking about transformer model numbers, in contrast to at least some rudimentary naming conventions used by most tube manufacturers, there have been no standards of naming audio and mains transformers used in tube amplifiers, so it is next to impossible to deduct a type of transformer from its model number only.

However, some manufacturers used their own naming systems. For example, Peerless used a naming convention where the prefix indicated the type of transformer and the suffix specified the type of housing (case). S-240-Q was an output transformer (S) in a "Q" type case. Peerless used suffix "A" for transformers with end bells.

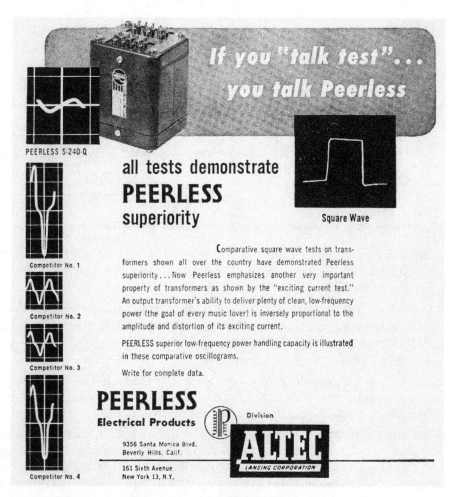

ABOVE: Peerless advertisement, March 1952, © Audio magazine

RIGHT: Peerless naming convention

PREFIX	TRANS. TYPE
C	FILTERING CHOKE
L	AUDIO CHOKE (PLATE CHOKE)
S	OUTPUT TRANSFORMER
K	INPUT TRANSFORMER
E	IMPEDANCE-MATCHING TRANSFORMER
R	POWER TRANSFORMER (PLATE +FILAMENT SUPPLY)

Estimating transformer's laminations quality by observing its hysteresis loop on an oscilloscope

The hysteresis curve of a particular transformer's core (lamination stack) has H, magnetic field strength or "magnetomotive force," on the X (horizontal) axis and B, magnetic flux density as a dependent variable on the vertical (Y) axis.

Since H is a direct function of the excitation current (primary current of the transformer without any load), if we connect a small value (1 ohm to 10 ohms) resistor in series with the primary winding, the voltage drop across such resistor will be proportional to the exciting current I1 and thus to H (just as in the previous test). This voltage should be applied to the horizontal input of the oscilloscope, which will work in X-Y mode (its time base generator will be disconnected).

Since the voltage induced in the secondary winding is proportional to the rate of change of magnetic flux Φ, this is expressed mathematically as a derivative ($V_2 = N_2 * d\Phi/dt$), where N_2, the number of secondary turns is a constant. This means the faster the rate of magnetic flux's change and the higher the number of secondary turns, the higher the induced secondary voltage!

B is proportional to magnetic flux Φ. To get B, we must perform the opposite mathematical operation on V_2. The opposite of differentiation is integration, so we need an integration of the secondary voltage V_2. A simple RC network can perform that task under the provision that its time constant t=RC is much larger than the period of the test signal.

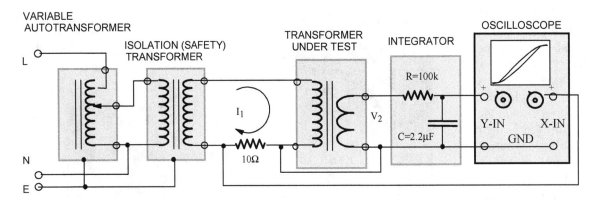

AOVE: The test setup for displaying the hysteresis curve on an oscilloscope

In other words, we must ensure that $\tau=RC \gg 1/\omega$, or that $\omega RC \gg 1$! Angular frequency ω (omega) is $\omega=2\pi f$! If this condition is not satisfied, the display will not accurately represent the hysteresis curve; the curve will be "folded" or warped as in the illustration below (far right).

If the mains voltage is used for this test, a frequency of 50 Hz (or 60 Hz in the USA and some other countries) needs to be used in the formulas above, so let's see what we get with our chosen values of R=100kΩ and C=2.2µF: $\omega RC = 2\pi fRC = 2*\pi*50*10^5*2.2*10^{-6} = 10*\pi*2.2 = 69$, which can be considered much larger than 1 (69$\gg$1), so the hysteresis curve will be displayed properly.

Going back to the examples below, even the one on the far left isn't bad; the curve is pretty narrow (low losses) and reasonably straight (low distortion). The example in the middle is even better; only the highest fidelity audio transformers using EI laminations or C-cores of superior quality will have such a linear hysteresis curve.

A few pointers regarding the test circuit. The capacitor used in the integrator network should be a low loss film type, polyester, or polypropylene. The test should start with the Variac in zero output position. Its dial should be slowly raised until the nominal primary voltage is reached. The vertical sensitivity control of the oscilloscope and the time base will need to be adjusted until a display such as those illustrated is obtained.

Due to the polarities of the signals involved, the display will be a mirror image unless the invert button on the scope is pressed (activated), which flips the signal brought to the scope's X-input.

If the scope's sensitivity cannot be lowered enough to display a large enough curve (budget scopes of low sensitivity or very low magnetizing current amplitudes), the 10-ohm series resistance will need to be increased.

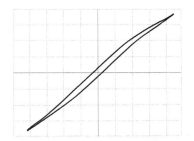
Hysteresis loop of a good quality output transformer with GOSS laminations

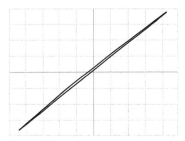
Hysteresis loop of a top quality output transformer with extremely small losses and very low distortion

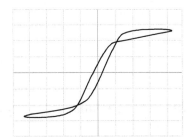
Distorted oscillogram of a hysteresis loop due to improperly selected RC integrator (ωRC product is too small).

Not all vintage audio transformers (or amplifiers) deserve their stellar reputation!

In April 1952, The Naval Research Laboratory of the US Navy prepared a report titled "Laboratory Tests of Some of the Popular Audio Amplifiers," in which they compared the design and performance of a dozen or so commercial tube amplifiers. Some amplifiers were constructed by their staff "in-house" using different output transformers available for purchase at the time, including Acrosound TO-300 and Peerless S-265-Q models.

They tested output transformers inside the same Williamson-type amplifier; two such oscillograms are shown on the next page, together with the A-f characteristics.

While its upper -3dB limit was around a respectable 70 kHz, at low power levels, Acrosound TO-300 exhibited an incredibly high resonant ultrasonic peak, which was detectable even at higher output levels. This manifested itself in a high overshoot at the rising edge of a square wave and low damping.

Peerless S-265-Q, on the other hand, showed no such ultrasonic instability and reproduced square wave signals much better, and also had a much higher upper -3dB limit!

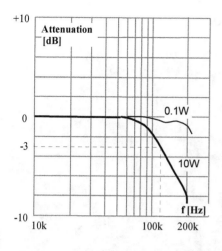
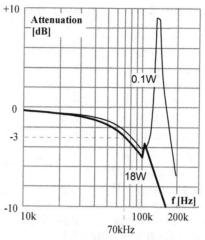

ABOVE LEFT: The A-f characteristics of a Williamson amplifier with Peerless S-265-Q output transformer and the fidelity of a 20 kHz signal at its output.

ABOVE RIGHT: The A-f characteristics of an ultralinear amplifier with Acrosound TO-300 output transformer and the fidelity of a 10 kHz signal at its output.

The moral of the story? While vintage output transformers generally fetch very high prices today, not all were created equal. As mentioned previously, the quality factor is the best single indicator of a transformer's quality.

For the Partridge UL2 model, the primary inductance is 140H, and since the leakage inductance is only 10mH, its quality factor is QF=LP/LL=14,000! Such a good result is most likely a consequence of their use of C-cores instead of inferior EI-laminations and advanced sectionalizing and winding techniques.

Notice very keenly competitive pricing of the Partridge UL2 model, which, despite being imported from the UK, was selling in the USA at around the same price ($25) as the smaller and inferior Acrosound TO-300 transformer.

PARTRIDGE UL2 MANUFACTURER'S SPECIFICATIONS:

- Z_{AA}=10kΩ
- BW: 30Hz-30kHz (-0.5dB)
- P_{OUT}: 50W @ 60Hz, 30W @ 30Hz
- THD: 0.5%
- L_P=140H, L_L=10 mH, QF= 14,000
- C_{PS}=0.5nF
- Dimensions: 85 x 85 x 90 mm

Partridge advertisement, December 1954, © Audio magazine

Lessons learned from the vintage German power transformer

This power transformer by Loewe Opta uses M-laminations, as do most of the smaller German and European power transformers of the 1950s and 60s. The normal marking practice has been to specify nominal voltages and full load currents for each winding except the primary, where the current is usually not specified. The manufacturer specified the voltages under full load but not the maximum allowed current levels. Instead, the number of turns and the wire diameters are marked.

The transformer weighs 1.4kg, and its core measures 74.3 x 74.3 x 34.7 mm, which indicates M74 laminations. The published dimensions (table below) specify a 2.3cm wide center leg, and with a 3.4cm stack the effective cross-sectional area is approximately 90% of the gross area, or $A_{EF}=0.9*A = 0.9*2.3*3.4 = 7.0$ cm^2.

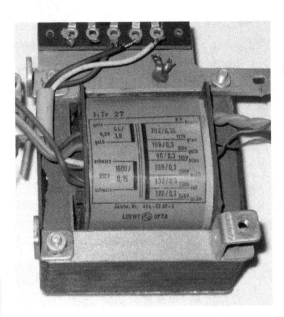

ABOVE: Vintage power transformer with M-laminations by Loewe Opta, salvaged from a tube amplifier

LEFT: The significant dimensions of "M-schnitten" and the table with standard laminations sizes, from the smallest (M30) to the largest (M102).

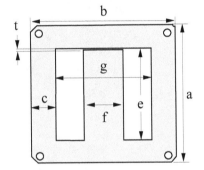

CORE	a = b	c	e = g	f
M30	30	5	20	7
M42	42	6	30	12
M55	55	8.5	38	17
M65	65	10	45	20
M74	74	11.5	51	23
M85	85	14.5	56	29
M102	102	17	68	34

Estimating the power rating and current capacities of the two secondary windings

Based on the given wire diameters, we can estimate the nominal current as $d=\sqrt{(1.27*I/J)}$, where d is wire diameter in mm, I is current in Amperes, and J is current density in A/mm^2. We can calculate current as $I=J*d^2/1.27$ [A].

We don't know the current density the designer chose 60 years ago; for small M cores J is usually between 2.7 and 3.5 A/mm2, so if we assume 3.2A/mm^2 for the heater winding with 1mm^2 wire, we get $I=J*d^2/1.27 = 3.2*1^2/1,27 = 2.2$ A! The power rating of this winding is $P_{S2}=6.3*2.5 = 15.75$ W.

Likewise, the HV winding can supply $I=J*d^2/1.27 = 3.2*0.15^2/1,27 = 57$mA, so its power rating is $P_{S1}=250*0.057 = 14.25$ W, and the total power capability of secondary windings is $P_S= P_{S1}+P_{S2}=14.25+15.75 = 30$W!

Vintage German literature and Din 41 302 standard specify that this particular core (3.4cm stack is one of the standard bobbin sizes) is approximately 84% efficient, which means the primary power rating must be $P_P=P_S/0.84 = 30/0.84 = 35.7$ Watts. The same literature lists this core as having a maximum power capacity of 50 Watts, which means the Loewe Opta transformer designer was very conservative in his approach, using only 36 watts out of the possible 50!

Determining the TPV (Turns-Per-Volt) figure

If you add up the individual number of turns in all the primary sections, up to the maximum of 240 Volts, you'll get 1,440 turns, so the Turns-Per-Volt figure is 1,440/240 = 6.0 TPV.

That figure is slightly higher for the high voltage secondary winding, 1,600/250 = 6.4, while the 6.3V secondary that supplies tube heaters is wound with TPV=44/6.3 = 7.0!

The explanation lies in the winding order. The primary is wound first in small power transformers, closest to the magnetic core, followed by the HV secondary. The secondaries supplying the heaters are wound last, furthest away from the core. This order is usually also the descending order of power ratings.

The highest power dissipation is in the primary, which is equal to the sum of power ratings of all the secondaries (plus transformer losses, a few percentages of the total), followed by the high voltage winding that needs to supply power to the output and all preamp and driver stages. Heater secondaries are usually of the lowest power rating, although in this case, their power is slightly higher than that of the high voltage winding.

Another reason heater winding(s) are wound last is their large wire size. If 6.3V heaters are used, that secondary winding usually has so few turns that it consists of one layer only, and when wound on top of all other windings, that layer pushes down on them and gives the whole coil mechanical strength and minimizes coil buzz.

Going back to the reason for the slightly increasing TPV figure, the clue lies, as always, in the fundamental transformer equation $TPV = N/V = 10^4/(4.44 B A_{EF} f)$.

As windings are wound on top of one another, the mean length of a turn increases, so their DC resistance increases. The turns also gradually encompass a larger and larger area, which includes the cross-section of the core's center leg and all the layers previously wound underneath. Since the magnetic flux (the number of magnetic lines) that couples all the windings is constant, the increase in the cross-sectional area encompassed by the upper windings means that B, the magnetic flux density, is dropping slightly, and since B is in the denominator of the TPV formula, to compensate for that slight reduction in B, a proportional increase in TPV is used!

A few important lessons from power supply filtering chokes

Vintage Motorola filtering chokes

An American eBay seller, electronic surplus parts company in Virginia, was selling half-a-dozen NOS chokes with "Motorola" printed on one and a number code on the other end-bell. The original price was US$49.99, and then they reduced it to US$4.99. However, they only showed a photo of the box and wrapping and not the actual chokes. No technical details were listed, not even the dimensions, so buyers were not interested.

We asked only one question - how many would fit in a Medium Flat Rate USPS Priority Mail International Box, which at the time cost US$66. The seller confirmed that they could ship four chokes in that box.

We could have asked for dimensions and a photo of one choke, some sellers would even be willing to perform basic ohmmeter tests, but the fact that only four would fit in a relatively large box was enough to give us a pretty good idea about their size and weight.

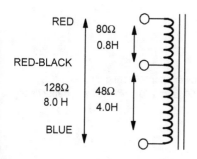

ABOVE: The results of the basic resistance and inductance measurements
BELOW: A typical application when both chokes are used to their maximum current-carrying capacity

We didn't know the diameter of the winding wire used (and thus the DC current capacity) or their inductance, but our logic was that if the chokes were unsuitable, we could always reuse the laminations and rewind them ourselves. With transport, each cost us around US$21.50.

The laminations are 76mm x 63mm with a 50mm stack thickness. The DC resistances are straightforward; however, the inductances cannot simply be added up. The two sections measure 0.8H and 4.0H, but the total winding measures 8.0 H!

If you go back to page 257, the two primary sections of our interstage transformer measure 9.0 Henry and 9.5 Henry, but their total inductance is 36 H in series!

When used as an 8H choke, the section wound with a thinner wire limits the maximum current. In this case, that is the high DC resistance section that measures 0.8H. Its winding wire was 0.22mm in diameter, suitable for up to 120 mA.

The topology illustrated below can be used to better utilize the 4H section, which is wound with 0.35mm wire and can pass up to 300mA. The filtered high voltage for the output stages is taken after the 4H choke section, while the DC supply voltage for output tubes' screens (if used as pentodes) and driver and preamp stages go through another filtering stage (another CLC filter).

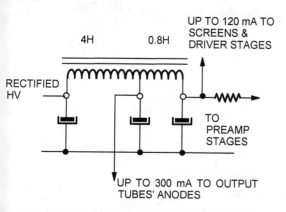

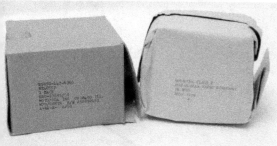

ABOVE: The only photo the ebay seller posted, not showing the actual choke at all.

ABOVE: The chokes are finished in matte black

How the inductance of a filtering choke varies with the magnitude of the DC current flowing through it

These vintage Halldorson chokes are rated at 8H @ 150mA, with 145Ω DC resistance. The printing on their boxes provides some valuable clues.

Notice how the inductance increases by 50% to 12H at a lower DC current of 100mA (point A) and reduces to half the nominal inductance, only 4H at 200mA (point C).

So, lesson #1 is that most chokes, and definitely ones made by reputable manufacturers, will tolerate higher DC current than that specified as nominal or maximum. However, the winding may get a bit warmer than usual (due to increased losses in copper), and the effective inductance will be reduced.

Likewise, the lower the load current below the nominal, the higher the inductance and the better the filtering action of the LC filter, resulting in reduced ripple!

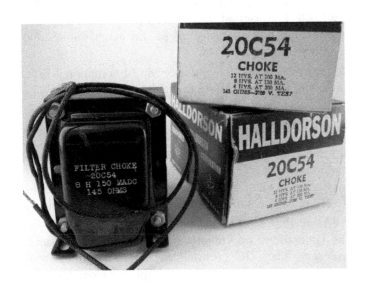
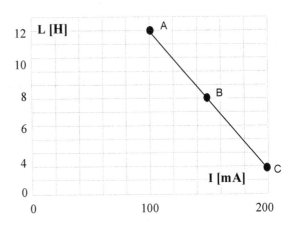

LEFT: A pair of HV power supply filtering chokes.
ABOVE: Between 100 and 200 mA their L-I dependency is almost linear!

Transformers For Tube Amplifiers: How to Design, Construct & Use Power, Output & Interstage Transformers and Chokes in Audiophile and Guitar Tube Amplifiers

FURTHER READING

- PHYSICAL FUNDAMENTALS OF MAGNETIC CIRCUITS AND TRANSFORMERS
- FILTERING CHOKES (INDUCTORS WITH DC CURRENT)
- TRANSFORMER MATERIALS, CONSTRUCTION METHODS AND ISSUES
- MAINS (POWER) TRANSFORMERS
- PHYSICAL FUNDAMENTALS OF AUDIO TRANSFORMERS
- SINGLE-ENDED OUTPUT TRANSFORMERS
- PUSH-PULL OUTPUT TRANSFORMERS
- SPECIAL MAGNETIC COMPONENTS: LOW POWER INPUT, PREAMP OUTPUT & DAC OUTPUT TRANSFORMERS, TRANSFORMER VOLUME CONTROL
- INTERSTAGE TRANSFORMERS, GRID & ANODE CHOKES
- OUTPUT AND INTERSTAGE TRANSFORMERS FOR TUBE GUITAR AMPS
- TRANSFORMER TESTS & MEASUREMENTS

MEASUREMENTS VERSUS LISTENING AND OTHER AUDIO DESIGN DILEMMAS

- What do we really listen for?
- Case study: Jadis SE300B monoblock amplifiers
- The crucial role of audio manufacturers, retailers and reviewers
- The impact of loudspeaker impedance on amplifier performance
- The whole brain approach to audio design

16

"Besides the noble art of getting things done, there is the noble art of leaving things undone."
Chinese proverb

What do we really listen for?

From the micro- to the macro-view

After detailed discussions of technical issues related to tube amplifiers, let's conclude with a few broader aspects of audio design based on the model below. Despite being relatively simple, the typical sound reproduction system model illustrates some important points.

First, every music system is a unidirectional chain of components; the signal propagates from the source through the amplifying chain to the final electroacoustic transducer, the speakers, and the listening room. Every link in the chain affects the sound, some more, others less.

Audiophiles disagree on most things audio. Some argue that the initial components, the turntable, the phono stage, the CD player, and the preamp are more important than the components down the line; others argue the exact opposite. Secondly, the chain does not end with the listening room; it includes the subject or the listener. Before we form the impression of a particular audio system how it sounds, the sound has to pass through another transducer, the human ear. That creates further problems since hearing is subjective; that is why the science that studies related phenomena is called psychoacoustics. Our psyche, including our moods, affects the way we hear things. The human ear is an imperfect transducer. Its sensitivity varies with the sound frequency, and the ear itself distorts, creating intermodulation components that were not present in the reproduced sound.

Some people hear more, some less. Some are sensitive to the bass frequencies, while others pay more attention to the high-frequency end of the audio range. Depending on their mood, even the same person will judge the same amplifier and system differently. I learned that lesson a long time ago.

When we build an amp and take it to the listening room for evaluations, I don't form my opinion just during the night listening, when the volumes are generally lower. The amp will sound different during the daytime when the output sound levels are generally higher. These are just some of the factors in play. Listening evaluations are like wine tasting, which means highly subjective.

I wish I had a dollar for every time a prospective buyer asked me, "Igor, which of your amplifiers will best match my XYZ speakers?" How a particular amplifier will sound with a specific loudspeaker cannot be predicted with any degree of certainty, just as the following link in the acoustical chain between the loudspeaker and the listening room cannot be established beforehand without the actual listening test.

This issue is so important because it also affects the projects and designs in this book. For a while, before finalizing this third volume, I entertained the thought of assigning a numerical score to each design, or at least of using a "star" system, assigning designs one to five or one to ten stars. However, that would be highly subjective and therefore next to useless. It could even be misleading and thus open to disputes and challenges.

There is only so much an author can do, and recommending some designs over others is not one of those things; you have to make that crucial decision yourself. I did express my preferences in many cases throughout this book, but that's all they are, impressions and preferences of a single individual. Since they result from only listening to a limited range of music on one system, they are not statistically valid, and your impressions could be the opposite.

However, as you listen to different sources, amps, and speakers, in different rooms and at different times, your preferences will gradually form if they haven't done so by now. You will have a pretty good idea of what kind of sound you like, what matters to you more, and what matters less.

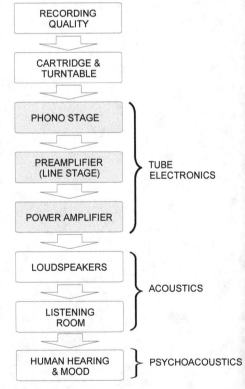

ABOVE: The simplest model of a turntable-based sound reproduction system has eight components and seven links. Volumes 1 to 3 of this book only considered three of them!

Three major aspects of the technical-musical gap

There are three significant issues in the technical-musical gap, the difference between design & measurements on one, and listening & other musical issues on the other side of the divide. They are not in any particular order, certainly not in the order of importance. Audiophiles and DIY constructors will disagree even on their relative significance!

We design amplifiers and test them with purely resistive loads, yet all loudspeakers are predominately inductive or capacitive (reactive), frequency-dependent loads. An amplifier that measures very well on a test bench may sound sterile and uninvolving with actual speaker loads. More on this issue soon.

Secondly, you have probably noticed that the "Measured Results" boxes in this book include only the output voltage/power, bandwidth (frequency range), and sometimes a damping factor. Similarly, only the voltage gain, frequency range, and output impedance are specified for preamplifiers. Why no harmonic distortion (THD) or intermodulation distortion (IM) figures? Simply because not all distortion is created equal!

The 2nd and higher-order even harmonics add body to the sound, make amps and preamps sound warmer, more "musical," and emotionally engaging. Odd harmonics are generally unpleasant and irritating, as is the discordant intermodulation distortion when the amplifier creates sidebands of unrelated frequencies. We have noticed that THD figures are not correlated with the listening evaluations. Many amps that measure very well sound ordinary and boring, while some or even most of the best sounding amps and preamps distort like hell!

Finally, the issue of negative feedback is probably the most contentious topic in audio design. The feedback-free puritan brigades don't even want to consider any use of NFB under any circumstances. However, we are not fanatics; this is not a religious but a technical book.

As with any other concept or intervention, we should understand its benefits and unwanted side effects. We have already covered most of them in volume 1 of this book, so we will only emphasize a few important points here.

NFB widens the linear region but makes clipping harsher and more abrupt

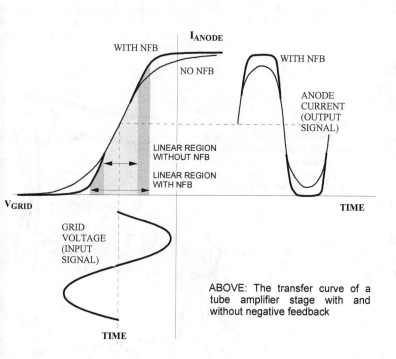

ABOVE: The transfer curve of a tube amplifier stage with and without negative feedback

When a global NFB is brought back to the input stage of a tube amplifier, the linear region of input (grid) voltages (signals) with NFB is wider than without it, so if you are driving that power amp with an active preamp, for instance, it will be able to accept larger input signals without clipping.

Notice how the input stage's transfer characteristics (far left) are "straightened" and how local negative feedback widens the linear region.

However, once clipping occurs, as indicated by the waveform of the anode current, it is more abrupt with NFB, and the output signal will resemble a square wave. This results in much higher and more unpleasant (odd harmonics) distortion, just as in solid-state amplifying stages.

When a local NFB is used in a tube amplification stage, typically through the un-bypassed cathode resistor, a similar effect happens; notice how the transfer characteristics of a triode and the cathode resistor (straight line) are combined (left). In effect, the un-bypassed RK "linearizes" the tube's transfer curve, straightens it, and makes it less steep.

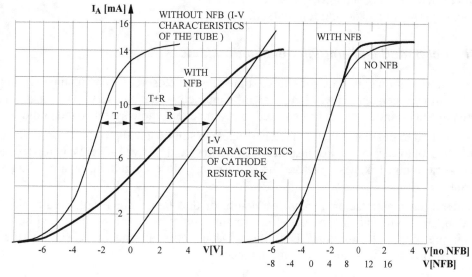

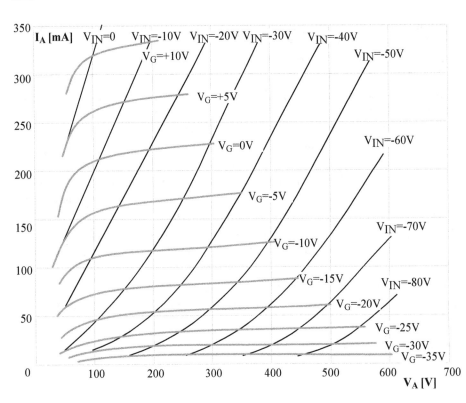

LEFT: Anode characteristics of a power pentode, original (thick gray lines) and after the application of negative voltage feedback (thinner black lines).

The graph on the left shows anode characteristics of a power pentode before (in gray) and after applying negative voltage feedback (in black). The original bias levels range from +10V to -35V, while after the application of NFB, the range is from 0 to -80V.

Thus, lower sensitivity, lower amplification factor, and lower internal resistance. The characteristics are steeper; the original ones were almost horizontal, indicating a very high internal resistance of the pentode, where anode current did not change much with changing anode voltage.

The curves now resemble those of a triode!

In conclusion, almost all audiophiles agree that very strong NFB is detrimental to sound quality, despite what vintage designers advocated in the 1960s and 1970s.

However, should NFB be used at all is another matter. We generally consider a very weak NFB beneficial. It widens the bandwidth and reduces output impedance, which increases the damping factor and improves the bass, among other benefits.

One strategy we can recommend is to install a variable NFB into an amplifier (either continuously variable with a potentiometer, or in steps, using a 3- or 4-position switch and fixed feedback resistors) and then ascertain for yourself, with your system, which way the amp sounds the best.

Never believe what self-proclaimed experts tell you; experiment and decide for yourself.

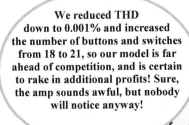

The octal model of amplifier evaluation factors

This short section aims only to achieve some sense of closure; it is by no means complete or scientifically validated. It is simply a personal view of eight significant factors that can serve as a basis for evaluating tube amplifiers and preamplifiers. Indeed, they can be used to evaluate the whole systems and audiophile setups.

I have tried to position the eight factors (boxes) in some logical proximity order, meaning that related factors should be adjacent to each other. For instance, "Dynamics" and "Drive" are closely related; some may even argue they mean the same thing. Likewise, "Drive" and "Bass" are close neighbors, as are "Transparency" and "Resolution," for instance.

Under each major factor, a few descriptors of that factor are listed. Again, there may be some repetition and redundancy here, and they overlap for sure; one cannot be certain where one descriptor stops and another starts. For example, under "Resolution," Definition, Detail retrieval, and Delicacy describe the same quality, although in slightly different ways and with slight nuances.

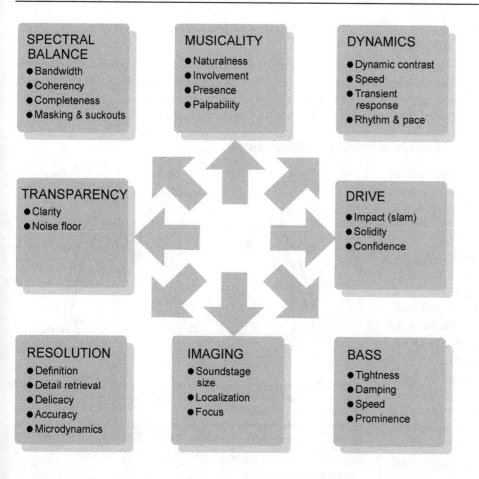

The principal aim of this discussion is to nudge you into thinking and, if possible, outlining your own model. What significant factors would you choose and why?

Don't worry if your model does not make complete sense or if there are overlaps, as mentioned. Not even a Ph.D. thesis in acoustical engineering would be sufficient to cover all these issues in a meaningful way. A much broader and thicker treatise is needed. After all is said and done, audio should be experienced through listening, not pontificated upon or studied in minute detail!

LEFT: One of the many possible models for evaluation of audio systems in general, and tube amplifiers and preamplifiers in particular

Case study: Jadis SE300B monoblock amplifiers

The decision to include this illuminating case study in this volume of "Audiophile Vacuum Tube Amplifiers" was made after reading the online evaluation and modification article of this amplifier by the late Australian valve expert Patrick Turner (http://www.turneraudio.com.au). His website is a treasure-trove of information and is highly recommended. Never one to blindly copy or follow the myths, Mr. Turner experiments for himself, measures, and builds, in contrast to the majority of online armchair "experts" who hypothesize, criticize and pontificate, instead of properly building, testing, and scrutinizing designs and commercial products.

Unfortunately, Mr. Turner did not publish the schematics of the original Jadis amplifier on his website, only the circuit after his modifications, so we couldn't include it here either.

The design of this amplifier is plain vanilla, two identical, capacitively-coupled 6SN7 triode stages in cascade, followed by a single-ended output stage with two 300B triodes in parallel. No negative feedback was used. Completing the tube complement, the two 5R4WGY rectifier tubes are in two high voltage supplies, one for the input stages and a separate one for the output stage.

The mains transformer has two 9.6 V_{AC} secondaries, but only one is used; after rectification and filtering, the DC voltage goes through a solid-state series regulator and supplies $5V_{DC}$ to both 300B tubes in parallel! Why wasn't the same circuit doubled using the other 9.6 V_{AC} secondary, so each 300B tube could have its own heater supply, a much better option? With a hefty US$13,000 price tag (in 1996!), one would not expect such a penny-pinching attitude from a supposedly reputable manufacturer.

Just as they share the heater supply, the two paralleled 300B lovelies share the cathode biasing circuit, a 300W resistance (actually nine 2k7 4W resistors in parallel), and 4x 47µF of capacitance (a total of 188µF). 300Ω in the cathode seems too low, equivalent to 600Ω for one 300B.

VERBATIM: Yes, they really said this!

"Like other Jadis units, the output transformers are handcrafted without any gap in the magnetic circuits, and are used well below their full possibilities (they can be employed on 250W amplifiers.)"
Specification sheet of Jadis amplifiers prepared by Jean-Paul Caffi

The time constant of the cathode RC biasing is τ=RC =300*188*10-6 = 0.0564 seconds, which means the -3dB frequency of the cathode circuit is 2.8 Hz. Fine.

The main issue with these amplifiers is that at low frequencies, the output transformers go into saturation very early, at only a couple of Watts of output power, so what does the claim that "they can be employed on 250W amplifiers" means if they can't even handle a measly couple of watts down to the required bass frequencies? Even with an air gap, these transformers cannot be used in a 250W amplifier; they simply aren't large enough!

How to reconcile measurement results and listening impressions

The two "Stereophile" reviewers (one from a few months earlier) raved about the amps' deep and punchy bass, "unlike anything single-ended" (!) and "treble extension without any treble hardness."

However, the measurements show that with a purely resistive load, the -3dB limits of this amp are 50Hz (1) and 23 kHz (2). A simulated speaker curve shows two major "suckout" regions, one in the bass-lower midrange (3), down up to -2dB, the other in lower treble (4), up to -3dB down from the midrange level.

Assuming these reviewers were honest in their glowing appraisals of this amplifier of questionable design and poor measurements, this leads to only one conclusion, that measurements don't matter one bit!

Frequency range does not matter, harmonic and intermodulation distortion figures don't matter, the amplifier's low damping factor doesn't matter; in conclusion, a tube amplifier can sound "sublime" despite its shocking measurements!

The illustration on the right shows the spectrum of the output signal from Jadis SE300B with a 50 Hz input signal. The power level was a low 1.6W into a 4W load. Notice H2 (2nd harmonic) only 14dB below the level of the 1st! Notice the nasties, H3, and other odd harmonics, how relatively high their amplitudes are.

As a side lesson, how can harmonic distortion in % be determined from these figures in dB?

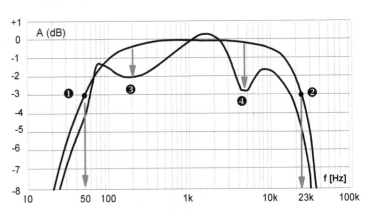

ABOVE: The amplitude vs. frequency response of Jadis SE300B at 1W into 8 ohm dummy load and into simulated speaker load SOURCE: redrawn from Stereophile, March 1996 review

BELOW: The spectrum of a 50Hz signal, 1.6 Watts into 4Ω for Jadis SE300B, SOURCE: redrawn from Stereophile, March 1996 review, both used with permission

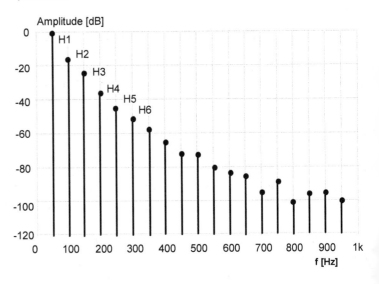

How to calculate harmonic distortion in % from individual spectral figures in dB?

The fundamental formula is dB = 20log(x), so "x" would be our harmonic distortion coefficient H (but not in % yet). Remember, if the base of the logarithm is not specified, it is assumed to be 10, so log(x) really means $log_{10}(x)$. Therefore $x=10^{(db/20)}$.

Here, the 2nd harmonic H_2, a 100 Hz signal, is −14.3dB below the zero level of the 1st harmonic H_1. We have $H_2 = 10^{(db/20)} = 10^{(-14.3/20)} = 10^{(-0.75)} = 0.1927$ To convert that figure into percentages, multiply it by 100 and get $H_2 = 19.27$ %!

This incredibly high distortion is due to the lack of the air gap in output transformers, which at such a low frequency would saturate even at very low power levels, let alone the relatively high 1.6 Watts!

The amplitude of the third harmonic (150Hz) of the distorted output signal is −23.4dB below that of the fundamental, which is a distortion of 6.8%. How could amplifiers with such horrendous distortion levels delight seasoned professional audio reviewers is mind-boggling.

The crucial role of audio manufacturers, retailers and reviewers

When manufacturers stretch the truth or ignorant dilettantes propagate myths and misconceptions on the web, that is one thing. Still, when professionals, people who are paid to create and disseminate relevant and accurate information and scrutinize products, such as audio reviewers, get their facts wrong, that is something unforgivable. An average reader thinks these guys are in-the-know and implicitly trusts their opinions.

In "Stereophile" magazine (Jan 2014, p27), the reviewer (Sam Tellig) wrote this about Unison Research "Simply Italy" SE amplifier: "Giovanni Sachetti designed the circuit to run the EL34B in pentode mode, but with a sound that closely emulates triode."

That amplifier operates in an ultra-linear mode, not in a pentode mode, and does not "emulate" the sound of triode! That sentence is not just inaccurate but also misleading, for the reader may think that the amp somehow processes the signal to emulate the triode sound - it doesn't!

It gets worse. In the Feb. 2012 issue of "Stereophile, on page 42, Art Dudley writes: "... Shindo Haut-Brion ($11,000), a stereo amplifier designed around complementary pairs of the famous 6L6 pentode tube." The reviewer managed to make two faux-passes in this short partial sentence. First, unlike bipolar transistors, there is no such thing as "complementary pairs" of tubes. Secondly, 6L6 is not a pentode but a beam power tube!

However, the title "Bull of the Century" must go to KR Enterprise, who say this in the user's manual for their VT6000BM amplifiers: "...the output stage uses the new, ultralinear power device called the VACUUM TRANSISTOR." The upper case is theirs. New power device? Vacuum transistor? Oh please!

Further down, the instructions read, "The red color on the L.E.D. front panel becomes bright green when preheating of the Vacuum Transistor units are (sic) completed." There is no such thing as a vacuum transistor, and if it needs preheating, it must be a vacuum tube!

The seven levels of truth in audio

Here is a simple 7-step model you can use whenever you are evaluating or choosing finished audio products, components, or anything else, for that matter, in any sphere of life. Similar "7 Levels of Truth" apply to real estate, white goods, cars, service providers (housebuilders, consultants, contractors, dentists, accountants, ...) pretty much all goods and services:

1. TECHNICAL FACTS: Accurate, measurable, and thus indisputable truth.
2. SIMPLIFICATIONS AND MODELS: More or less accurate representations of the reality, but only under certain conditions, and within clearly specified and verified limitations.
3. EDUCATED GUESSES: An intuitive feeling of an expert, possibly valid, albeit without any conclusive proof or test results that would verify such respected opinion(s).
4. OPINIONS: Without much value, unless the source is experienced and knowledgeable, in which case those opinions could serve as a starting point for thinking and further investigation.
5. MYTHS: Started either innocently, out of ignorance, or deliberately, due to one's vested interests, and then propagated by the more-or-less ignorant majority of audiophiles and bystanders ("tube kickers").
6. MISREPRESENTATIONS: Opinions, myths and half-truths presented as facts and "distinguishing features" by manufacturers, importers, and retailers with vested interests.
7. DELIBERATE LIES: Untruths that are clear and obvious to experienced and educated experts but are often taken as gospel by audiophiles.

As an example of an audio product, let's take this book. Even though you cannot listen to it (many audio products don't produce or process sound), it is an audio product. At what level of truth is this book?

Well, it certainly deals with facts, so most of it is or should be at level 1. When we built and tested a particular circuit and then included it in this book, that constitutes a verifiable experiment. You or anyone else should be able to reproduce it, and if you do, you should get the same or very close results. The same applies to our tests & evaluations of commercial amplifiers and components.

However, to minimize the maths, simplify things and make them more user-friendly, there are many models, simplifications, and rules-of-thumb, as in any technical book, so level 2 is also used. This is not an admission of inadequacy. Every university course of study uses models and simplifications; otherwise, it would take 15-20 years to fully comprehend things and eventually earn a university degree!

Occasionally, I voiced my opinions and hypothesized about the causal or correlational links between certain issues. However, no matter how experienced and insightful the author may be, educated guesses and personal opinions should be minimized in a book of this kind, something I tried to do as much as possible.

Finally, I believe there are no myths, misrepresentations, or deliberate lies propagated in this book. The keyword is deliberate. As in any book, there may be inconsistencies, omissions, and repetitions, there may even be factual errors, but they are by no means deliberate. As an author, I have no interest in making you buy an E86C tube instead of 12AU7, for instance, so why would I aggrandize one and denigrate the other? As always, we come to the underlying issue, which is someone's motive!

Case study: Psvane "Teflon®" capacitors

Psvane film capacitors come in three values. As the values go up, so does the (very high) price. In 2015 the pricing was 0.1mF/600V (US$99 a pair), 0.22mF/600V (US$149 a pair), and 0.47mF/600V (US$199 a pair).

Printed on the capacitors' bodies are the words "Reference Cap," which is meaningless marketing hype, and "Teflon Copper Foil." These capacitors do use copper foil, but the insulating film is polyester. The only Teflon is the sleeving around the outside leads, one black, the other red.

Does the printing of "Teflon Copper Foil" constitute misleading advertising, and which of the seven levels of truth would it fall under? Is it a misrepresentation or misleading advertising (Level 6) or a deliberate lie, Level 7?

Had "Teflon Film Copper Foil" been printed, that would be a deliberate lie, no doubt. However, by omitting the word "film," the manufacturer can argue that these are Teflon caps since they contain Teflon, albeit only on the outside, where such insulation of the leads does not make any difference to the sound.

However, an unsuspecting DIY constructor or audiophile "upgrading" their capacitors are likely to automatically assume that Teflon refers to the most crucial aspect of a capacitor (sonically), its insulating film; therefore, it can be argued that this constitutes misleading advertising!

The impact of loudspeaker impedance on amplifier performance

A speaker as a complex impedance

A typical dynamic (moving coil) loudspeaker, both drivers by themselves, and especially a combination of drivers with crossover networks in a speaker box, are complex RCL networks whose impedance modulus and phase angle vary widely with frequency. This typical impedance vs. frequency curve of a dynamic loudspeaker shows peaks and dips. Notice a sharp increase of overall impedance at $f_R=100Hz$, the peak of around 20Ω. The rise of impedance at higher frequencies is due to the dominant inductance of the voice coil and the inductors in the crossover.

The "nominal" speaker impedance Z_{NOM}, usually 4, 6, or 8 ohms, "proclaimed" by manufacturers is meaningless, but for simplicity's sake, we still assume such a fixed load when designing power amps!

The illustrated impedance curve is very benign; many speakers exhibit not just one but two or even three resonant peaks (as here at 100 Hz) and dips. Some dip way below 2Ω or even 1Ω, meaning an amplifier will struggle to push signal currents through such a low impedance, practically a short circuit!

To understand the sonics of a particular amplifier, you have to understand the loudspeakers they are driving in those listening tests, and that primarily means their impedance curve.

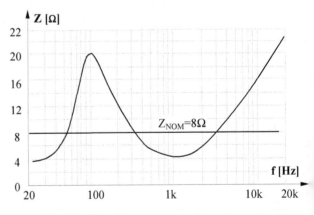

ABOVE: Typical impedance curve of a dynamic speaker driver

A simple way to plot the loudspeaker impedance curve

Since the 220Ω series resistor has a much higher DC resistance than the impedance of the measured loudspeaker, this simple circuit approximates a CCS (Constant Current Source). That same current flows through both the series resistor and the speaker.

You are measuring the difference between the readings of two AC voltmeters: the output voltage of a function generator (sinewave) and the AC voltage drop across the loudspeaker's impedance at various frequencies.

Had the speaker's impedance been constant with changing signal frequency, the ratio of the two measured voltages would remain constant. However, since the speaker's impedance is highly frequency-dependent, the ratio of the measured voltages will also change with frequency.

MEASUREMENTS VERSUS LISTENING AND OTHER AUDIO DESIGN DILEMMAS

The voltmeters must be capable of operation up to at least 20 kHz, so cheap multimeters are out of the question. If you don't have two AC voltmeters or multimeters, you can use a two-channel oscilloscope, but estimating signals' amplitude from a CRO screen is much slower and less precise than using AC voltmeters.

From 10 to 150 Hz we took measurements every 10 Hz, then from 200 to 1,000 Hz every 100 Hz, and then every 1kHz. After punching in all the figures into a spreadsheet, we got a recognizable dynamic speaker impedance curve, shown on the next page.

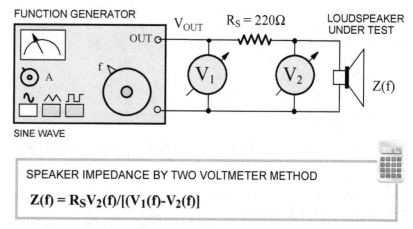

SPEAKER IMPEDANCE BY TWO VOLTMETER METHOD

$$Z(f) = R_S V_2(f)/[(V_1(f) - V_2(f)]$$

ABOVE: Test setup for manual (point-by-point) plotting of a loudspeaker impedance versus frequency curve
LEFT: Opera III (Terza) speakers, tested using this method

EXPERIMENT: Loudspeaker impedance curve for Opera III (Terza) loudspeaker

These babies have served us faithfully for many years as one of our test speakers. Technically, these should be unsuitable for low-power SET amplifiers due to their complex crossovers (a 3-way design with high-order filters) and relatively low sensitivity (86-87 dB/W). Yet, even a lowly 3W amp can drive them relatively easily.

The bass drivers are small, so don't expect ceiling-collapsing bass levels, and they are not the most accurate or transparent speakers around. But, they are musical, great-looking (solid mahogany, slim and elegant), and serve their purpose well.

We kept V1 constant and only had to measure V2. Once the test results were plotted, we got the impedance curve illustrated below, and a few things became clear. Opera Terza is a very easy load to drive! Its impedance never dropped below 6.2W; in the bandwidth between 3 and 20 kHz, it stayed in the 12-16W range. The three resonant peaks are self-evident, at around 30Hz, 80Hz, and 2,500Hz.

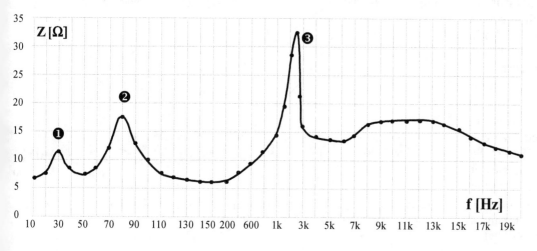

LEFT: The impedance modulus versus frequency for Opera III (Terza) loudspeaker

Amplitude-frequency response of SET amplifier with a dummy load and Opera III speakers

With a dummy (purely resistive) 8Ω load, the frequency range of this amplifier (SET with EL156 tubes) was flat, dropping to -3dB at 15 Hz and 41 kHz. The curve is shown below and marked (4). With Opera III speakers connected and at the same power level (1 Watt at 1 kHz, the 0 dB level), the speakers' impedance changes heavily influenced the amplitude-frequency characteristics. This is due to the relatively low damping factor of around 3.6 in triode mode and only 2.5 in the ultra-linear mode!

The three resonant peaks of the speaker result in three amplification peaks (marked 5, 6, and 7), but the overshoots are not severe in any of the cases.

The biggest drawback is two dips, one in lower bass, around 45Hz, and the other in the upper bass, with a minimum of -3.5 dB at around 160 Hz. These two "suckout" regions are the main reason for the relatively weak bass of these speakers when driven by most SET amplifiers.

The midrange frequencies between 1 and 3 kHz will be slightly emphasized (7), up to +1.2 dB, which could make the bass seem even weaker in comparison.

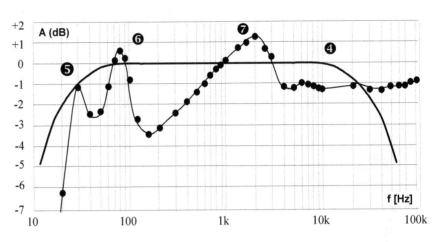

ABOVE: The amplitude-versus-frequency response of EL156 SET amplifier with an 8Ω dummy load (solid line) and with Opera Terza speakers. Dots indicate measured values.

Reactive loads

An audio output stage with a purely resistive load is a designer's dream that never comes true. The path of operation for a partially inductive load such as a typical dynamic loudspeaker or a capacitive load which as an electrostatic speaker is an ellipse.

The illustration below shows the anode characteristics of a 2A3 SET output stage with a purely resistive load line (2,500Ω anode resistance) and with a loudspeaker load. The operating point does not move along the A-B load line anymore, but it rotates following an elliptic path.

The only difference between the inductive and capacitive loads is in the direction of such rotation, counterclockwise or clockwise.

For positive anode current swings (50 to 120mA), the tube operates in class A. For negative swings, down from 50 mA towards zero (where the tube would still be conducting current with a purely resistive load), the elliptic operating path takes it into a cutoff mode where the anode current is zero. This will result in significant distortion.

Luckily, at lower power levels, the ellipse stays undistorted. We can also conclude that the "fatter" the ellipse (the more inductive or capacitive the load), the earlier such cutoff situation will happen, and the lower the maximum attainable output power will be!

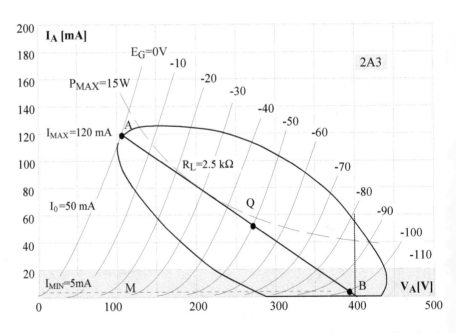

ABOVE: One load ellipsis for a SET stage with a 2A3 power tube. Notice how the ellipse is cut off at its bottom, resulting in significant distortion.

The whole brain approach to audio design

In this final section of the book, after some raving and ranting, and some serious evaluation and discussion of the most crucial interface of all, that between the power amplifier and loudspeaker, let's conclude with a few aspects of audio design.

First, books on engineering design methodology are relatively rare, and secondly, they mostly talk about relatively obscure issues and specialist areas. They cover topics that are easy to cover, not those that matter the most! Some educators even claim that it is impossible to teach design in a holistic and general manner. So, until we get our hands on that ultimate, still unwritten design book, a few issues to ponder about.

The Yin and Yang of audio design and construction

To design and build a world-class tube amplifier or anything else for that matter, requires input from both sides of the creative divide, the right and the left brain.

The left sides of human brains deal predominately with quantitative issues, facts and figures, and does so in a logical and orderly manner. Even the order of its progression makes sense. Firstly, one has to master mathematics, the "language" or the "tool" for explaining and expressing scientific and engineering principles and methods. Mathematics helps us with models, simplified approximations of the real life, for instance, a model of vacuum tube as a linear amplifier.

Then, one must understand the scientific principles that are the foundation of engineering, concepts such as the conservation of energy, efficiency and effectiveness, and many others.

The right side of the brain involves qualitative issues, those that cannot easily (or at all) be expressed in numbers or even outlined or explained, issues that are felt and imagined. This side of design starts with an inspiration, which eventually starts forming into an idea or vision, and finally progresses into the artistic side of the design, the looks and the feel of the finished product, together with all the emotions it evokes in the viewer, listener or user.

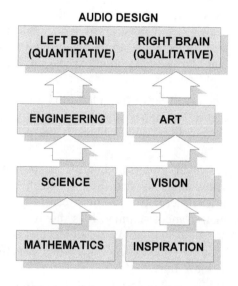

ABOVE: A simple model of audio design's dual nature

Ideally, both aspects of a certain design should be balanced. For instance, a tube amplifier should be of sound design (no pun intended), properly built and reliable, which are all left brain attributes, but should also be pleasing to the eye and ear, it should look good and be pleasurable to use, all right-brain qualities.

Technical or left brain aspects of design are easier to judge and evaluate. Technical performance, build quality and reliability are still open to a certain degree of interpretation, but not nearly as much as the right brain aspects are. The beauty, after all, is in the eyes of a beholder! Thus, what seems to be a well balanced piece of audio engineering to one audiophile may seem a butt-ugly product to another.

That illusive and contentious point of diminishing returns

Audiophiles have a notorious reputation for constant experimentation, tube rolling, perpetual modifications, component upgrading, and all kinds of minor and major system changes.

Some of those make no sense at all. For instance, quite a few posts on online audio fora are of the "I bought this amp, can you guys suggest what components I should replace?" kind. This is silly. The owner has neither listened critically to his new amp in his system nor identified its strengths and weaknesses from the sonic, design, construction, or reliability point of view. He has no clue about electronics yet wants to start changing things.

On the other hand, this constant experimentation quest for that elusive ultimate sound is understandable. The question, however, is how far is far enough, and how far is too far?

The S-curve is well known in project management and management in general. Just like Murphy's law, it applies to most aspects of life.

For instance, when you start improving something, often the results aren't immediately noticeable. You have to persist with your efforts until you overcome the initial period of "inertia" (for the lack of a better word) and reach the linear portion of the curve, where further efforts result in noticeable achievement or improvement.

Common examples in audio would be interstage and output transformers, which, when replaced, take some time to be broken in and may sound terrible initially. So, patience is often required, and hasty decisions should not be made too early. Likewise, silver interconnects and speaker cables also take quite a few hours to "settle down." Initially, they sound detailed but bright, even harsh on some systems.

Finally, going back to our S-curve, there usually comes the point where the further investment of time, money, or effort (or in audio "further upgrades") are not resulting in any meaningful improvements. You have reached the point of "saturation," the plateau, the point of diminishing returns. You are investing a lot but getting little or nothing in return.

The concept of diminishing returns is closely related to a bottleneck - both feature prominently in audio. For instance, ignorant audiophiles buy "hi-end" film coupling capacitors at $600 a pop, retube the amp at $1,500 a pop, or rewire the whole amp using silver wire with Teflon insulation, or "all of the above"!

However, the amp is a $900 Made-in-China concoction, poorly designed and constructed, with anode and heater voltages 20% higher than expected, so those expensive tubes will only last 3-6 months if that. To use a farming expression, this is like putting lipstick on a pig!

The amplifier or the "platform" is clearly a bottleneck or the limiting factor. No matter what you do to it, you will not improve it beyond a certain level and will reach the point of diminishing returns very soon.

However, understanding a concept is one thing, while applying that knowledge in real life is much more difficult. Life is not a well-structured book or a carefully rehearsed and smooth-flowing management seminar. It is messy, problems are not neatly defined, solutions may not work, and results are far from guaranteed. It is next to impossible to determine where exactly on the S-curve we are at any point in time.

However, don't let that discourage you. It would be unwise to end this (or any) book on a sour note, so keep learning, keep experimenting, and, most importantly, keep having fun!

I wish you every success on your journey!

ABOVE: The S-curve applies to most human endeavors, and audio is no different

THE END MATTER OF VOLUME 3

- Reference data
- Index
- Further reading

17

"It's not how you start, but how you finish!'
Chef Didier, played by Gérard Depardieu, in movie "Last Holiday"

Reference data

Transformer lamination sizes - EI type

TYPE	f	e	a	c	b	w	d
EI 19	19	12.5	5	2.5	4.5	10	2.5
EI 24	24	15	6	3	6	12	3
EI 25.4	25.4	16.1	6.4	3.1	6.4	13.1	3
EI 28	28	21	8	4	6	17	4
EI 35	35	24.5	9.6	5	7.7	19.5	5
EI 41	41	27	13	6	8	21	6
EI 48	48	32	16	8	8	24	8
EI 54	54	36	18	9	9	27	9
EI 57	57	38	19	9.5	9.5	28.5	9.5
EI 60	60	40	20	10	10	30	10
EI 66	66	44	22	11	11	33	11
EI 73	73	49	23	11.5	13.5	34.5	11.5
EI 74	74	51	23	11.5	14	34	14
EI 75	75	51.5	23	12	14	37.5	14
EI 76.2	76.2	50.8	25.4	12.7	12.7	38.1	12.7
EI 84	84	56	28	14	14	42	14
EI 85.8	85.8	57.2	28.6	14.3	14.3	42.9	14.3
EI 86	86	59	26	13	17	46	13
EI 88	88	60	29.4	13.3	16	44	16
EI 90	90	60	30	15	15	45	15
EI 96	96	64	32	16	16	48	16
EI 100	100	67	32	16	18	51	16
EI 105	105	70	35	17.5	17.5	52.5	17.5
EI 111	111	76	35	17.5	20.5	55.5	20.5
EI 114	114	76	38	19	19	57	19
EI 120	120	80	40	20	20	60	20
EI 133.2	133.2	88.8	44.4	22.2	22.2	66.6	22.2
EI 152.4	152.4	101.6	50.8	25.4	25.4	76.2	25.4
EI 181.2	181.2	120.8	60.4	30.2	30.2	90.6	30.2
EI 192	192	128	64	32	32	96	32

ABOVE: EI lamination sizes. Wastless sizes are highlighted in gray.

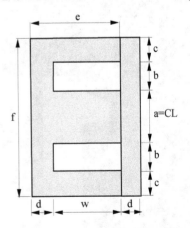

Dimensions and magnetic path length for all EI sizes. a = center leg width (CL), b = window height, w = window width, S = stack thickness (not shown), W = Window area W= b*w, A = gross core area A = a*S
For scrapless EI laminations (gray shaded in tables) b = c = g and a = 2b = 2c = 2g

Resistor color chart and standard values

Resistor values follow geometric progressions, such as E12, E24 and E48 (10%, 5% and 2% tolerance). The values between 1 and 10 are shown below. All other values are obtained by multiplying those with 10, 100, 1000, and so on. E12 series (10%): 1.00 1.20 1.50 1.80 2.20 2.70 3.30 3.90 4.70 5.60 6.80 8.20

E24 series (5%): 1.00 1.10 1.20 1.30 1.50 1.60 1.80 2.00 2.20 2.40 2.70 3.00 3.30 3.60 3.90 4.30 4.70 5.10 5.60 6.20 6.80 7.50 8.20 9.10

Example: Red - Violet - Red - Silver: 27 x 100 = 2k7 (2,700 ohms or 2.7 kohms), tolerance +/-10%

You need an 82k anode resistor. 8 is Gray, 2 is Red, the multiplier is 1,000 (to get from 82 to 82,000), which is Orange. So you need Gray-Red-Orange-Silver (from E12 series, with 10% tolerance) or Gray-Red-Orange-Gold (from E24 series, 5% tolerance) or Gray-Red-Orange-Red (from E48 series, 2% tolerance)

Color	Band 1 1st digit	Band 2 2nd digit	Band 3 Multiplier	Band 4 Tolerance
Black	0	0	1	
Brown	1	1	10	+/- 1%
Red	2	2	100	+/- 2%
Orange	3	3	1000	
Yellow	4	4	10,000	
Green	5	5	100k	
Blue	6	6	1M	
Violet	7	7	10M	
Gray	8	8	100M	
White	9	9	1G	
Gold			0.1	+/- 5%
Silver			0.01	+/-10%
None				+/- 20%

BY THE SAME AUTHOR:

Sound Improvement Secrets For Audiophiles: Get Better Sound Without Spending Big

Publisher: Career Professionals
Year published: 2021
Language: English
Paperback: 328 pages
ISBN: 978-0648298205

Avoid the hit-and-miss approach and stop wasting money on overpriced high-end products in the blind hope of sonic improvement. Achieve the ultimate audio synergy and get more enjoyment from your audio system by making it as good sounding as possible.

"Sound Improvement Secrets for Audiophiles" will teach you how things work, why some circuits, designs, and technologies sound the way they do, and how to make them sound even better through simple modifications and improvements. It is like having an audio and acoustic consultant by your side to guide you through optimizing and voicing your audio system and your listening room.

While relatively technical and in-depth, this practical manual goes way beyond "a dozen quick tips" and the simplistic advice you read elsewhere. Instead, the focus is on dozens of DIY projects, case studies, and examples of commercial audio components – turntables, preamplifiers, amplifiers, loudspeakers, power supplies, and acoustic treatments.

With over 400 photographs, diagrams, and illustrations, "Sound Improvement Secrets for Audiophiles" makes it easy for you to understand and comprehend complex technical concepts and issues.

The author does not shy away from many controversial and hotly debated topics. Tubes vs transistors, objectivists vs subjectivists, measurements vs listening, and digital vs analog: all of these are discussed in detail.

The money invested in this book would not even buy you a budget-priced pair of cables: it will prove to be one of the best financial investments you ever make. Even if you implement only a few improvements from the hundreds described within its pages – you will never look back!

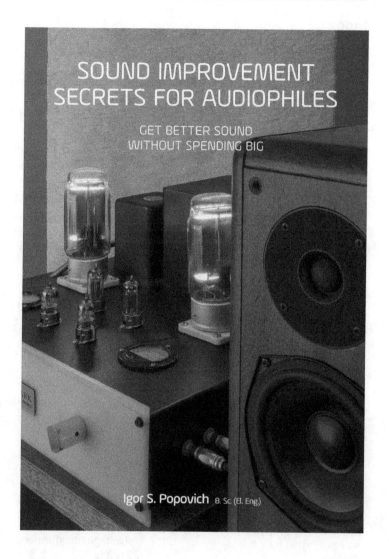

BOOK CONTENTS:

1. WHY YOU SHOULD READ THIS BOOK AND HOW YOU WILL BENEFIT FROM IT
2. BEFORE YOU BUY AN AUDIO SYSTEM OR COMPONENT - THINGS TO DO & MISTAKES TO AVOID
3. WHAT DO WE LISTEN FOR AND WHAT DO WE ACTUALLY HEAR?
4. CLEANING UP THE POWER SUPPLY TO REDUCE NOISE, HUM, AND INTERFERENCE
5. CABLES, FUSES, CONTACTS, AND CONNECTIONS
6. UPGRADING & FINE-TUNING THE SOURCES: OPEN REEL RECORDERS, TURNTABLES, PHONO STAGES AND CD PLAYERS
7. AUDIO AMPLIFIERS - HOW THEY WORK AND HOW TO IMPROVE THEIR SOUND
8. HEADPHONES AND HEADPHONE AMPLIFIERS
9. LOUDSPEAKER TYPES, TESTS, AND IMPROVEMENTS
10. COMPONENT MATCHING AND AUDIO SYSTEM INTEGRATION ISSUES
11. LOUDSPEAKER POSITIONING
12. OPTIMIZING THE ACOUSTIC PERFORMANCE OF YOUR LISTENING ROOM
13. ACOUSTIC TREATMENTS
14. MINIMIZING UNWANTED VIBRATIONS & OSCILLATIONS
15. TROUBLESHOOTING YOUR AUDIO SYSTEM

Index

A

Anode choke, 14, 54-55, 60, 152, 226
Antique Sound Lab MG-Power 805 amplifier, 154-155
Audio Note Kit 1, 11-113
ARC LS7 and LS8 preamplifiers, 26-28
Audio chain, 270
Audio design, 279-280
AudioRomy 838 amplifier, 144-145
Audion
    Sterling EL34 amplifier, 141-142
    Super Sterling amplifier, 129
Auto-transformer, 226
Automatic Bias Control, 220-221

B

Back-to-back transformers, 188, 238
Balanced input, 185, 227, 230
Balanced output, 31, 49
BAT Rex II line stage, 48
Battery bias, 23, 52
Battery powered amplifier, 137-139
Bel Canto SRPP stage, 19-22, 89-90
Bell 2145-A amplifier, 186-187
BEZ Q3B-1 line stage,
BEZ Q6A line stage, 2930
BEZ T10B amplifier, 149-153
Bias line, 45
Bogen DB130 amplifier, 88
Bridged mono operation, 227-228
Brook 10C amplifier, 220

C

Cary 805C amplifier, 145-147
Cascode, 136, 161, 227
Cash drawers as chassis, 159, 247
Cathode bypass capacitor, 49
Cathode-coupled inverter, 195, 202
Cathode loading, 182, 192
Cathode follower, 26-28, 31, 33, 41, 109, 136, 150, 155, 161, 182, 187
CCS, 33, 37, 277
Cheater plugs, 147

D

Damping factor, 64, 168
Dared SL2000-A line stage, 45
Darling amplifiers, 64-65
D&Y A-300 amplifier, 100-106
DC coupling, 14, 41, 101, 123, 125, 128, 140, 149, 155, 191, 195, 209, 210
DC heating and filament wear, 113
DHT (directly heated triodes), 34-38, 58-63, 69-72, 100-116, 140, 144-164
Differential amplifier, 182, 185, 215, 227
Driver stage for Class A2 or Class B, 136, 149
Dynaco MkII amplifier, 178-180

E

E.A.R. 509 amplifier, 215
E.A.R. Yoshino V20 amplifier, 214
Eico ST40 and ST70 amplifiers, 194-196
Enhanced triode mode, 135-137, 160-161, 214

F

Fisher SA-1000 amplifier, 181-183
Floating paraphase inverter, 187, 192, 204

G

Grid stopper resistors, 213
Grid choke, 126, 150

H

Harmonic distortion, 274
Heater-cathode voltage limit, 176
Heathkit UA2 amplifier, 167-168
HF compensation, 167, 175, 182, 190-191, 197-198
Hybrid rectifier, 36, 38, 62, 104

I

Inductors, 267-268
Interstage transformers, 114-117, 146, 231, 255

J

Jadis SE300B monoblocks, 273-274
JFET CCS, 28

L

LC-coupling, 14, 55, 107, 115, 152, 155, 226
Leben CS300 amplifier, 204-206
LED bias, 73
Levels of truth in audio, 275
Line output transformers, 254
Line stage
    transformer coupled, 30, 48-49, 53
Long tailed pair, see: Cathode-coupled inverter
Loudspeaker impedance curve, 277-278
Luxman MQ60 amplifier, 183

M

Marantz T-1, 229-233
Mercury-vapor rectifiers, 127
Miller's capacitance, 81, 190, 218
Mingda MC-7R line stage, 42-44
Minimal Signal Path Length (MSPL), 63
μ (mju) follower, 29, 33
MOSFET CCS, 26
Motor start capacitors, 63, 238
Mr. Liang LS845 amplifier, 154

N

Negative feedback, 271-272
NTC thermistor, 131, 159

O

Opera Consonance Cyber 222 line stage, 32-33

P

Paco SA40 amplifier, 169
Paralleled tubes, 29, 68, 90, 95, 107, 116, 128, 131, 135-136, 140, 174, 217-218
Paraphase inverter, 187
Partial cathode coupling, 180-181
Payback Rule, 148
Positive ground amplifier, 225-226
Psvane "Teflon®" capacitors, 276

Q

Quad II amplifier, 192-193
Quicksilver 8417 amplifier, 173-175

R

RC-filter, 96
Reactive loudspeaker loads, 279
Regulated power supply, 127, 203
RimLock sockets, 94, 247

S

Schottky diodes, 74
Screen grid, 180
Screen driven tubes, 19-20
Selenium rectifier, 178
Series voltage regulator, 55, 57, 136-139
Sheer Audio DHT-01A line stage, 34-39

S

Shindo Cortese amplifier, 120-121
Shindo Sinhonia amplifier, 210
Shunt regulators, 53
Snubber (spark killer), 159, 206
Split-load phase inverter, 169, 174, 180, 182, 191, 195, 209, 220
SRPP, 22, 33, 112, 115, 120, 123, 128, 138, 149, 158, 163, 202, 213
Stromberg Carlson ASR-444 amplifier, 170-172
Sun Audio SV-300BE amplifier, 110-111
Suppressor grid, 97, 132

T

Transformer
    hysteresis loop, 263
    magnetizing current, 262
    TPV (Turns-Per-Volt), 266
Transformer-coupled stages, 53, 73, 114-116, 124-125, 146, 155, 210, 213, 220, 230
Triplett 3444 tube tester, 84-86, 92
TV damper diodes, 120
TVC-transformer volume control, 257-258

U

Ultra fast rectifiers, 131
Ultralinear connection, 90, 96, 126, 129, 131, 135, 142, 202, 209-213

V

VAL MP-211 amplifier, 148-149
Voltage doubler, 162, 172, 206, 243
Voltage regulator
    solid state, 38, 203, 223
    tube, 53, 127, 238, 241-243
VR tubes, 32
VU-meter, 210

W

White Follower, 41-42
Williamson amplifier, 190-191

X

Xiang Sheng 708B headphone amplifier & line stage, 40-41

Y

Yarland FVII amplifier, 139-141
Yarland PM7 line preamplifier, 39-40

Z

Zobel network, 50, 203

FURTHER READING

IN ENGLISH

Amplifiers, H. Lewis York, 1964, 254 pages

Amplifiers (The Why and How of Good Amplification), G. A. Briggs, 1952, 216 pages

Analysis and Design of Electronic Circuits, Paul M. Chirlian, 1965, 570 pages

Applied Electronics, Truman S. Gray, 1955, 882 pages

An Approach to Audio Frequency Amplifier Design, G.E.C. Valve and Electronics Department, 1957, 126 pages

Audio Cyclopedia, Howard M. Tremaine, 1st edition, 1959, 1269 pages

Audio Design Handbook, H. A. Hartley, 1958, 224 pages

Audio Transformer Design Manual, Robert G. Wolpert, 1989, 108 pages

Basic Audio, Norman Crowhurst, 1959, Volume 1 (114 pages), Volume 2 (122 pages), Volume 3 (113 pages)

Capacitors, Magnetic Circuits, and Transformers, Leander Matsch, 1964, 350 pages

Circuit Theory of Electron Devices, E. Milton Boone, 1959, 483 pages

Designing Power Supplies for Tube Amplifiers, Merlin Blencowe, 2010, 246 pages

Designing Tube Preamps for Guitar and Bass, Merlin Blencowe, 2009, 294 pages

Electron Tubes and Semiconductors, Joseph J DeFrance, 1958, 288 pages

Electronic Amplifier Circuits, Joseph Petit and Malcolm McWhorter, 1961, 325 pages

Electronic and Radio Engineering, Frederick E. Terman, 1955, 1078 pages

Electronic Circuits, Thomas L. Martin. 1959, 708 pages

Electronic Circuits and Tubes, Harry E. & Wing, Alexander H. Clifford, 1947, 996 pages

Electronic Circuits: A Unified Treatment of Vacuum Tubes and Transistors, Ernest Angelo, 1964, 652 pages

Electronic Designers' Handbook, Landee, Davis and Albrecht, 1957

Electronic Engineering, C. Alley and K. Atwood, 1973, 838 pages

Electronic Transformers and Circuits, Reuben Lee, 1955, 349 pages

Electronics and Electron Devices, A. L. Albert, 1956

Electron-Tube Circuits, Samuel Seely, 1950, 530 pages

Engineering Electronics, George Happell and Wilfred Hesselberth, 1953, 508 pages

Essentials of Radio-Electronics, M. Slurzberg and W. Osterheld, 1961, 716 pages

Engineering Electronics With Industrial Applications, John D. Ryder, 1957, 666 pages

Fundamental Amplifier Techniques with Electron Tubes, Rudolf Moers, 2011, 834 pages

Fundamentals of Semiconductor and Tube Electronics, H. Alex Romanowitz, 1962, 620 pages

General Electronics Circuits, Joseph J. DeFrance, 1963, 526 pages

High-End Valve Amplifiers 2, Menno van der Veen, 2011, 416 pages

High-Fidelity Circuit Design, Norman Crowhurst and George Cooper, 1957, 296 pages

How to Gain Gain, A Reference Book on Triodes in Audio Pre-Amps, Vogel Burkhard, 2013

Inside the Vacuum Tube, John F. Rider, 1945, 407 pages

Instruments and Measurements For Electronics, Clyde N. Herrick, 1972, 542 pages

Maintaining Hi-fi Equipment, Joseph Marshall, 1956, 224 pages

Magnetic Circuits and Transformers, M.I.T. Electrical Engineering Staff, 1943, 718 pages

Modern High-end Valve Amplifiers Based on Toroidal Output Transformers, M. van der Veen, 1999, 250 pages

Network analysis and feedback amplifier design, Hendrik W. Bode, 1952, 551 pages

Practical Transformer Design Handbook, Eric Lowdon, 1989, 389 pages

Principles and Applications of Electron Devices, Paul D. Ankrum, 1959, 667 pages

Principles of Electron Tubes, Gewartowski and Watson, 1965, 655 pages

Principles of Electron Tubes, Herbert Reich, 1941, 398 pages

Radio Engineer's Handbook, F. Terman, 1943, 1,021 pages

Radiotron Designer's Handbook, F. Langford Smith, 4th edition, 1952, 1,482 pages

IN ENGLISH, cont.

The Sound of Silence: Lowest-Noise RIAA Phono-Amps: Designer's Guide, Burkhard Vobel, 2011
The Thermionic Vacuum Tube and Its Applications, H. J. Van Der Bijl, 1920, 391 pages
The Tube Preamp Cookbook, Allen Wright, 1997
The Ultimate Tone, Kevin O'Connor, 1995
Thermionic Valve Circuits, Emrys Williams, 1961, 427 pages
Tu-be or Not Tu-be: Modification Manual For Vacuum Tube Electronics, H. I. Eisenson, 1977, 198 pages
Theory and Applications of Electron Tubes, Herbert Reich, 2nd edition 1941, 716 pages
Theory of Thermionic Vacuum Tubes, E. Leon Chafee, 1933, 652 pages
Vacuum Tube and Semiconductor Electronics, Jacob Millman, 1958, 644 pages
Vacuum Tube Circuits, Lawrence Baker Arguimbau, 1948, 668 pages
Vacuum Tubes, Karl A. Spangenberg, 1948, 860 pages
Valve and Transistor Audio Amplifiers, John Linsley Hood, 1997, 250 pages
Valve Amplifiers, Morgan Jones, 4th edition, 2012, 700 pages

IN FRENCH

Les Tubes Electroniques et leurs Applications. Tome 1 : Principes Généraux, H. Barkhausen, 1949, 228 pages
Les tubes à vide et leurs applications, tome 2 : Les amplificateurs, H. Barkhousen, 1942, 301 pages
Physique et technique des tubes electroniques, R. Champeix, 1958, 214 pages
Théorie et pratique des circuits de l'électronique et des amplificateurs, Tome 1-2, J. Quinet, 1962
Traité moderne des amplificateurs haute fidélité à tubes, Lallie/Fiderspi, 2008, 331 pages

IN GERMAN

Audio-Röhrenverstärker von 0,3 - 10 W erfolgreich selbst bauen, Wilfried Frohn, 2005, 128 pages
Elektronische Speisegeräte, Karl Steimel, 1957, 246 pages
High-End mit Röhren, Gerhard Haas, 2007, 319 pages
Hören mit Röhren, Friedrich Hunold, 1999,
Niederfrequenzverstärker-Praktikum, Otto Diciol, 1959, 393 pages
Röhren NF Verstärker Praktikum, Otto Diciol, 2003, 393 pages
Röhrenprojekte von 6 bis 60 Volt, Burkhard Kainka, 2003, 153 pages
Röhrentechnik ganz modern, Winfried Knobloch, 1991, 103 pages
Röhrenverstärker selber bauen, Richard Zierl, 2011, 264 pages
Telefunken Laborbuch, Band I, 1970, 404 pages
Theorie der Spulen und Übertrager, R. Feldtkeller, 1957, 186 pages
Theorie und Praxis des Röhrenverstärkers, Peter Dieleman, 2007, 253 pages

IN ITALIAN

Amplificatori valvolari tecnica e pratica di autocostruzione, Ivano Incerti, 2012, 144 pages
I piccoli trasformatori : calcolo e costruzione ad uso degli elettricisti, Mario Pierazzuoli, 1949, 146 pages
I trasformatori tipo radio e simili, Dr. Ing. Enrico Baldoni, 1955, 182 pages
La construzione e il calcolo dei piccoli trasformatori, Ernesto Carbone, 1966, 80 pages
Manuale hi-fi a valvole, Luciano Macrì and Riccardo Gardini, 1994, 348 pages
Teoria e calcolo dei piccoli trasformatori, Giacomo Giuliani, 1948, 80 pages
Valvole e dintorni Hi-Fi, Atto E. Rinaldo, 2009, 228 pages
Valvole e trasformatori per Hi-Fi, Gieffe, 2004, 280 pages

IN SERBIAN

Elektronika, Dr Slavoljub Marjanovic, 1981, 580 pages
Hi-Fi Lampaška Pojacala, N.Vukušic, 2011, 726 pages